IMMUNE MECHANISMS AND DISEASE

ANNALS OF THE NEW YORK ACADEMY OF SCIENCES
Volume 987

IMMUNE MECHANISMS AND DISEASE

Edited by Nicholas Chiorazzi, Robert G. Lahita, J. Donald Capra, Manlio Ferrarini, and John B. Zabriskie

The New York Academy of Sciences
New York, New York
2003

Library of Congress Cataloging-in-Publication Data

Immune mechanisms and disease / edited by Nicholas Chiorazzi ... [et al.]
p. ; cm. — (Annals of the New York Academy of Sciences, ISSN 0077-8923 ; v.987)
Result of a conference held April 14-17, 2002 in St. George's, Grenada, West Indies.
Includes index.
ISBN 1-57331-434-X (cloth : alk. paper) — ISBN 1-57331-435-8 (paper: alk. paper)
1. Immune response—Congresses.
[DNLM: 1. Immunity—physiology—Congresses. 2. Autoimmune Diseases—immunology—Congresses. QW 540 I321 2003] I. Chiorazzi, Nicholas. II. Series.
Q11.N5 vol. 987
[QR186]
500 s—dc21
[616.07/
987 2003004730

GYAT / PCP

Printed in the United States of America

ISBN 1-57331-434-X (cloth)

ISBN 1-57331-435-8 (paper)

ISSN 0077-8923

ANNALS OF THE NEW YORK ACADEMY OF SCIENCES

Volume 987
April 2003

IMMUNE MECHANISMS AND DISEASE

Editors and Conference Organizers
NICHOLAS CHIORAZZI, ROBERT G. LAHITA, J. DONALD CAPRA, MANLIO FERRARINI, AND JOHN B. ZABRISKIE

This volume is the result of a conference entitled Immune Mechanisms and Disease, sponsored jointly by the New York Academy of Sciences and The Henry Kunkel Society, held April 14–17, 2002 in St. George's, Grenada, West Indies.

CONTENTS

Preface. *By* NICHOLAS CHIORAZZI, ROBERT G. LAHITA, J. DONALD CAPRA, MANLIO FERRARINI, AND JOHN B. ZABRISKIE xi

X^th Henry Kunkel Society Lecture

Activation-Induced Cytidine Deaminase Links Class Switch Recombination and Somatic Hypermutation. *By* ILMI OKAZAKI, KIYOTSUGU YOSHIKAWA, KAZUO KINOSHITA, MASAMICHI MURAMATSU, HITOSHI NAGAOKA, AND TASUKU HONJO 1

Part I. Antigen Presentation

Creation versus Destruction of T Cell Epitopes in the Class II MHC Pathway. *By* C. WATTS, C.X. MOSS, D. MAZZEO, M.A. WEST, S.P. MATTHEWS, D.N. LI, AND B. MANOURY 9

Dendritic Cell Function *in Vivo* during the Steady State: A Role in Peripheral Tolerance. *By* RALPH M. STEINMAN, DANIEL HAWIGER, KANG LIU, LAURA BONIFAZ, DAVID BONNYAY, KARSTEN MAHNKE, TOMONORI IYODA, JEFFREY RAVETCH, MADHAV DHODAPKAR, KAYO INABA, AND MICHEL NUSSENZWEIG 15

Molecular Mechanisms of B Cell Antigen Receptor Trafficking. *By* MARCUS R. CLARK, DON MASSENBURG, MIAO ZHANG, AND KARYN SIEMASKO . 26

Part II. Induction of Adaptive and Autoimmune Responses

Activation of Rheumatoid Factor (RF) B Cells and Somatic Hypermutation Outside of Germinal Centers in Autoimmune-Prone MRL/*lpr* Mice. *By* MARK J. SHLOMCHIK, CHAD W. EULER, SEAN C. CHRISTENSEN, AND JACQUELINE WILLIAM 38

Coordination of T Cell Activation and Migration through Formation of the Immunological Synapse. *By* MICHAEL L. DUSTIN 51

Intrinsic T Cell Defects in Systemic Autoimmunity. *By* PHILIP L. KONG, JARED M. ODEGARD, FARIDA BOUZAHZAH, JIN-YOUNG CHOI, LEAH D. EARDLEY, CHRISTINA E. ZIELINSKI, AND JOSEPH E. CRAFT 60

Opsonization of Apoptotic Cells and Its Effect on Macrophage and T Cell Immune Responses. *By* SUN JUNG KIM, DEBRA GERSHOV, XIAOJING MA, NATHAN BROT, AND KEITH B. ELKON 68

Major Peptide Autoepitopes for Nucleosome-Centered T and B Cell Interaction in Human and Murine Lupus. *By* SYAMAL K. DATTA 79

Mechanisms of Autoantibody Diversification to SLE-Related Autoantigens. *By* UMESH S. DESHMUKH, FELICIA GASKIN, JANET E. LEWIS, CAROL C. KANNAPELL, AND SHU MAN FU 91

Mechanisms Inducing or Controlling $CD8^+$ T Cell Responses against Self- or Non-Self-Antigens. *By* DANIELE ACCAPEZZATO, VITTORIO FRANCAVILLA, ANTONELLA PROPATO, MARINO PAROLI, AND VINCENZO BARNABA 99

Part III. Germinal Center Reaction

Role of Homeostatic Chemokine and Sphingosine-1-Phosphate Receptors in the Organization of Lymphoid Tissue. *By* GERD MÜLLER, PHILLIP REITERER, UTA E. HÖPKEN, SVEN GOLFIER, AND MARTIN LIPP 107

The Human Marginal Zone B Cell. *By* MARIELLA DONO, SIMONA ZUPO, MONICA COLOMBO, ROSANNA MASSARA, GIANLUCA GAIDANO, GIUSEPPE TABORELLI, PAOLA CEPPA, VITO L. BURGIO, NICHOLAS CHIORAZZI, AND MANLIO FERRARINI 117

B Cell Immunity Regulated by the Protein Kinase C Family. *By* KAORU SAIJO, INGRID MECKLENBRÄUKER, CHRISTIAN SCHMEDT, AND ALEXANDER TARAKHOVSKY 125

A Modified Digestion-Circularization PCR (DC-PCR) Approach to Detect Hypermutation-Associated DNA Double-Strand Breaks. *By* SARAH K. DICKERSON AND F. NINA PAPAVASILIOU 135

Part IV. Abnormalities of the Germinal Center Reaction Associated with Disease

Ectopic Germinal Center Formation in Rheumatoid Synovitis. *By* CORNELIA M. WEYAND AND JÖRG J. GORONZY 140

The V(D)J Recombination/DNA Repair Factor Artemis Belongs to the Metallo-β-Lactamase Family and Constitutes a Critical Developmental Checkpoint of the Lymphoid System. *By* DESPINA MOSHOUS, ISABELLE CALLEBAUT, RÉGINA DE CHASSEVAL, CATHERINE POINSIGNON, ISABELLE VILLEY, ALAIN FISCHER, AND JEAN-PIERRE DE VILLARTAY . . 150

Hypermutation in Human B Cells *in Vivo* and *in Vitro*. *By* SANDRA WELLER, AHMAD FAILI, SAID AOUFOUCHI, QUENTIN GUÉRANGER, MORITZ BRAUN, CLAUDE-AGNÈS REYNAUD, AND JEAN-CLAUDE WEILL 158

Gene Expression Dynamics during Germinal Center Transit in B Cells. *By* ULF KLEIN, YUHAI TU, GUSTAVO A. STOLOVITZKY, JEFFREY L. KELLER, JOSEPH HADDAD, JR., VLADAN MILJKOVIC, GIORGIO CATTORETTI, ANDREA CALIFANO, AND RICCARDO DALLA-FAVERA 166

Somatic Hypermutation and B Cell Receptor Selection in Normal and Transformed Human B Cells. *By* RALF KÜPPERS 173

Part V. Novel Therapies

Dendritic Cells: Controllers of the Immune System and a New Promise for Immunotherapy. *By* JACQUES BANCHEREAU, SOPHIE PACZESNY, PATRICK BLANCO, LYNDA BENNETT, VIRGINIA PASCUAL, JOSEPH FAY, AND A. KAROLINA PALUCKA 180

Co-Stimulatory Blockade in the Treatment of Murine Systemic Lupus Erythematosus (SLE). *By* ANNE DAVIDSON, XIAOBO WANG, MASAHIKO MIHARA, MEERA RAMANUJAM, WEIQING HUANG, LENA SCHIFFER, AND JAYASHREE SINHA 188

The Transfer of Immunity from Mother to Child. *By* LARS Å. HANSON, MARINA KOROTKOVA, SAMUEL LUNDIN, LILJANA HÅVERSEN, SVEN-ARNE SILFVERDAL, INGER MATTSBY-BALTZER, BIRGITTA STRANDVIK, AND ESBJÖRN TELEMO 199

Novel Method to Control Pathogenic Bacteria on Human Mucous Membranes. *By* VINCENT A. FISCHETTI 207

Workshop I. Induction of Normal and Maladaptive Immune Responses

Peptide Mimics of a Major Lupus Epitope of SmB/B′. *By* KENNETH M. KAUFMAN, MONICA Y. KIRBY, JOHN B. HARLEY, AND JUDITH A. JAMES 215

CD137-Mediated T Cell Co-Stimulation Terminates Existing Autoimmune Disease in SLE-Prone NZB/NZW F1 Mice. *By* JUERGEN FOELL, MEGAN MCCAUSLAND, JENNIFER BURCH, NICHOLAS CHIORAZZI, XIAO-JIE YAN, CAROLYN SUWYN, SHAWN P. O'NEIL, MICHAEL K. HOFFMANN, AND ROBERT S. MITTLER 230

Using Single-Gene Deletions to Identify Checkpoints in the Progression of Systemic Autoimmunity. *By* K. MICHAEL POLLARD, PER HULTMAN, AND DWIGHT H. KONO 236

Activation of the Ets Transcription Factor Elf-1 Requires Phosphorylation and Glycosylation: Defective Expression of Activated Elf-1 Is Involved in the Decreased TCR ζ Chain Gene Expression in Patients with Systemic Lupus Erythematosus. *By* GEORGE C. TSOKOS, MADHUSOODANA P. NAMBIAR, AND YUANG-TAUNG JUANG 240

Aberrant Activation of B Cells in Patients with Rheumatoid Arthritis. *By* STEFFI LINDENAU, SUSANN SCHOLZE, MARCUS ODENDAHL, THOMAS DÖRNER, ANDREAS RADBRUCH, GERD-RÜDIGER BURMESTER, AND CLAUDIA BEREK 246

Workshop II. Innate and Adaptive Immune Mechanisms in Health and Disease

The Role of *rel*B in Regulating the Adaptive Immune Response. *By* MAURIZIO ZANETTI, PAOLA CASTIGLIONI, STEPHEN SCHOENBERGER, AND MARA GERLONI 249

Tolerogenic Dendritic Cells Induced by Vitamin D Receptor Ligands Enhance Regulatory T Cells Inhibiting Autoimmune Diabetes. *By* LUCIANO ADORINI 258

Antibody Repertoire in a Mouse with a Simplified D_H Locus: The D-Limited Mouse. *By* GREGORY C. IPPOLITO, JUKKA PELKONEN, LARS NITSCHKE, KLAUS RAJEWSKY, AND HARRY W. SCHROEDER, JR. 262

Long-Lived Plasma Cells in Immunity and Inflammation. *By* A.E. HAUSER, G. MUEHLINGHAUS, R.A. MANZ, G. CASSESE, S. ARCE, G.F. DEBES, A. HAMANN, C. BEREK, S. LINDENAU, T. DOERNER, F. HIEPE, M. ODENDAHL, G. RIEMEKASTEN, V. KRENN, AND A. RADBRUCH 266

Molecular Mechanism of Serial VH Gene Replacement. *By* ZHIXIN ZHANG, YUI-HSI WANG, MICHAEL ZEMLIN, HARRY W. FINDLEY, S. LOUIS BRIDGES, PETER D. BURROWS, AND MAX D. COOPER 270

Oligoclonal Expansions of Antigen-Specific $CD8^+$ T Cells in Aged Mice. *By* DAVID N. POSNETT, DMITRY YARILIN, JENNIFER R. VALIANDO, FANG LI, FOO Y LIEW, MARC E. WEKSLER, AND PAUL SZABO 274

A Genome-Wide Analysis of the Acute-Phase Response and Its Regulation by Stat3β. *By* STEPHEN DESIDERIO AND JOO-YEON YOO 280

Poster Papers

Ii-CS on Gastric Epithelial Cells Interacts with CD44 on T Cells and Induces Their Proliferation. *By* C.A. BARRERA, T. CHAN, S.E. CROWE, P.B. ERNST, AND V.E. REYES 285

Tracking CD40 Signaling during Normal Germinal Center Development by Gene Expression Profiling. *By* KATIA BASSO, ULF KLEIN, HUIFENG NIU, GUSTAVO A. STOLOVITZKY, YUHAI TU, ANDREA CALIFANO, GIORGIO CATTORETTI, AND RICCARDO DALLA-FAVERA 288

Elevated Levels of Soluble CD27 in Sera of Primary Sjögren's Syndrome Patients Correlate Positively with Serum Concentration of IgG: A Result of Abnormal B Cell Differentiation? *By* JANNE Ø. BOHNHORST, JØRN E. THOEN, ROLAND JONSSON, JACOB B. NATVIG, AND KEITH M. THOMPSON 291

IL-1β and TNF-α Produce Divergent Acute Inflammatory and Skeletal Lesions in the Knees of Lewis Rats. By GIUSEPPE CAMPAGNUOLO, BRAD BOLON, LI ZHU, DIANE DURYEA, ULRICH FEIGE, AND DEBRA ZACK 295

Correlation of Innate Immune Response with IgA against *Gardnerella vaginalis* Cytolysin in Women with Bacterial Vaginosis. *By* SABINA CAUCI, SECONDO GUASCHINO, DOMENICO DE ALOYSIO, SILVIA DRIUSSI, DAVIDE DE SANTO, PAOLA PENACCHIONI, ALINE BELLONI, PAOLO LANZAFAME, AND FRANCO QUADRIFOGLIO 299

B Cell Chronic Lymphocytic Leukemia: Distinct Phenotypic Features and Replicative Histories among Subgroups. *By* RAJENDRA N. DAMLE, FABIO GHIOTTO, ANGELO VALETTO, EMILIA ALBESIANO, STEVEN L. ALLEN, PHILIP SCHULMAN, VINCENT P. VINCIGUERRA, KANTI R. RAI, MANLIO FERRARINI, AND NICHOLAS CHIORAZZI 302

New Structural Model for HLA-Restricted Presentation of Nickel to T Cells. *By* KATHARINA GAMERDINGER, CORINNE MOULON, AND HANS ULRICH WELTZIEN 305

Why Are Mice with Targeted Mutation of Co-Stimulatory Molecules Prone to Autoimmune Disease? *By* YANG LIU, JIAN-XIN GAO, HUIMING ZHANG, XUE-FENG BAI, JING WEN, XINCHENG ZHENG, JINQUING LIU, AND PAN ZHENG 307

Facets of Dendritic Cell Maturation: Environmental Instructions Lead to Distinct Transcriptional Pattern in Myeloid Dendritic Cells. *By* DAVORKA MESSMER, BRADLEY MESSMER, AND NICHOLAS CHIORAZZI . 309

IRTA Family Proteins: Transmembrane Receptors Differentially Expressed in Normal B Cells and Involved in Lymphomagenesis. *By* IRA MILLER, GEORGIA HATZIVASSILIOU, GIORGIO CATTORETTI, CATHY MENDELSOHN, AND RICCARDO DALLA-FAVERA 312

Transcriptional Deregulation of Mutated BCL6 Alleles by Loss of Negative Autoregulation in Diffuse Large B Cell Lymphoma. *By* LAURA PASQUALUCCI, ANNA MIGLIAZZA, B. HILDA YE, AND RICCARDO DALLA-FAVERA 314

Analysis of the Structure of Proteasome-Proliferating Cell Nuclear Antigen (PCNA) Multiprotein Complex and Its Autoimmune Response in Lupus Patients. *By* YOSHINARI TAKASAKI, KAZUHIKO KANEDA, KEN TAKEUCHI, RAN MATSUDAIRA, MASAKAZU MATSUSHITA, HIROFUMI YAMADA, MASUYUKI NAWATA, KEIGO IKEDA, AND HIROSHI HASHIMOTO 316

Ethyl Pyruvate Protects against Lethal Systemic Inflammation by Preventing HMGB1 Release. *By* LUIS ULLOA, MITCHELL P. FINK, AND KEVIN J. TRACEY 319

Central Tolerance in a Prostate Cancer Model TRAMP Mouse. *By* PAN ZHENG, XIN-CHENG ZHENG, JIAN-XIN GAO, HUIMING ZHANG, TERRENCE GEIGER, AND YANG LIU 322

Long-Lasting Bioactive Substance Mimicking Basic Fibroblast Growth Factor Was Associated with the Heavy Chain(s) of Immunoglobulin G in Serum from Three Patients with Breast Cancer. *By* M.B. ZIMERING AND S. THAKKER-VARIA 324

Index of Contributors 329

Financial assistance was received from:

Co-Sponsor

- THE HENRY KUNKEL SOCIETY

Major Funders

- CARE PLUS HEALTH PLAN
- MUSHETT FAMILY FOUNDATION INC.
- WINDREF RESEARCH INSTITUTE, ST. GEORGE'S UNIVERSITY

Supporter

- THE JEFFREY MODELL FOUNDATION

Contributors

- ABBOTT LABORATORIES
- AMERICAN AUTOIMMUNE RELATED DISEASES ASSOCIATION, INC.
- BIOGEN
- BOEHRINGER INGELHEIM PHARMACEUTICALS INC.
- COLEY PHARMACEUTICAL GROUP
- IMMUNEX CORPORATION
- PAN AMERICAN LEAGUE AGAINST RHEUMATISM
- PFIZER INC.

Preface

Over the past thirty years, our understanding about the basic mechanisms underlying the immune response has been consistently and progressively evolving. This understanding has been propelled by several seminal observations, including, among others, the demonstration of specialized antigen processing and presenting cells that initiate the first stages of an immune response, the definition of the molecules and mechanisms whereby immune cells interact in an adaptive immune response, and the identification of the molecules and processes that permit the generation of antigen-binding diversity in the immune system after productive cellular interactions. More recently there has been a rebirth of interest and understanding of innate immune responses, how these responses augment adaptive immune responses, and how the two responses combine to afford protective immunity.

Of key importance is the relationship between these critical immune mechanisms and disease. The Henry Kunkel Society is dedicated to fostering investigation that addresses this relationship. The Society also recognizes the diverse yet interconnected ways that knowledge about the immune system evolves and, in particular, its evolution from studying both normal and disease states. Professor Henry G. Kunkel of The Rockefeller University demonstrated that important immunologic principles can be discerned by studying disease (in his case, liver diseases such as autoimmune hepatitis, B cell lymphoproliferative disorders such as multiple myeloma, and autoimmune disorders such as rheumatoid arthritis and systemic lupus erythematosus), and that once defined, from whatever context, these principles should be aggressively applied to more precisely understand and treat the diseases from which some of the information was born. The Society has striven to promote this paradigm by creating an environment whereby immunologists who focus on both human and animal systems can meet and share their knowledge, with the eventual goal of applying it to disease states.

In this spirit, The Henry Kunkel Society and the New York Academy of Sciences partnered to sponsor the conference entitled "Immune Mechanisms and Disease." This meeting, which also was the Xth annual scientific session of The Henry Kunkel Society, was held in April 2002 in St. George's, Grenada, in the West Indies. The conference began with the presentation of the Xth Henry Kunkel Society Lecture by Professor Tasuku Honjo of Kyoto University, Japan. This was followed over the next three days by four major sessions that dealt with important immune mechanisms, including antigen presentation, induction of adaptive and autoimmune immune responses, the germinal center reaction, and clinical abnormalities related to the germinal center reaction; a fifth major session was devoted to novel therapies for immunologically relevant disorders and emerging immune-based therapies for various illnesses. These topics were discussed by a series of invited, internationally recognized experts in these fields. In the workshop sessions, shorter presentations dealing with these issues were delivered by attendees chosen to speak on the basis of abstracts submitted in advance of the meeting. Most of these presentations follow in this volume.

Ann. N.Y. Acad. Sci. 987: xi–xii (2003). © 2003 New York Academy of Sciences.

We hope that these papers will be a valuable tool to understand the specific immune mechanisms presented and the diseases to which they relate. We also hope that they indicate the commitment of The Henry Kunkel Society and the New York Academy of Sciences to nourishing and facilitating the ongoing productive interactions of basic and clinical investigators who have interests in immunology and other biomedical sciences.

—NICHOLAS CHIORAZZI
—ROBERT G. LAHITA
—J. DONALD CAPRA
—MANLIO FERRARINI
—JOHN B. ZABRISKIE

Activation-Induced Cytidine Deaminase Links Class Switch Recombination and Somatic Hypermutation

ILMI OKAZAKI,[a] KIYOTSUGU YOSHIKAWA,[a,b] KAZUO KINOSHITA,[a] MASAMICHI MURAMATSU,[a] HITOSHI NAGAOKA,[a] AND TASUKU HONJO[a]

[a]*Department of Medical Chemistry and Molecular Biology, Graduate School of Medicine, and* [b]*Radioisotope Research Center, Kyoto University, Yoshida Konoe-cho, Sakyo-ku, Kyoto, 606-8501, Japan*

ABSTRACT: Activation-induced cytidine deaminase (AID), a putative RNA-editing cytidine deaiminase that is expressed strictly in activated B cells, is indispensable in three apparently distinct genetic alterations of immunoglobulin genes—namely, class switch recombination, somatic hypermutation, and gene conversion. Recent findings led us to propose a common DNA cleaving mechanism, in which the transient secondary structure of the S and V region DNA is recognized by a nicking enzyme regulated by the putative RNA-editing activity of AID.

KEYWORDS: activation-induced cytidine deaminase; class switch recombination; somatic hypermutation; gene conversion; S regions

INTRODUCTION

Immunoglobulin (Ig) genes are highly diversified through a series of genetic alterations. During the early stages of development, B lymphocytes acquire the primary repertoire of their antigen receptors by V(D)J recombination, which assembles variable (V), diversity (D), and joining (J) segments of the V exon of Ig genes. This process is well characterized and known to be mediated by the RAG-1 and RAG-2 proteins.[1] Mature B lymphocytes activated by antigens further diversify their Ig genes through somatic hypermutation (SHM)[2] or gene conversion (GC)[3] in the V region gene and class switch recombination (CSR)[4,5] in the heavy-chain constant (C_H) region gene. SHM introduces untemplated point mutations in the rearranged V region genes, while GC introduces pseudogene templated nucleotide replacements into the functional V genes, both giving rise to antibodies with higher affinity to antigens when coupled with selection by limited amounts of antigens. On the other hand, CSR changes the Ig heavy-chain constant (C_H) region gene from Cμ to one of the other C_H genes, resulting in the generation of antibodies of different isotypes and

Address for correspondence: T. Honjo, Department of Medical Chemistry and Molecular Biology, Graduate School of Medicine, Kyoto University, Yoshida Konoe-cho, Sakyo-ku, Kyoto, 606-8501, Japan. Voice: 81-75-753-4371; fax: 81-75-753-4388.
honjo@mfour.med.kyoto-u.ac.jp

Ann. N.Y. Acad. Sci. 987: 1–8 (2003). © 2003 New York Academy of Sciences.

effector functions with the same antigen specificity. In contrast to V(D)J recombination, the molecular mechanisms of SHM, GC, and CSR have been poorly understood and considered to be distinct from each other. The recent finding that activation-induced cytidine deaminase (AID) is indispensable for all three processes[6–9] has led us to propose an AID-regulated common mechanism to recognize and cleave target DNAs in these events. In this paper, we summarize our recent findings on CSR and SHM, and discuss how AID is involved in these DNA modifications.

IN VITRO MODEL SYSTEM FOR CSR

CSR is accompanied by looping-out deletion of DNA segments between μ switch (S) region and one of downstream S regions located 5′ to each C_H gene.[4,5] CSR can be divided into three steps: (1) selection of a target S region; (2) recognition and cleavage of the targeted S and Sμ regions by a putative recombinase; and (3) repair and ligation of the broken DNA ends by the non-homologous end-joining (NHEJ) repair system.[10–12] Signaling through CD40 and cytokine receptors induces two critical events in CSR: (a) transcription from an intronic (I) promoter located 5′ to each S region (so-called germline transcription)[13-15] and (b) expression of AID. The former determines a target S region to be recombined with Sμ. Expression of AID is restricted to stimulated B lymphocytes,[16] while the NHEJ repair system is constitutively expressed in almost all cell types. AID is likely to be involved in the cleaving step of CSR,[17,18] although it is unknown whether AID itself cleaves DNA or regulates the putative DNA cleaving enzyme.

To dissect the molecular mechanism of CSR, we have developed a series of artificial switch substrates and utilized them in combination with a murine B lymphoma CH12F3-2,[15,19–22] which can switch efficiently from IgM to IgA when stimulated with CD40L and cytokines *in vitro.*[23] Our artificial switch substrates contain the Sμ and Sα regions directed by separate promoters (FIG. 1a). The upstream promoter is constitutively active, whereas the downstream promoter is either constitutive or regulatable by tetracycline. Both S sequences are removed by splicing from each transcript. The coding sequences for the extracellular (EC) domain of CD8α and the sequence for the transmembrane (TM) domain of CD8α fused with green fluorescent protein (GFP) are separated into two transcription units. The EC domain of CD8α could be anchored on the cell surface only after its fusion with the TM domain by recombination between the two S regions, which allows detection of switched cells by fluorescence-activated cell sorting (FACS). CSR in this substrate in CH12F3-2 cells is dependent on both transcription of S regions[15] and stimulation with cytokines,[19] as shown in the endogenous immunoglobulin locus in B cells. Cytokine stimulation can be replaced by exogenous AID expression,[22] indicating that our artificial substrates fulfill another criterion of CSR, that is, dependency on AID. Thus, this system well reflects the properties of physiological CSR.

SELECTION OF TARGET S REGIONS

Specificity of the CSR target S region is determined by cytokines provided by T cells.[24] Each cytokine stimulation activates a specific I promoter and induces germline transcription of a target S region. Such transcription is supposed to be required to open the chromatin structure of a specific S region, rendering it accessible to a putative recombinase.[13,14] However, this is not the only role of germline transcription. Using an artificial switch substrate containing a tetracyucline-inducible promoter as the downstream promoter, we have recently shown that germline transcription levels quantitatively correlate with the CSR efficiency.[15] Germline transcripts per se do not appear to be required for CSR, because CSR is not affected by expression of either germline transcripts in *trans* or excess amounts of *E.coli* RNase H. A plausible explanation for these results is that the secondary structure of the denatured S region, which is transiently formed during transcription, may play important roles.

RECOGNITION AND CLEAVAGE OF S REGIONS

Although CH12F3-2 cells switch almost exclusively to IgA in the endogenous locus, CSR can be induced efficiently in an artificial substrate containing the Sγ or Sε sequences instead of the Sα sequence.[19] In addition, an inverted Sα sequence is also efficient for CSR in CH12F3-2 cells. Thus, the CSR recombinase might recognize other structures of the S region rather than the primary sequence. S regions are composed of abundant palindromic sequences and tandem repetitions of G-rich sequences. Palindromic but not G-rich nor repetitive sequences appeared to be most important for functional S regions. We have shown that palindromic but AT-rich sequences of the *Xenopus* S region can support CSR in the artificial substrate.[20] While vertebrate telomere sequences containing repetitive sequences of G-rich motifs cannot replace the S region for CSR, the multiple cloning sequence composed of many short palindromes can be functional for CSR. These results suggest that the stem-loop structure that can be formed by transient denaturation during active transcription might serve as a recognition target of the CSR recombinase. Furthermore, the distribution of CSR is biased to the proximity of the stem-loop structure in single-stranded S sequences, predicted by a single-stranded DNA folding program.[20,25]

To investigate the nature of cleavage by CSR recombinase, we have developed another artificial substrate, in which inversion-type CSR takes place to retain both recombination breakpoints on the substrate DNA.[21] Sequencing analysis of CSR junctions in this substrate revealed that deletions and duplications of variable lengths exist exclusively at junction sites. This suggests that the initial cleavage of the S region is most likely to be the staggered cleavage and that single-stranded DNA tails generated by such cleavage may be converted to blunt-end by either exonuclease cleavage and/or filling by DNA polymerase(s) so that cleaved ends are repaired by the NHEJ system.

AID IS NOT ONLY REQUIRED BUT SUFFICIENT FOR CSR AND SHM

Not only the sequence homology of AID to APOBEC-1, a catalytic subunit of the RNA-editing complex for the apolipoprotein B messenger RNA precursor, but the *in vitro* cytidine deaminase activity of recombinant AID suggests that it might function as an RNA-editing enzyme.[16] However, the exact function of AID is unknown. Because AID deficiency abolishes CSR and SHM in mice[6] and human,[7] AID is re-

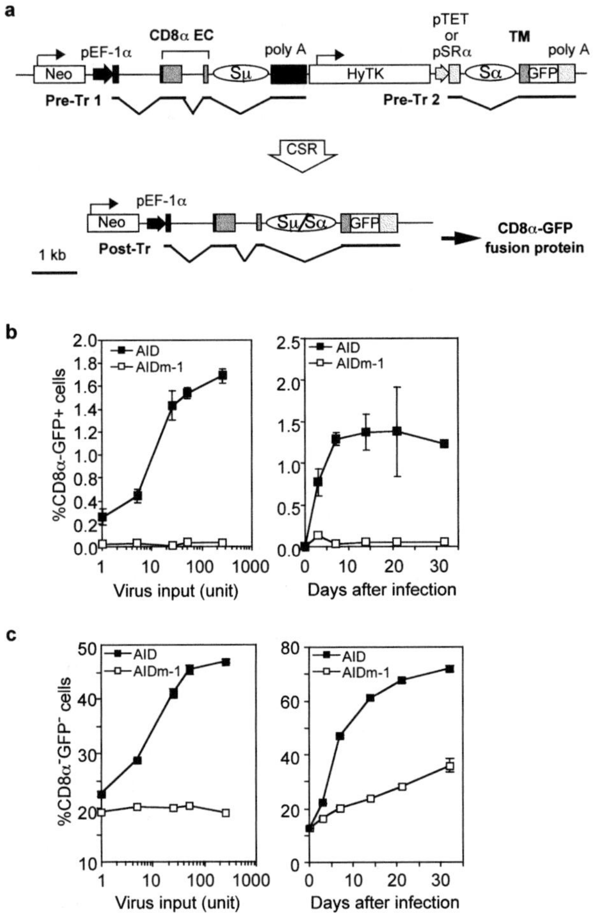

FIGURE 1. Induction of CSR in the artificial substrate in NIH3T3 cells by expression of AID. (**a**) Structure of the CSR substrate. pEF-1α, elongation factor 1α promoter; pTET, tetracycline-responsive promoter; pSRα, SRα promoter; Neo, neomycin resistance gene with the TK promoter; HyTK, hygromycin phosphotransferase-thymidine kinase fusion protein gene with the CMV promoter; Pre-Tr, pre-switch transcripts; Post-Tr, post-switch transcripts; rectangles, exons; arrows, promoters. EC and TM indicate extracellular and transmembrane domains of CD8α, respectively. When deletion occurs between the two S regions by CSR, the CD8α-GFP fusion protein is expressed on the cell surface. (**b, c**) AID-induced accumulation of CD8α-GFP$^+$ (**b**) and CD8α$^-$GFP$^-$ (**c**) at day 7 with different virus amounts (*left*) or at a fixed virus titer (50 units) at different days as indicated (*right*). AIDm-1, a loss-of-function mutant of AID. AID was introduced by retroviral infection into NIH3T3 cells with a single copy of the substrate. The CSR efficiency increased dose dependently on the amount of AID viruses used for infection. CD8α-GFP$^+$ cells were detectable as early as day 3 and reached the maximum level at day 7 after infection (**b**). CD8α$^-$GFP$^-$ cells also increased in an AID-dependent manner during culture of NIH3T3 cells (**c**). TM-GFP fragment sequences in non-rearranged substrate that were amplified from AID-infected cells had frequent mutations (3.1×10^{-3} per bp), whereas those in AIDm-1-infected and uninfected cells had point mutations at the same level with PCR error frequency (3.4×10^{-4} per bp and 5.6×10^{-4} per bp, respectively). (From Okazaki *et al.*[22] Reprinted, with permission from Macmillan Magazines Ltd.)

quired for CSR and SHM. Now the question is whether AID is sufficient for these events. The fact that CSR can be induced in the transcribed substrate in CH12F3-2 cells without any cytokine stimulation[22] suggests that all the other *trans*-acting molecules may be constitutively expressed in B cells; this prompted us to examine whether AID can function in non-B cells such as fibroblasts. Surprisingly, ectopic expression of AID is sufficient to activate CSR in the CSR substrate even in NIH3T3 fibroblasts at a level comparable to that in stimulated CH12F3-2 cells[22] (FIG. 1b). AID-induced CSR in fibroblasts was dependent on target transcription, and their

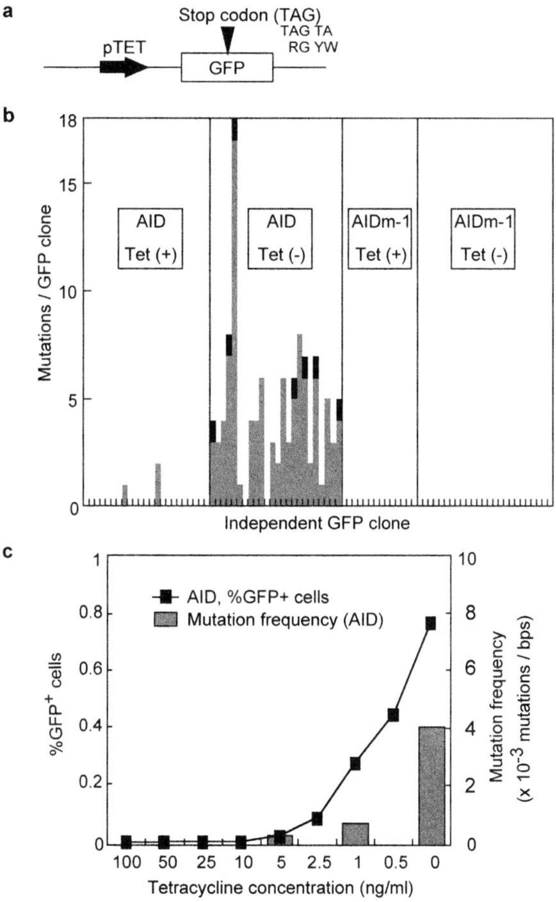

FIGURE 2. Induction of SHM in the pI substrate in NIH3T3 cells by expression of AID. (**a**) Structure of the pI substrate. (**b**) Mutations accumulated in the GFP sequence AID dependently at day 10 after infection. AID was introduced by retroviral infection into NIH3T3 cells with a single copy of the substrate. Gray, point mutations; black, deletions or duplications. (**c**) Frequencies of GFP+ cells (*line*) and mutations in the GFP sequence (*bar*) at day 10 after infection at different concentrations of tetracycline at a fixed virus titer (500 units). (From Yoshikawa *et al.*[27] Reprinted, with permission, from The American Association for the Advancement of Science.)

CSR products had features similar to CSR in the endogenous immunoglobulin loci in B cells. In addition, expression of AID induced accumulation of a large number of point mutations in the actively transcribed GFP sequence of the CSR substrate, suggesting that AID is also sufficient to induce SHM in fibroblasts (FIG. 1c).

Recently, we have shown that ectopic expression of AID can induce SHM in an SHM assay construct pI[26] which carries a mutant GFP sequence driven by the tetracycline-responsive promoter[27] (FIG. 2a, b). The mutant GFP has a premature stop codon (TAG) in an RGYW [R, purine (A/G); Y, pyrimidine (C/T); W, A/T] motif, a known SHM hot spot.[28] The appearance of GFP-positive revertant cells, which is an indication of mutations in this hot spot, as well as the frequency of mutations in the whole GFP sequence was closely correlated with the transcription level of the target (FIG. 2c). Furthermore, AID-induced mutations were biased not only to the RGYW/WRCY motif but also to the stem-loop structure in the GFP sequence as shown in the V region of the immunoglobulin gene.[29] These results imply that all the *trans*-acting factors required for the CSR and SHM reaction, except AID, are expressed constitutively and ubiquitously. Although it is still unclear whether AID itself cleaves DNA or regulates a putative DNA-cleaving enzyme, this finding supports the idea that AID is involved in the cleaving step of CSR and SHM.

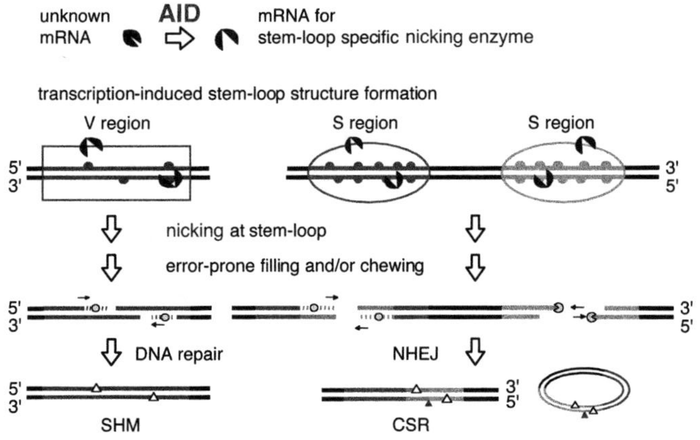

FIGURE 3. Model for the regulation of class-switch recombination and somatic hypermutation by AID. V and S regions are indicated by a *rectangle* and *ovals*, respectively, superimposed with their predicted secondary structures indicated by half circles. AID probably edits pre-mRNA for an unknown but ubiquitously expressed enzyme (*scissors*), converting it to mRNA for a nickase specific to the stem-loop structure (*open-mouth scissors*). Nicking of S-region DNA occurs frequently at different positions on both strands of DNA, giving rise to double-stranded cleavages with staggered ends. Nicking of V-region DNA is less frequent, generating mostly single-stranded breaks. Single-stranded tails are either digested by exonuclease(s) (*white pacman*) or double-stranded by error-prone DNA polymerases (*white circle*), giving rise to deletion (*black triangles*), duplication and/or mutations (*white triangles*) near cleavage sites. Processed ends in S regions are ligated by NHEJ pathway, whereas those in V regions are repaired by an unknown mechanism other than NHEJ. (Modified from Honjo *et al.*[4])

MODEL FOR CSR AND SHM

CSR and SHM share many common features although their targets and products are quite distinct: (a) AID dependency, (b) positive correlation of target transcription with genetic alterations, and (c) association of DNA cleavage with mutations.[4] Furthermore, the stem-loop structure in the V region is also speculated to be the recognition target for SHM.[29] However, the joining system of cleaved DNA ends is known to be different in CSR and SHM. All of these features support the following scenario (FIG. 3): AID is likely to modify an unknown but ubiquitously expressed pre-mRNA thorough its RNA-editing activity to generate mRNA encoding a nickase that specifically recognizes the stem-loop structure. AID can function in concert with putative cofactors, which are expressed constitutively and ubiquitously. Transcription of the S and V regions, which contain palindromic sequences, leads to transient denaturation, forming the stem-loop structure that is cleaved by the AID-regulated nickase. Single-strand tails of cleaved ends of S regions will be processed by error-prone DNA polymerase-mediated gap-filling or exonuclease-mediated reaction. CSR ends are then ligated by the NHEJ system. SHM nicked ends provide entry sites for exonuclease and error-prone DNA polymerase, followed by a ligation system. Mismatched bases will be corrected or fixed by mismatch repair enzymes.

ACKNOWLEDGMENTS

We thank Y. Shiraki for secretarial help. This work was supported by a Center of Excellence Grant from the Ministry of Education, Culture, Sports, Science and Technology of Japan. I.O. is a research fellow of the Japan Society for the Promotion of Science.

REFERENCES

1. HESSLEIN, D.G. & D.G. SCHATZ. 2001. Factors and forces controlling V(D)J recombination. Adv. Immunol. **78:** 169–232.
2. NEUBERGER, M.S. & C. MILSTEIN. 1995. Somatic hypermutation. Curr. Opin. Immunol. **7:** 248–254.
3. REYNAUD, C.A. *et al.* 1987. A hyperconversion mechanism generates the chicken light chain preimmune repertoire. Cell **48:** 379–388.
4. HONJO, T., K. KINOSHITA & M. MURAMATSU. 2002. Molecular mechanism of class switch recombination: linkage with somatic hypermutation. Annu. Rev. Immunol. **20:** 165–196.
5. KINOSHITA, K. & T. HONJO. 2001. Linking class-switch recombination with somatic hypermutation. Nat. Rev. Mol. Cell Biol. **2:** 493–503.
6. MURAMATSU, M. *et al.* 2000. Class switch recombination and hypermutation require activation-induced cytidine deaminase (AID), a potential RNA editing enzyme. Cell **102:** 553–563.
7. REVY, P. *et al.* 2000. Activation-induced cytidine deaminase (AID) deficiency causes the autosomal recessive form of the hyper-IgM syndrome (HIGM2). Cell **102:** 565–575.
8. ARAKAWA, H., J. HAUSCHILD & J.M. BUERSTEDDE. 2002. Requirement of the activation-induced deaminase (AID) gene for immunoglobulin gene conversion. Science **295:** 1301–1306.
9. HARRIS, R.S. *et al.* 2002. AID is essential for immunoglobulin V gene conversion in a cultured B cell line. Curr. Biol. **12:** 435–438.

10. ROLINK, A., F. MELCHERS & J. ANDERSSON. 1996. The SCID but not the RAG-2 gene product is required for Sμ-Sε heavy chain class switching. Immunity **5:** 319–330.
11. CASELLAS, R. *et al.* 1998. Ku80 is required for immunoglobulin isotype switching. EMBO J. **17:** 2404–2411.
12. MANIS, J.P. *et al.* 1998. Ku70 is required for late B cell development and immunoglobulin heavy chain class switching. J. Exp. Med. **187:** 2081–2089.
13. STAVNEZER, N.J. & S. SIRLIN. 1986. Specificity of immunoglobulin heavy chain switch correlates with activity of germline heavy chain genes prior to switching. EMBO J. **5:** 95–102.
14. YANCOPOULOS, G.D. *et al.* 1986. Secondary genomic rearrangement events in pre-B cells: VHDJH replacement by a LINE-1 sequence and directed class switching. EMBO J. **5:** 3259–3266.
15. LEE, C.G. *et al.* 2001. Quantitative regulation of class switch recombination by switch region transcription. J. Exp. Med. **194:** 365–374.
16. MURAMATSU, M. *et al.* 1999. Specific expression of activation-induced cytidine deaminase (AID), a novel member of the RNA-editing deaminase family in germinal center B cells. J. Biol. Chem. **274:** 18470–18476.
17. PETERSEN, S. *et al.* 2001. AID is required to initiate Nbs1/γ-H2AX focus formation and mutations at sites of class switching. Nature **414:** 660–665.
18. NAGAOKA, H. *et al.* 2002. Activation-induced deaminase (AID)-directed hypermutation in the immunoglobulin Smu region: implication of AID involvement in a common step of class switch recombination and somatic hypermutation. J. Exp. Med. **195:** 529–534.
19. KINOSHITA, K. *et al.* 1998. Target specificity of immunoglobulin class switch recombination is not determined by nucleotide sequences of S regions. Immunity **9:** 849–858.
20. TASHIRO, J., K. KINOSHITA. & T. HONJO. 2001. Palindromic but not G-rich sequences are targets of class switch recombination. Int. Immunol. **13:** 495–505.
21. CHEN, X., K. KINOSHITA & T. HONJO. 2001. Variable deletion and duplication at recombination junction ends: implication for staggered double-strand cleavage in class-switch recombination. Proc. Natl. Acad. Sci. USA **98:** 13860–13865.
22. OKAZAKI, I. *et al.* 2002. The AID enzyme induces class switch recombination in fibroblasts. Nature **416:** 340–345.
23. NAKAMURA, M. *et al.* 1996. High frequency class switching of an IgM+ B lymphoma clone CH12F3 to IgA+ cells. Int. Immunol. **8:** 193–201.
24. STAVNEZER, J. 1996. Antibody class switching. Adv. Immunol. **61:** 79–146.
25. MUSSMANN, R. *et al.* 1997. Microsites for immunoglobulin switch recombination breakpoints from Xenopus to mammals. Eur. J. Immunol. **27:** 2610–2619.
26. BACHL, J. *et al.* 2001. Increased transcription levels induce higher mutation rates in a hypermutating cell line. J. Immunol. **166:** 5051–5057.
27. YOSHIKAWA, K. *et al.* 2002. AID enzyme-induced hypermutation in an actively transcribed gene in fibroblasts. Science **296:** 2033–2036.
28. ROGOZIN, I.B. & N.A. KOLCHANOV. 1992. Somatic hypermutagenesis in immunoglobulin genes. II. Influence of neighbouring base sequences on mutagenesis. Biochim. Biophys. Acta **1171:** 11–18.
29. KOLCHANOV, N.A., V.V. SOLOVYOV & I.B. ROGOZIN. 1987. Peculiarities of immunoglobulin gene structures as a basis for somatic mutation emergence. FEBS Lett. **214:** 87–91.

Creation versus Destruction of T Cell Epitopes in the Class II MHC Pathway

C. WATTS, C.X. MOSS, D. MAZZEO, M.A. WEST, S.P. MATTHEWS, D.N. LI, AND B. MANOURY

Division of Cell Biology & Immunology, School of Life Sciences, University of Dundee, DD1 5EH, United Kingdom

ABSTRACT: Proteases perform two key roles in the class II MHC antigen processing pathway. They initiate removal of the invariant chain chaperone for class II MHC and they generate peptides from foreign and self proteins for eventual capture and display to T cells. How a balance is achieved between generation of suitable peptides versus their complete destruction in an aggressive proteolytic environment is not known. Nor is it known in most cases which proteases are actually involved in antigen processing. Our recent studies have identified asparagine endopeptidase (AEP or legumain) as an enzyme that contributes to both productive and destructive antigen processing in the class II MHC pathway. The emerging consensus seems to be that individual proteolytic enzymes make clear and non-redundant contributions to antigen processing.

KEYWORDS: antigen processing; class II MHC; proteases; AEP

INTRODUCTION

In the class I MHC loading pathway, the proteolytic machinery that generates peptides, primarily the proteasome, is separated from the machinery that catalyzes the loading of peptides onto nascent class I MHC molecules. The key events of peptide generation and peptide loading are separated by the membrane of the endoplasmic reticulum (ER) and rely on the TAP transporter to link them. Although recent studies show that N-terminal trimming of peptides for final class I MHC "fitting" is performed by ER localized aminopeptidases,[1,2] the situation is quite different to that which exists in the class II MHC pathway. Here the entire process of antigen processing and peptide capture by class II MHC takes place within the same compartmental system.[3,4] Moreover, many of the enzymes present within the endosomes and lysosomes of antigen presenting cells are rather non-specific in terms of their cleavage specificity.[5,6] The class II MHC loading system therefore faces challenges not faced by the class I MHC system. In particular, how can T cell epitopes survive and

Address for correspondence: C. Watts, Division of Cell Biology & Immunology, School of Life Sciences, University of Dundee, DD1 5EH, UK. Voice: 44-1382-344233; fax: 44-1382-345783.
c.watts@dundee.ac.uk

Ann. N.Y. Acad. Sci. 987: 9–14 (2003).

be loaded onto class II MHC in such an aggressive proteolytic environment? Other questions also can be posed: are specific enzymes required for optimal processing and loading of T cell epitopes; in other words, how redundant is antigen processing in the class II MHC pathway? Also, are the same or different enzymes used to process antigens and remove the invariant chain? Recently we have made progress in answering some of these questions in studies that have also identified a novel cysteine protease found in antigen presenting cells.

Processing of the Tetanus Toxin Antigen

For a number of years we have used the tetanus toxin antigen as a model system to analyze processing in the class II MHC pathway (see Ref. 7 for review). Surprisingly, *in vitro* lysosomal digestion of the 47kDa C-terminal domain of the tetanus toxin antigen (TTCF) revealed that one processing enzyme dominated the digestion pattern in spite of the presence of many other lysosomal enzymes.[8,9] This enzyme turned out to be a novel cysteine protease with unusually strict cleavage specificity, in this case for asparagine residues.[8] Although it had not been observed before in mammalian cells, a homologous enzyme had been previously described in plant cells and named legumain.[10] We refer to the mammalian enzyme simply as AEP for asparagine endopeptidase. Inhibition of AEP activity or mutagenesis of AEP cleavage sites in tetanus toxin slowed down the kinetics of antigen presentation, demonstrating that at least for this antigen, efficient presentation required the action of this specific enzyme.[9] Because presentation of several T cell epitopes in TTCF appeared to be dependent on cleavage of one or two AEP sites,[11] we have suggested that the initial action of AEP on the TTCF substrate may be to "unlock" the native antigen structure.[4] AEP activity may or may not be important for processing other antigens but the same idea could apply to other processing enzymes. In other words, different substrates may be preferentially cleaved by different members of the repertoire of enzymes available in antigen presenting cells.

How much processing is needed before binding to class II MHC occurs? Earlier studies have shown that class II MHC can bind long peptides and even intact proteins.[12–14] However, it has been more difficult to determine whether this actually happens during normal antigen processing; intriguing functional[15] and biochemical data[16,17] exist to support the notion that class II MHC can capture long peptides. Our more recent data (C.X. Moss *et al.*, in preparation) indicates that in the case of the TTCF substrate, very limited processing by AEP is sufficient for generation of a form of antigen able to bind to available class II MHC molecules. In other words, capture by class II MHC can be an early event involving long, unfolded stretches of the antigen substrate. A key feature of the class II MHC protein is the open-ended peptide binding groove that allows extended peptides to bind.[18] This may have evolved to allow early capture of extended peptides and thereby provide protection from further proteolysis. Further exemplification of this result for other antigen substrates will be needed, but we can begin to see how the problem of epitope destruction can be solved in the class II MHC pathway. This is illustrated diagrammatically in FIGURE 1. It may be useful to think of three types of proteases: (1) initiating or unlocking proteases; (2) endoproteases that preferentially act on partially processed and unfolded antigen; and (3) exopeptidases that trim long MHC-bound peptides to their normal size range (12–25 residues).

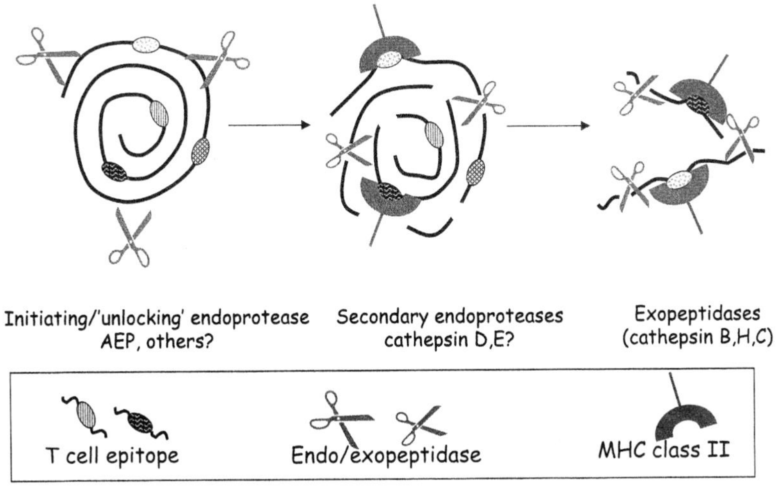

FIGURE 1. Model for antigen processing. The cartoon emphasizes that capture of processed antigen by class II MHC can and may need to be an early event to minimize T cell epitope destruction. It is suggested that a limited number of enzymes may be able to initiate antigen processing leading to unfolding and binding to the open-ended groove of class II MHC. Secondary enzymes may act on the partially denatured substrate to create other sequences for class II binding and to initiate trimming, which is then completed by amino and carboxyexopeptidases.

Destructive Processing and Escape of Autoreactive T Cells from Tolerance Induction

Inevitably, some potential T cell epitopes will contain protease cleavage sites and will be destroyed during antigen processing. However, the immunological consequences of this have not been established. In the case of foreign antigens, it is simply the price paid for antigen processing: in order to generate T cell epitopes, a few may have to be destroyed. However, in the case of self-proteins, this "destructive processing" may have other more serious consequences, especially if other factors come into play later on which modulate this "crypticity" of a self T cell determinant.[19] We have recently identified a clear example of a self-epitope whose level of presentation correlates inversely with the activity of the AEP protease. Inhibition of AEP results in enhanced presentation and overexpression of AEP in diminished presentation as shown in FIGURE 2 (see also Ref. 20). The reason for this is that the epitope in question, MBP 85-99, contains a very good AEP processing site at Asn 94,[20,21] which effectively targets its destruction. MBP is expressed in the thymus,[22,23] leading to tolerance induction of some MBP-specific T cells.[24,25] However, tolerance is not established to the MBP 85-99 region, and we have suggested this is because AEP, which is also expressed in the thymus,[20] destroys it. Perhaps this is a general mechanism that allows some autoreactive T cells escape thymic tolerance induction. How then can this epitope, rendered "cryptic" through destructive processing, be presented as a target for autoreactive T cells? We do not yet know, and important ques-

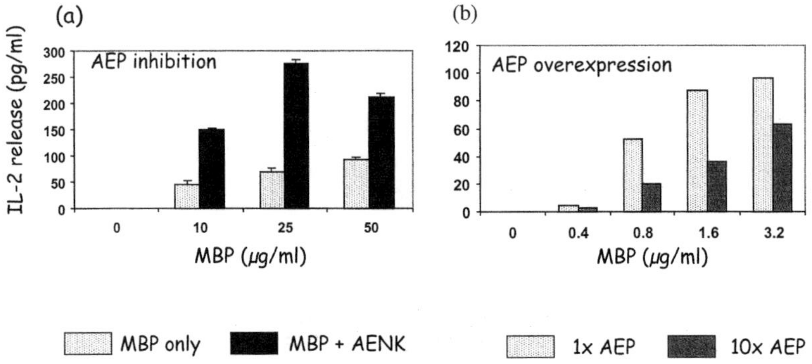

FIGURE 2. Presentation of MBP 85-99 is inversely proportional to AEP activity. (**a**) Prior to pulsing with MBP, the cell line MGAR was treated with or without the AEP inhibitor AENK. Presentation of the peptide MBP 85-99 was detected using the T cell hybridoma Ob15. (**b**) The cell line MelJuSO.SD1 with doxycycline-regulatable AEP expression was transfected with the human class II MHC β-chain DRB1*1501 and cultured with (low AEP) or without (high AEP) doxycycline prior to pulsing with MBP. Presentation of the MBP 85-99 was detected as in (**a**). (Modified from Manoury *et al.*,[20] with permission.)

tions remain to be answered concerning the level of AEP expression, for example, in the central nervous system, where MBP 85-99 reactive T cells are thought be responsible for immunopathology.

Pathogen Interference with the Class II MHC Pathway

There are well-documented examples of pathogen interference with the class I MHC pathway, with some of the best examples provided by viruses.[26] Fewer examples exist in the class II MHC pathway although it is well-documented that certain organisms such as *Mycobacteria* perturb the normal maturation of the phagocytic compartments in which they reside and by doing so interfere with class II MHC presentation.[27] The proteolytic enzymes that are responsible for antigen processing and invariant chain removal would be reasonable targets for pathogen-encoded gene products seeking to interfere with antigen processing and class II MHC function. Indeed, there is evidence that under normal conditions the antigen presenting activity of immature dendritic cells is regulated in part by expression of natural cysteine protease inhibitors called cystatins.[28] We have recently shown in collaboration with R.M. Maizels's group that cystatin homologues of the nematode parasite *Brugia malayi* are able to block antigen processing in the class II MHC pathway *in vitro.* [29] Interestingly, the Brugia cystatin named *Brugia malayi* cysteine protease inhibitor-2 (Bm-CPI-2), inhibits both papain family cysteine proteases such as cathepsins B and S, and AEP. Like mammalian cystatin C, AEP inhibition is mediated by a distinct site located distant from the cathepsin binding site.[30] Whether Bm CPI-2 perturbs immune responses against this pathogen *in vivo* remains to be shown.

CONCLUSIONS

Redundancy of protease function in the class II MHC pathway appears to be less than might have been suspected. This conclusion is not only based on our work with AEP but on a number of other recent reports on the effects of chemical or genetic ablation of other proteolytic enzymes in the class II MHC pathway such as cathepsins B, S, and L.[31–33] Individual enzymes can apparently control both the production and destruction of specific T cell epitopes, raising the possibility that they may be targets for therapeutic intervention, for example, in allergy or autoimmunity. Indeed, inhibitors against cathepsin S are already being tested for their immunosuppressive capacity.[34,35] Future studies will examine whether other enzymes in this pathway can be selectively ablated to influence immunological outcomes.

ACKNOWLEDGMENTS

This work is supported by The Wellcome Trust, the European Union, and by Medivir UK Ltd. The authors thank collaborators R. Maizels, D. Wraith, and L. Fugger for reagents and stimulating discussions.

REFERENCES

1. SERWOLD, T. *et al.* 2002. ERAAP customizes peptides for MHC class I molecules in the endoplasmic reticulum. Nature **419:** 480–483.
2. York, I.A. *et al.* 2002. The ER aminopeptidase ERAP1 enhances or limits antigen presentation by trimming epitopes to 8–9 residues. Nat. Immunol. **3:** 1177–1184.
3. LENNON-DUMENIL, A.M. *et al.* 2002. A closer look at proteolysis and MHC-class-II-restricted antigen presentation. Curr. Opin. Immunol. **14**: 15-21.
4. WATTS, C. 2001. Antigen processing in the endocytic compartment. Curr. Opin. Immunol. **13:** 26–31.
5. RIESE, R.J. & H.A. CHAPMAN. 2000. Cathepsins and compartmentalization in antigen presentation. Curr. Opin. Immunol. **12:** 102–113.
6. TURK, B., D. TURK & V. TURK. 2000. Lysosomal cysteine proteases: more than scavengers. Biochim. Biophys. Acta **1477:** 98–111.
7. WATTS, C. *et al.* 1998. Modulation by epitope-specific antibodies of class II MHC-restricted presentation of the tetanus toxin antigen. Immunol. Rev. **164:** 11–16.
8. CHEN, J.M. *et al.* 1997. Cloning, isolation, and characterization of mammalian legumain, an asparaginyl endopeptidase. J. Biol. Chem. **272:** 8090–8098.
9. MANOURY, B. *et al.* 1998. An asparaginyl endopeptidase processes a microbial antigen for class II MHC presentation [see comments]. Nature **396:** 695–699.
10. KEMBHAVI, A.A. *et al.* 1993. The two cysteine endopeptidases of legume seeds: purification and characterization by use of specific fluorometric assays. Arch. Biochem. Biophys. **303:** 208–213.
11. ANTONIOU, A.N. *et al.* 2000. Control of antigen presentation by a single protease cleavage site. Immunity **12:** 391–398.
12. LEE, P., G.R. MATSUEDA & P.M. ALLEN. 1988. T cell recognition of fibrinogen. A determinant on the A alpha-chain does not require processing. J. Immunol. **140:** 1063–1068.
13. SETTE, A. *et al.* 1989. Capacity of intact proteins to bind to MHC class II molecules. J. Immunol. **143:** 1265–1267.
14. BUSCH, R. *et al.* 1996. Invariant chain protects class II histocompatibility antigens from binding intact polypeptides in the endoplasmic reticulum. EMBO J. **15:** 418–428.

15. DENG, H. *et al.* 1993. Determinant capture as a possible mechanism of protection afforded by major histocompatibility complex class II molecules in autoimmune disease. J. Exp. Med. **178:** 1675–1680.
16. DAVIDSON, H.W. *et al.* 1991. Processed antigen binds to newly synthesized MHC class II molecules in antigen-specific B lymphocytes. Cell **67:** 105–116.
17. CASTELLINO, F. *et al.* 1998. Large protein fragments as substrates for endocytic antigen capture by MHC class II molecules. J. Immunol. **161:** 4048–4057.
18. STERN, L.J. & D.C. WILEY. 1994. Antigenic peptide binding by class I and class II histocompatibility proteins. Structure **2:** 245–251.
19. SERCARZ, E.E. *et al.* 1993. Dominance and crypticity of T cell antigenic determinants. Annu. Rev. Immunol. **11:** 729–766.
20. MANOURY, B. *et al.* 2002. Destructive processing by asparagine endopeptidase limits presentation of a dominant T cell epitope in MBP. Nat. Immunol. **3:** 169–174.
21. BECK, H. *et al.* 2001. Cathepsin S and an asparagine-specific endoprotease dominate the proteolytic processing of human myelin basic protein in vitro. Eur. J. Immunol. **31:** 3726–3736.
22. HEATH, V.L. *et al.* 1998. Intrathymic expression of genes involved in organ specific autoimmune disease. J. Autoimmun. **11:** 309–318.
23. PRIBYL, T.M. *et al.* 1993. The human myelin basic protein gene is included within a 179-kilobase transcription unit: expression in the immune and central nervous systems. Proc. Natl. Acad. Sci. USA **90:** 10695–10699.
24. TARGONI, O.S. & P.V. LEHMANN. 1998. Endogenous myelin basic protein inactivates the high avidity T cell repertoire. J. Exp. Med. **187:** 2055–2063.
25. HARRINGTON, C.J. *et al.* 1998. Differential tolerance is induced in T cells recognizing distinct epitopes of myelin basic protein. Immunity **8:** 571–580.
26. PLOEGH, H.L. 1998. Viral strategies of immune evasion. Science **280:** 248–253.
27. MERESSE, S. *et al.* 1999. Controlling the maturation of pathogen-containing vacuoles: a matter of life and death. Nat. Cell Biol. **1:** E183-8. java/Propub/cellbio/ncb1199_E183.fulltext java/Propub/cellbio/ncb1199_E183.abstract.
28. PIERRE, P. & R. MELLMAN. 1998. Development regulation of invariant chain proteolysis controls MHC class II trafficking in mouse dendritic cells. Cell **93:** 1135–1145.
29. MANOURY, B. *et al.* 2001. Bm-CPI-2, a cystatin homolog secreted by the filarial parasite *Brugia malayi*, inhibits class II MHC-restricted antigen processing. Curr. Biol. **11:** 447–451.
30. ALVAREZ-FERNANDEZ, M. *et al.* 1999. Inhibition of mammalian legumain by some cystatins is due to a novel second reactive site. J. Biol. Chem. **272:** 19195–19203.
31. DRIESSEN, C., A.M. LENNON-DUMENIL & H.L. PLOEGH. 2001. Individual cathepsins degrade immune complexes internalized by antigen-presenting cells via Fcgamma receptors. Eur. J. Immunol. **31:** 1592–1601.
32. PLUGER, E.B. *et al.* 2002. Specific role for cathepsin S in the generation of antigenic peptides in vivo. Eur. J. Immunol. **32:** 467–476.
33. HSIEH, C.S. *et al.* 2002. A role for cathepsin L and cathepsin S in peptide generation for MHC class II presentation. J. Immunol. **168:** 2618–2625.
34. RIESE, R.J. *et al.* 1998. Cathepsin S activity regulates antigen presentation and immunity. J. Clin. Invest. **101:** 2351–2363.
35. SAEGUSA, K. *et al.* 2002. Cathepsin S inhibitor prevents autoantigen presentation and autoimmunity. J. Clin. Invest. **110:** 361–369.

Dendritic Cell Function *in Vivo* during the Steady State: A Role in Peripheral Tolerance

RALPH M. STEINMAN,[a] DANIEL HAWIGER,[b] KANG LIU,[a] LAURA BONIFAZ,[a] DAVID BONNYAY,[a] KARSTEN MAHNKE,[a] TOMONORI IYODA,[e] JEFFREY RAVETCH,[c] MADHAV DHODAPKAR,[d] KAYO INABA,[e] AND MICHEL NUSSENZWEIG[e]

The Rockefeller University, Laboratories of [a]Cellular Physiology and Immunology, [b]Molecular Immunology, [c]Molecular Genetics and Immunology, and [d]Human Immunology and Immune Therapy, New York, New York 10021-6399, USA

[e]Department of Animal Development and Physiology, Graduate School of Biostudies, and Department of Zoology, Graduate School of Science Kyoto University, Kyoto 606-8502, Japan

ABSTRACT: The avoidance of autoimmunity requires mechanisms to actively silence or tolerize self reactive T cells in the periphery. During infection, dendritic cells are not only capturing microbial antigens, but also are processing self antigens from dying cells as well as innocuous environmental proteins. Since the dendritic cells are maturing in response to microbial and other stimuli, peptides will be presented from both noxious and innocuous antigens. Therefore it would be valuable to have mechanisms whereby dendritic cells, prior to infection, establish tolerance to those self and environmental antigens that can be processed upon pathogen encounter. In the steady state, prior to acute infection and inflammation, dendritic cells are in an immature state and not fully differentiated to carry out their known roles as inducers of immunity. These immature cells are not inactive, however. They continuously circulate through tissues and into lymphoid organs, capturing self antigens as well as innocuous environmental proteins. Recent experiments have provided direct evidence that antigen-loaded immature dendritic *in vivo* silence T cells either by deleting them or by expanding regulatory T cells. In this way, it is proposed that the immune system overcomes at least some of the risk of developing autoimmunity and chronic inflammation. It is proposed that dendritic cells play a major role in defining immunologic self, not only centrally in the thymus but also in the periphery.

KEYWORDS: dendritic cells; DEC-205; endocytosis; antigen presentation; peripheral tolerance

Address for correspondence: Ralph M. Steinman, The Rockefeller University, Laboratory of Cellular Physiology and Immunology, New York, NY 10021-6399. Voice: 212-327-8106; fax: 212-327-8875.

Steinma@mail.rockefeller.edu

Ann. N.Y. Acad. Sci. 987: 15–25 (2003).

A ROLE FOR DENDRITIC CELLS IN ANTIGEN-SPECIFIC TOLERANCE

The immunoglobulin gene rearrangements that allow lymphocytes to generate a diverse array of antigen receptors also yield lymphocytes with reactivity to self and with harmless environmental proteins. In the case of T cells, self-reactive clones can be silenced during their development in the thymus. Dendritic cells (DCs) likely play a major role in this developmental or central tolerance.[1–3] Central tolerance is nevertheless incomplete. Some self-reactive lymphocytes escape deletion.[4] For other T cells, such as those specific for many innocuous environmental antigens, the thymus may not be given a chance to exert its purgatory role.[5] Peripheral tolerance is therefore a vital component of immune function.

Achieving peripheral tolerance is challenging. For one thing, it is difficult to induce experimentally. The main approach has been to inject mice with large amounts of preprocessed antigenic peptides, typically 100 μg or more.[6–8] If antigen processing only allows 1 in 100–1000 protein molecules to provide peptides for presentation on MHC products,[9] then it would seem that current strategies for achieving tolerance require exposure to 10–100 mg of protein, even for mice. Clinically, there is a need to develop the means to silence the immune response in an antigen-specific fashion, as in the settings of transplantation, allergy, and autoimmunity. Current suppressive agents appear to silence immunity in an antigen-nonspecific fashion, without inducing true immunological tolerance. In a parallel vein, in chronic infections and tumors, ineffective host resistance may reflect in part the exploitation of natural tolerogenic pathways.[10]

The induction of tolerance is emerging as a new theme in the biology of DCs, one that is intrinsically connected to their better-known roles in inducing immunity. These two functions can now both be analyzed by targeting antigens selectively to DCs *in situ*. In the steady state, in the absence of acute infection, antigen-targeted DCs are able to silence the corresponding peripheral T cells. In contrast, if the DCs are induced to differentiate further or mature, through stimulation of Toll-like and TNF-receptor family members, immunity develops. The work summarized here emanates from the collaborations of five laboratories.

DENDRITIC CELL MATURATION AS A PIVOTAL STEP IN THE CONTROL OF IMMUNITY

DCs are located in peripheral tissues where their turnover (except in the case of the epidermis) can be quite rapid, for example, a half-life of less than two days.[11,12] This turnover is associated with the traffic of DCs through tissues into lymphatics and then to the T cell areas of lymphoid tissues, whereupon most DCs seem to die, also with a half-life of less than two days.[12] During this journey, a critical control point of DC function is termed "maturation." It allows the DCs to change from an antigen-capturing but poorly immunogenic cell to a poorly endocytic but highly immunogenic one.[13–15] By the term "immunogenic," one means the induction of effector T cells (such as $CD4^+$ Th1 type helpers, producing IFNγ, and $CD8^+$ cytolytic T lymphocytes) as well as high-affinity memory cells. All these outcomes have been achieved now using mature DCs to immunize humans, either healthy individuals or patients with advanced melanoma.[16–19]

Maturation also creates a dilemma from the perspective of immune tolerance. The risk is as follows. When DCs are capturing microbes and also maturing in response to signals delivered through Toll-like and TNF-receptor family members, they also capture dying self-tissues, for example, cells dying as a result of infection as is characteristic for agents like influenza and Salmonella,[20,21] as well as innocuous environmental proteins.[22] How do maturing DCs focus the immune response on the microbe and not induce immunity to peptides in self-tissues and harmless proteins in the air or intestine (which are essentially part of ourselves). The risk of inducing inappropriate immune response is especially great when one considers that DCs require very small amounts of a TCR stimulus to activate naïve T cells.[15,23] Therefore, there is a considerable risk that the induction of immunity to pathogens could be accompanied by the induction of immunity to self and environmental antigens captured by the DCs.

To solve the dilemma posed by DC maturation, it has been proposed that in the steady state, prior to an encounter with pathogens, DCs are defining immunologic self and inducing tolerance to those proteins (self and environmental) that would later be processed in the setting of infection.[10,24] This field is in its early stages but, currently, tolerance can take place by either a deletional route,[25–27] or through the induction of antigen-dependent regulatory T cells.[28,29] These distinct tolerogenic pathways may reflect the type of DC used to induce tolerance. Deletional tolerance occurs with immature DCs in lymphoid tissues in the steady state, while regulatory tolerance is observed with DCs that are also classified as immature or semi-mature but are generated *ex vivo*. In either case, the induction of peripheral tolerance to self in the steady state allows DCs to carry out their better-known roles in adaptive immunity. For example, in the case of influenza infection of the lung, DCs will mediate resistance to the virus but not leave the recovered individual with autoimmunity to lung tissue or with chronic reactivity to inhaled proteins. Self-tolerance by DCs is therefore a valuable and essential counterpart of immunity by DCs.

RECEPTOR-MEDIATED UPTAKE OF DYING CELLS AND PROTEINS BY DCS IN THE STEADY STATE

In earlier literature, the capture of proteins *in vivo* by dendritic cells in different animals (mice, rats, sheep) in the steady state is described.[30–34] When one isolates from antigen-injected animals the cells capable of stimulating the corresponding antigen-specific T cells, DCs are the main source of activity.[34] However, the animals are not ostensibly immunized. More recent studies have visualized directly the capture by DCs of proteins administered into the airway of mice[22] and of dying epithelial cells from the intestine of rats.[35] These beautiful new findings stimulate research on uptake mechanisms and immunologic consequences. To approach these questions, one requires methods to selectively target experimental antigens to DCs *in situ*.

DEC-205 is an endocytic receptor expressed at high levels by DCs *in situ*, primarily in the lymphoid tissues of mice. This molecule was discovered with a monoclonal antibody developed by George Kraal. It was called NLDC-145, to describe a nonlymphoid DC product of 145 kDa.[36] The molecule recognized by NLDC-145 is expressed at low levels by nonlymphoid DCs, for example, the Langerhans cells of the epidermis; however, the DCs in lymphoid tissues (and also certain epithelia such as the thymic cortical epithelium) are actually the major site for expression. When the corresponding antigen was purified [37] and cloned,[38] it proved to have a molecular weight of 205 kDa, with 10 contiguous C-type lectin domains. Hence, the molecule was renamed DEC-205, a decalectin of molecular weight 205 kDa. Cloning also revealed homology in sequence with a known endocytic receptor, the macrophage mannose receptor; indeed, DEC-205 localized to coated pits and trafficked into the endocytic system of DCs.[38]

The natural ligands for DEC-205 are currently unknown. Therefore, antibodies to DEC-205 have been used as surrogate ligands to present peptides from the Ig[38,39] or peptides from antigens linked to the antibody.[25,26] The use of antibodies as surrogate ligands was pioneered by Chesnut and Grey, who showed that mouse B cells efficiently presented peptides from rabbit anti-mouse Ig to rabbit Ig-specific T cells.[40]

Studies with anti-DEC-205 are showing that this receptor is functionally quite different from its macrophage mannose receptor cousin. First, the cytosolic domain of DEC-205 mediates efficient presentation on MHC class II products, 30–100 times more efficiently than presentation via the cytosolic domain of the MMR or the complete MMR protein.[39] This may be related to an unusual feature of DEC-205, its capacity to recycle into and out of cells through late endosomal, MHC class II–rich compartments;[39] classically, most recycling endocytic receptors traffic through nonproteolytic and peripheral early endosomes. Second, when antibodies to DEC-205 are injected subcutaneously into mice, the Ig targets efficiently and selectively to DCs and leads to presentation of the antibody associated peptides on MHC class II products.[25] In sections through the lymph node, the injected antibody can be visualized in DCs in the T cell areas, while by FACS analysis the antibody is taken up by most DCs but not other cell types.[25] When isolated, the DCs also selectively present antigens to T cells that are specific for the anti-DEC-205 associated peptides.[25] Third, anti-DEC-205 delivers proteins efficiently for presentation on MHC class I.[26] This is the so-called exogenous pathway, whereby endocytosed proteins are presented by a TAP-dependent process, without the standard requirement for endogenous biosynthesis in the DCs.

These special features of DEC-205 are all evident in lymph node DCs in the steady state. DCs in the T cell areas continually take up the surrogate anti-DEC-205 ligands and present the associated antigens, without any ostensible change in DC phenotype, even in the presence of the corresponding antigen specific T cells.[25,26] It remains to be seen whether other receptors are able to function *in vivo* to bring about selective and efficient antigen presentation by DCs. Some other candidate DC receptors for antigen targeting *in vivo* include Langerin,[41] an asialoglycoprotein receptor,[42] DC-SIGN,[43,44] and BDCA-2.[45]

Another uptake pathway, which is likely to be receptor mediated, has just become evident with DCs *in vivo*. We are referring to the uptake of dying cells.[46] Most $CD11c^+$ DCs *in vivo* are able to take up particulates,[12] but in the case of dying cells, only the $CD8^+$ subset is active.[46] The apoptotic cells include self splenocytes, allogeneic cells killed by NK cells, and tumor cells. The selective uptake by $CD8^+$ DCs implies the existence of a receptor pathway. At the moment, CD36 is the only candidate receptor (among 12 tested) that is more abundant at the mRNA level in $CD8^+$ DCs.[46] Importantly, the uptake of dying cells takes place continuously in the steady state, as observed directly by injecting dying self (splenocytes) or tumor cells labeled with the stable fluorochrome, CSFE. The uptake of dying cells is followed by efficient presentation to MHC class I– and class II–restricted T cells.[46]

In summary, immature DCs in the steady state can continuously take up dying cells and soluble proteins. At least one receptor, DEC-205, has been pinpointed for uptake and presentation with high efficiency, including the exogenous pathway to MHC class I. What then is the evidence that this uptake results in peripheral tolerance in the steady state, or reciprocally to immunity, if DC maturation is simultaneously induced along with antigen uptake?

DELETIONAL TOLERANCE INDUCED BY DCs PRESENTING ANTIGENS IN THE STEADY STATE

In the first study of antigens targeted selectively to DCs *in situ*, Hawiger *et al.* engineered the heavy chain of the anti-DEC-205 monoclonal antibody to encode an immunogenic peptide from hen egg lysozyme (HEL); the heavy chain was further modified to comprise mouse constant regions which in turn carried mutations to obviate FcγR binding.[25] The anti-DEC-205 HEL antibody presented HEL efficiently *in vivo*, as assessed with 3A9, HEL-specific and MHC class II–restricted, TCR transgenic T cells. Only 200 ng of antibody, or a few nanograms of peptide, saturated the response of 2×10^6 T cells. It is estimated that $< 2 \times 10^4$ T cells are present for any one specific antigen, so that this antigen-presenting pathway is both efficient and high capacity *in vivo*. The T cells began to proliferate, but within a week most were deleted. Importantly, the animals by days 7 and 21 could no longer be immunized with antigen (HEL peptide) delivered in complete Freund's adjuvant. Therefore, a few nanograms of peptide delivered within an intact protein were able to tolerize a large number of antigen-specific T cells, whereas prior studies without DC targeting have had to use 100 μg or more to silence the immune response.

Two more recent studies have extended the results of Hawiger *et al.* In the work of Bonifaz *et al.*, ovalbumin protein (OVA) was chemically conjugated to rat anti-mouse DEC-205. The antibody:OVA conjugate was presented several hundred times more efficiently than free OVA.[26] Presentation took place on both MHC class I and II products in the steady state. The former was TAP dependent, so that DEC-205 was able to mediate the exogenous pathway to MHC class I *in vivo*. When OT-I, OVA-specific and MHC class I–restricted, TCR transgenic T cells were followed, deletion again took place after an initial burst of proliferation. Furthermore, the mice became solidly tolerant to an injection of OVA in complete Freund's adjuvant, even though only 0.1 μg of OVA conjugated to antibody had been injected.

In a second recent study, Liu *et al.* documented tolerance to OVA delivered within dying cells to DCs in mouse spleen in the steady state.[27] The OVA was taken up by applying protein along with an osmotic shock. This treatment was found to induce apoptosis and as a result efficient capture by $CD8^+$ DCs *in vivo* (see above). When the immunologic consequences of this capture were followed, again the T cells proliferated, followed by deletion and establishment of profound peripheral tolerance. The induction of tolerance was efficient. Approximately one microgram of OVA had been injected intravenously in association with the dying cells, and $>10^6$ T cells would undergo deletional tolerance.

The tolerance outcome in all three examples of antigen targeting to DCs described above could be converted to immunity by delivering the antigen together with the agonistic anti-CD40 monoclonal antibody, developed and provided by Dr. T. Rolink. The T cells differentiated into IFNγ secreting effectors and were primed to respond rapidly to rechallenge with specific antigens. In addition, T cells emigrated into the lung in mice given a CD40 stimulus, whereas in the steady state, the T cells appeared to remain within the circulation and peripheral lymphoid organs.

In summary, the targeting of different antigens to DCs *in vivo* reveals that these cells induce deletional tolerance in the steady state and immunity following maturation. In contrast to prior work, tolerance is achieved with small amounts of protein and is manifest as profound unresponsiveness to challenge *in vivo* with antigen in strong adjuvant. We believe this physiology will help to explain the enigma of peripheral tolerance discussed in the introduction. DCs in the steady state are able to define immunological self, modulating the reactivity of the T cell repertoire; when microbial infection later matures the DCs, the repertoire and the immune attack are selectively focused on the pathogen.

INDUCTION OF REGULATORY (SUPPRESSOR) T CELLS VIA IMMATURE DCs

In the above examples, deletional tolerance was induced by DCs in lymphoid tissues, following selective antigen uptake in the steady state. Another pathway to tolerance has emerged in studies of DCs that are generated *ex vivo* and reinfused. These DCs are either called "immature" or "semi-mature"; this nomenclature will be discussed further below. Nevertheless, the DCs induce antigen-dependent regulatory T cells, able to silence other antigen-dependent effector T cells. This pathway to tolerance is an exciting one, because it has the potential to silence effector cells, as would be valuable in transplantation, allergy, and autoimmunity.

Regulatory or suppressor T cells were discovered in two settings, although their relationship—that is, one or two types of regulatory T (T-reg) cells—is currently unclear. The setting studied by Sakaguchi and by Mason (reviewed in Refs. 47 and 48), originates in the thymus. In addition to its better-known roles in positive and negative selection, the thymus generates $CD25^+$ $CD4^+$ T cells that regulate autoimmunity. Such T cells are readily evident in human peripheral blood.[49–52] $CD25^+$ cells turn off responses to anti-CD3 antibodies in culture, but the $CD25^+$ cells are themselves anergic, poorly responsive (in terms of cell proliferation) to anti-CD3. $CD25^+$ suppressors seem to be generated initially through the presenting function of thymic cortical epithelium.[53] Another setting, first noted by Roncarolo during allogeneic bone marrow transplantation,[54] involves the peripheral induction of T-reg cells (reviewed in Ref. 55). These T-reg produce IL-10 and are anergic (in terms of cell proliferation) even though they suppress responses of other effector T cells. Cells with the properties of T-reg can be induced by immature DCs, generated *ex vivo* from monocytes cultured in GM-CSF and IL-4.[56]

Dhodapkar studied the immune responses in volunteers injected with DCs that had been generated *ex vivo* with GM-CSF and IL-4 and then pulsed with an MHC class I binding peptide from the influenza matrix protein. The hypothesis was that these immature DCs would lack immunogenicity, previously observed with DCs matured by inflammatory monocyte products.[16,17] In the first two volunteers studied, the immature DCs were not "ignored" as predicted. Instead, the DCs specifically silenced the matrix peptide-specific, IFNγ-producing cells, and induced in their place peptide-specific IL-10 producing cells.[57] The T cells from the volunteers also could silence influenza-specific, CD8 IFNγ-producing effectors in the preimmunization or recovery blood samples,[28] whereas CMV- and EBV-specific effectors were not regulated. These human studies were stopped because, ostensibly, influenza immunity in the volunteers was being compromised. Nevertheless, it became evident that immature DCs could somehow induce regulatory cells that blocked effectors in an antigen-dependent way.

In the studies of Menges *et al.* in mice, DCs were generated from bone marrow precursors, pulsed with an encephalogeneic peptide, and injected into naïve mice. These DCs expanded peptide-specific, IL-10–producing, $CD4^+$ T cells *in vivo*, and were able to block the induction of EAE with peptide in strong adjuvants.[29] The authors emphasized that the developing DCs had to be exposed to TNF during culture to be become active in this type of tolerance. As a result, the DCs expressed certain markers found on mature DCs.[29] They proposed that an intermediate or semi-mature cell induces T-reg cells *in vivo*.

The work of Akbari *et al.* has lead to the proposal that DCs isolated from the lung also are able to induce regulatory T cells. The pulmonary DCs capture OVA protein and produce IL-10.[58] The production of IL-10 may be one requirement for the induction of IL-10–producing T-reg cells.[59] This work brings up the possibility that DCs at exposed body surfaces, like the lung and intestine, are the physiologic counterparts to the DCs being generated *ex vivo* in the other experiments sited above.[28,29,56] The concept is that the uptake of environmental proteins at exposed body surfaces is so risky, in terms of inducing chronic cell-mediated immunity to environmental proteins, that DCs at these surfaces induce regulatory T cells.

DISCUSSION: TOLEROGENIC DCs AND SOME IMPLICATIONS FOR CLINICAL IMMUNOLOGY

These emerging data on peripheral tolerance leave a lot to be desired in terms of mechanism, both at the levels of the DC and the T cell. For example, how do DCs in lymphoid organs, which delete T cells, relate to the DCs generated from human monocytes and mouse bone marrow precursors *ex vivo*, which induce regulatory T cells? Are there different subsets of DCs for deletional and regulatory T cell tolerance, or do these represent different phenotypes produced by the environment within a common DC lineage? The terms "immature" or "mature" are by themselves incomplete to describe the different properties of the DCs that have been studied. Nevertheless, the terms helpfully distinguish the antigen uptake and immunizing roles of immature and mature cells, respectively, as originally revealed in studies of Langerhans cells from the epidermis.[13–15]

In our introduction, we reasoned that DC-based tolerance is an essential prerequisite for the function of mature DCs as adjuvants for protective immunity. The information also seems pertinent to antigen-specific immunosuppression clinically. A few examples follow. One pathway to autoimmunity would entail the chronic activation of immature DCs in the steady state, obviating their tolerance role. Evidence for chronic interferon-α–based maturation of DCs in the setting of systemic lupus erythematosus has been reported.[60] Likewise, chronic CD40-based stimulation of epidermal Langerhans cells in mice is associated with profound autoimmunity.[61]

The tolerogenic function of immature DCs might be bypassed in another way in the setting of allergy. Allergy is characterized by the induction of Th2 cells, whereas DCs characteristically induce Th1 type responses (partly on the basis of their capacity to produce IL-12 and related Th1-inducing cytokines). Allergy may reflect the use of non-DCs as antigen-presenting cells, such as activated B cells. If the latter capture and present antigens *in vivo*, the tolerogenic function of DCs may be bypassed and a Th2 response initiated.

In infectious diseases, the tolerogenic role of immature DCs could be exploited by pathogens that replicate chronically and gain access to immature DCs. One example is *Leishmania major,* which is DC-tropic[62,63] and known to induce the formation of suppressor T cells.[64,65] Another example is HIV-1. This virus is produced in large amounts but targets to receptors that are expressed primarily by immature DCs, for example, CD4, CCR5, and DC-SIGN. We have suggested that a major obstacle to HIV-1 resistance is the capacity of the virus to use immature DCs to induce T cell deletion and regulatory T cells.[10]

DCs are proving to be valuable for assessing responsiveness to tumor antigens *ex vivo* and for inducing immunity *in situ.* It has been of major concern that tumors pose as "self-tissues" and tolerize specific immunity. Interestingly, this need not be the case, as illustrated by some of the first studies with DCs as antigen-presenting cells for human tumor immunity. *In vivo*, it is possible to induce melanoma immunity in patients who receive autologous DCs pulsed with melanoma peptides.[19] *In vitro*, it is possible to induce strong tumor-specific killer T cell responses in patients with progressive multiple myeloma.[66] The tumor antigen-specific T cells can be generated from lymphocytes in the involved marrow, and they kill authentic myeloma targets.[66] In other words, tumor-bearing patients are not completely tolerant but have tumor-reactive T cells that can be expanded by maturing DCs. Perhaps the results are less surprising if it is considered that the anti-apoptotic nature of tumors prevents their capture by immature DCs and thereby the induction of tolerance.

This discussion is, of course, speculative. Nonetheless, it should be valuable to continue to pursue DC function *in situ*, to analyze and induce antigen-specific tolerance.

REFERENCES

1. MATZINGER, P. & S. GUERDER. 1989. Does T-cell tolerance require a dedicated antigen-presenting cell? Nature **338:** 74–76.
2. ZAL, T., A. VOLKMANN & B. STOCKINGER. 1994. Mechanisms of tolerance induction in major histocompatibility complex class II-restricted T cells specific for a blood-borne self-antigen. J. Exp. Med. **180:** 2089–2099.
3. VOLKMANN, A., T. ZAL & B. STOCKINGER. 1997. Antigen-presenting cells in thymus that can negatively select MHC class II-restricted T cells recognizing a circulating self antigen. J. Immunol. **158:** 693–706.
4. BOUNEAUD, C., P. KOURILSKY & P. BOUSSO. 2000. Impact of negative selection on the T cell repertoire reactive to a self-peptide: a large fraction of T cell clones escapes clonal detection. Immunity **13:** 829–840.
5. NOSSAL, G.J.V. 2001. A purgative mastery. Nature **412:** 685–686.
6. KEARNEY, E.R., K.A. PAPE, D.Y. LOH, *et al.* 1994. Visualization of peptide-specific T cell immunity and peripheral tolerance induction in vivo. Immunity **1:** 327–339.

7. Parijs, L.V., A. Ibraghimov & A.K. Abbas. 1996. The roles of costimulation and fas in T cell apoptosis and peripheral tolerance. Immunity **4:** 321–328.
8. Aichele, P., K. Brduscha-Riem, R.M. Zinkernagel, *et al.* 1995. T cell priming versus T cell tolerance induced by synthetic peptides. J. Exp. Med. **182:** 261–266.
9. Dadaglio, G., C.A. Nelson, M.B. Deck, *et al.* 1997. Characterization and quantitation of peptide-MHC complexes produced from hen egg lysozyme using a monoclonal antibody. Immunity **6:** 727–738.
10. Steinman, R.M. & M.C. Nussenzweig. 2002. Avoiding horror autotoxicus: the importance of dendritic cells in peripheral T cell tolerance. Proc. Natl. Acad. Sci. USA **99:** 351–358.
11. Holt, P.G., S. Haining, D.J. Nelson, *et al.* 1994. Origin and steady-state turnover of class II MHC-bearing dendritic cells in the epithelium of the conducting airways. J. Immunol. **153:** 256–261.
12. Kamath, A.T., J. Pooley, M.A. O'Keeffe, *et al.* 2000. The development, maturation, and turnover rate of mouse spleen dendritic cell populations. J. Immunol. **165:** 6762–6770.
13. Inaba, K., G. Schuler, M.D. Witmer, *et al.* 1986. The immunologic properties of purified Langerhans cells: distinct requirements for the stimulation of unprimed and sensitized T lymphocytes. J. Exp. Med. **164:** 605–613.
14. Romani, N., S. Koide, M. Crowley, *et al.* 1989. Presentation of exogenous protein antigens by dendritic cells to T cell clones: intact protein is presented best by immature, epidermal Langerhans cells. J. Exp. Med. **169:** 1169–1178.
15. Romani, N., K. Inaba, E. Pure, *et al.* 1989. A small number of anti-CD3 molecules on dendritic cells stimulate DNA synthesis in mouse T lymphocytes. J. Exp. Med. **169:** 1153–1168.
16. Dhodapkar, M., R.M. Steinman, M. Sapp, *et al.* 1999. Rapid generation of broad T-cell immunity in humans after single injection of mature dendritic cells. J. Clin. Invest. **104:** 173–180.
17. Dhodapkar, M.V., J. Krasovsky, R.M. Steinman, *et al.* 2000. Mature dendritic cells boost functionally superior T cells in humans without foreign helper epitopes. J. Clin. Invest. **105:** R9–R14.
18. Schuler-Thurner, B., D. Dieckmann, P. Keikavoussi, *et al.* 2000. Mage-3 and influenza-matrix peptide-specific cytotoxic T cells are inducible in terminal stage HLA-A2.1+ melanoma patients by mature monocyte-derived dendritic cells. J. Immunol. **165:** 3492–3496.
19. Banchereau, J., A.K. Palucka, M. Dhodapkar, *et al.* 2001. Immune and clinical responses in patients with metastatic melanoma to CD34(+) progenitor-derived dendritic cell vaccine. Cancer Res. **61:** 6451–6458.
20. Albert, M.L., S.F.A. Pearce, L.M. Francisco, *et al.* 1998. Immature dendritic cells phagocytose apoptotic cells via $\alpha_v\beta_5$ and CD36, and cross-present antigens to cytotoxic T lymphocytes. J. Exp. Med. **188:** 1359–1368.
21. Yrlid, U. & M.J. Wick. 2000. *Salmonella*-induced apoptosis of infected macrophages results in presentation of a bacteria-encoded antigen after uptake by bystander dendritic cells. J. Exp. Med. **191:** 613–623.
22. Vermaelen, K.Y., I. Carro-Muino, B.N. Lambrecht, *et al.* 2001. Specific migratory dendritic cells rapidly transport antigen from the airways to the thoracic lymph nodes. J. Exp. Med. **193:** 51–60.
23. Bhardwaj, N., J.W. Young, A.J. Nisanian, *et al.* 1993. Small amounts of superantigen, when presented on dendritic cells, are sufficient to initiate T cell responses. J. Exp. Med. **178:** 633–642.
24. Steinman, R.M., S. Turley, I. Mellman, *et al.* 2000. The induction of tolerance by dendritic cells that have captured apoptotic cells. J. Exp. Med. **191:** 411–416.
25. Hawiger, D., K. Inaba, Y. Dorsett, *et al.* 2001. Dendritic cells induce peripheral T cell unresponsiveness under steady state conditions in vivo. J. Exp. Med. **194:** 769–780.
26. Bonifaz, L., D. Bonnyay, K. Mahnke, *et al.* 2002. Efficient targeting of protein antigen to the dendritic cell receptor DEC-205 in the steady state leads to antigen presentation on MHC class I products and peripheral $CD8^+$ T cell tolerance. J. Exp. Med. **196:** 1627–1638.
27. Liu, K., T. Iyoda, M. Saternus, *et al.* 2002. Immune tolerance after delivery of dying cells to dendritic cells in situ. J. Exp. Med. **196:** 1091–1097.

28. DHODAPKAR, M.V. & R.M. STEINMAN. 2002. Antigen-bearing, immature dendritic cells induce peptide-specific, CD8+ regulatory T cells *in vivo* in humans. Blood **100:** 174–177.
29. MENGES, M., S. ROSSNER, C. VOIGTLANDER, *et al.* 2002. Repetitive injections of dendritic cells matured with tumor necrosis factor α induce antigen-specific protection of mice from autoimmunity. J. Exp. Med. **195:** 15–22.
30. KYEWSKI, B.A., C.G. FATHMAN & R.V. ROUSE. 1986. Intrathymic presentation of circulating non-MHC antigens by medullary dendritic cells. An antigen-dependent microenviroment for T cell differentiation. J. Exp. Med. **163:** 231–246.
31. HOLT, P.G., M.A. SCHON-HEGRAD & J. OLIVER. 1987. MHC class II antigen-bearing dendritic cells in pulmonary tissues of the rat. Regulation of antigen presentation activity by endogenous macrophage populations. J. Exp. Med. **167:** 262–274.
32. BUJDOSO, R., J. HOPKINS, B.M. DUTIA, *et al.* 1989. Characterization of sheep afferent lymph dendritic cells and their role in antigen carriage. J. Exp. Med. **170:** 1285–1302.
33. LIU, L.M. & G.G. MACPHERSON. 1993. Antigen acquisition by dendritic cells: intestinal dendritic cells acquire antigen administered orally and can prime naive T cells in vivo. J. Exp. Med. **177:** 1299–1307.
34. CROWLEY, M., K. INABA & R.M. STEINMAN. 1990. Dendritic cells are the principal cells in mouse spleen bearing immunogenic fragments of foreign proteins. J. Exp. Med. **172:** 383–386.
35. HUANG, F.-P., N. PLATT, M. WYKES, *et al.* 2000. A discrete subpopulation of dendritic cells transports apoptotic intestinal epithelial cells to T cell areas of mesenteric lymph nodes. J. Exp. Med. **191:** 435–442.
36. KRAAL, G., M. BREEL, M. JANSE, *et al.* 1986. Langerhans cells, veiled cells, and interdigitating cells in the mouse recognized by a monoclonal antibody. J. Exp. Med. **163:** 981–997.
37. SWIGGARD, W.J., A. MIRZA, M.C. NUSSENZWEIG, *et al.* 1995. DEC-205, a 205 kDa protein abundant on mouse dendritic cells and thymic epithelium that is detected by the monoclonal antibody NLDC-145: purification, characterization and N-terminal amino acid sequence. Cell. Immunol. **165:** 302–311.
38. JIANG, W., W.J. SWIGGARD, C. HEUFLER, *et al.* 1995. The receptor DEC-205 expressed by dendritic cells and thymic epithelial cells is involved in antigen processing. Nature **375:** 151–155.
39. MAHNKE, K., M. GUO, S. LEE, *et al.* 2000. The dendritic cell receptor for endocytosis, DEC-205, can recycle and enhance antigen presentation via major histocompatibility complex class II-positive lysosomal compartments. J. Cell Biol. **151:** 673–683.
40. CHESNUT, R.W. & H.M. GREY. 1981. Studies on the capacity of B cells to serve as antigen-presenting cells. J. Immunol. **126:** 1075–1079.
41. VALLADEAU, J., O. RAVEL, C. DEZUTTER-DAMBUYANT, *et al.* 2000. Langerin, a novel C-type lectin specific to Langerhans cells, is an endocytic receptor that induces the formation of Birbeck granules. Immunity **12:** 71–81.
42. VALLADEAU, J., V. DUVERT-FRANCES, J.-J. PIN, *et al.* 2001. Immature human dendritic cells express asialoglycoprotein receptor isoforms for efficient receptor-mediated endocytosis. J. Immunol. **167:** 5767–5774.
43. KWON, D.S., G. GREGARIO, N. BITTON, *et al.* 2002. DC-SIGN mediated internalization of HIV is required for *trans*-enhancement of T cell infection. Immunity **16:** 135–144.
44. ENGERING, A., T.B. GEIJTENBEEK, S.J. VAN VLIET, *et al.* 2002. The dendritic cell-specific adhesion receptor DC-SIGN internalizes antigen for presentation to T cells. J. Immunol. **168:** 2118–2126.
45. DZIONEK, A., Y. SOHMA, J. NAGAFUNE, *et al.* 2001. BDCA-2, a novel plasmacytoid dendritic cell-specific type II C-type lectin, mediates antigen-capture and is a potent inhibitor of interferon-α/β induction. J. Exp. Med. **194:** 1823–1834.
46. IYODA, T., S. SHIMOYAMA, K. LIU, *et al.* 2002. The CD8+ dendritic cell subset selectively endocytoses dying cells in culture and in vivo. J. Exp. Med. **195:** 1289–1302.
47. SAKAGUCHI, S. 2000. Regulatory T cells: key controllers of immunologic self-tolerance. Cell **101:** 455–458.

48. SAOUDI, A., B. SEDDON, V. HEATH, *et al.* 1996. The physiological role of regulatory T cells in the prevention of autoimmunity: the function of the thymus in the generation of the regulatory T cell subset. Immunol. Rev. **149:** 195–216.
49. JONULEIT, H., E. SCHMITT, M. STASSEN, *et al.* 2001. Identification and functional characterization of human CD4(+)CD25(+) T cells with regulatory properties isolated from peripheral blood. J. Exp. Med. **193:** 1285–1294.
50. LEVINGS, M.K., R. SANGREGORIO & M.G. RONCAROLO. 2001. Human CD25(+) CD4(+) T regulatory cells suppress naive and memory T cell proliferation and can be expanded in vitro without loss of function. J. Exp. Med. **193:** 1295–1302.
51. DIECKMANN, D., H. PLOTTNER, S. BERCHTOLD, *et al.* 2001. Ex vivo isolation and characterization of CD4(+) CD25(+) T cells with regulatory properties from human blood. J. Exp. Med. **193:** 1303–1310.
52. NG, W.F., P.J. DUGGAN, F. PONCHEL, *et al.* 2001. Human CD4(+) CD25(+) cells: a naturally occurring population of regulatory T cells. Blood **98:** 2736–2744.
53. BENSINGER, S.J., A. BANDEIRA, M.S. JORDAN, *et al.* 2001. Major histocompatibility complex class II-positive cortical epithelium mediates the selection of $CD4^{+}25^{+}$ immunoregulatory T cells. J. Exp. Med. **194:** 427–438.
54. BACCHETTA, R., M. BIGLER, J.-L. TOURAINE, *et al.* 1994. High levels of interleukin 10 production in vivo are associated with tolerance in SCID patients transplanted with HLA mismatched hematopoietic stem cells. J. Exp. Med. **179:** 493–502.
55. RONCAROLO, M.G., R. BACCHETTA, C. BORDIGNON, *et al.* 2001. Type 1 T regulatory cells. Immunol. Rev. **182:** 68–79.
56. JONULEIT, H., E. SCHMITT, G. SCHULER, *et al.* 2000. Induction of human IL-10-producing, non-proliferating CD4+ T cells with regulatory properties by repetitive stimulation with allogeneic immature dendritic cells. J. Exp. Med. **192:** 1213–1222.
57. DHODAPKAR, M.V., R.M. STEINMAN, J. KRASOVSKY, *et al.* 2001. Antigen specific inhibition of effector T cell function in humans after injection of immature dendritic cells. J. Exp. Med. **193:** 233–238.
58. AKBARI, O., R.H. DEKRUYFF & D.T. UMETSU. 2001. Pulmonary dendritic cells producing IL-10 mediate tolerance induced by respiratory exposure to antigen. Nat. Immunol. **2:** 725–731.
59. LEVINGS, M.K., R. SANGREGORIO, F. GALBIATI, *et al.* 2001. IFN-α and IL-10 induce the differentiation of human type 1 T regulatory cells. J. Immunol. **166:** 5530–5539.
60. BLANCO, P., A.K. PALUCKA, M. GILL, *et al.* 2001. Induction of dendritic cell differentiation by IFN-α in systemic lupus erythematosus. Science **294:** 1540–1543.
61. MEHLING, A., K. LOSER, G. VARGA, *et al.* 2001. Overexpression of CD40 ligand in murine epidermis results in chronic skin inflammation and systemic autoimmunity. J. Exp. Med. **194:** 615–628.
62. MOLL, H., H. FUCHS, C. BLANK, *et al.* 1993. Langerhans cells transport Leishmania major from the infected skin to the draining lymph node for presentation to antigen-specific T cells. Eur. J. Immunol. **23:** 1595–1601.
63. VON STEBUT, E., Y. BELKAID, T. JAKOB, *et al.* 1998. Uptake of Leishmania major amastigotes results in activation and interleukin 12 release from murine skin-derived dendritic cells: implications for the initiation of anti-Leishmania immunity. J. Exp. Med. **188:** 1547–1552.
64. HOWARD, J.G., C. HALE & F.Y. LIEW. 1980. Immunological regulation of experimental cutaneous leishmaniasis. III. Nature and significance of specific suppression of cell-mediated immunity in mice highly susceptible to Leishmania tropica. J. Exp. Med. **152:** 594–607.
65. LIEW, F.Y., A. SINGLETON, E. CILLARI, *et al.* 1985. Prophylactic immunization against experimental leishmaniasis. V. Mechanism of the anti-protective blocking effect induced by subcutaneous immunization against Leishmania major infection. J. Immunol. **135:** 2102–2107.
66. DHODAPKAR, M.V., J. KRASOVSKY & K. OLSON. 2002. T cells from the tumor microenvironment of patients with progressive myeloma can generate strong, tumor-specific cytolytic responses to autologous, tumor-loaded dendritic cells. Proc. Natl. Acad. Sci. USA **99:** 13009–13013.

Molecular Mechanisms of B Cell Antigen Receptor Trafficking

MARCUS R. CLARK, DON MASSENBURG, MIAO ZHANG, AND KARYN SIEMASKO

University of Chicago, Section of Rheumatology, 5841 South Maryland Avenue, Chicago, Illinois 60637, USA

ABSTRACT: B lymphocytes are among the most efficient cells of the immune system in capturing, processing, and presenting MHC class II restricted peptides to T cells. Antigen capture is essentially restricted by the specificity of the clonotypic antigen receptor expressed on each B lymphocyte. However, receptor recognition is only one factor determining whether an antigen is processed and presented. The context of antigen encounter is crucial. In particular, polyvalent arrays of repetitive epitopes, indicative of infection, accelerate the delivery of antigen to specialized processing compartments, and up-regulate the surface expression of MHC class II and co-stimulatory molecules such as B7. Recent studies have demonstrated that receptor-mediated signaling and receptor-facilitated peptide presentation to T cells are intimately related. For example, rapid sorting of endocytosed receptor complexes through early endosomes requires the activation of the tyrosine Syk. This proximal kinase initiates all BCR-dependent signaling pathways. Subsequent entry into the antigen-processing compartment requires the tyrosine phosphorylation of the BCR constituent Igα and direct recruitment of the linker protein BLNK. Signals from the BCR also regulate the biophysical and biochemical properties of the targeted antigen-processing compartments. These observations indicate that the activation and recruitment of signaling molecules by the BCR orchestrate a complex series of cellular responses that favor the presentation of even rare or low-affinity antigens if encountered in contexts indicative of infection. The requirement for BCR signaling provides possible mechanisms by which cognate B:T cell interactions can be controlled by the milieu in which antigen engagement occurs.

KEYWORDS: antigen receptor trafficking; signaling; MIIC; Igα; Igβ; BLNK; Syk

RECOGNITION AND CAPTURE OF ANTIGEN

Most adaptive immune responses require the processing and presentation of antigens to T cells. At the usual portals of entry for infectious agents, resident macrophages, dendritic cells, and other professional antigen presenting cells (APCs)

Address for correspondence: Marcus R. Clark, University of Chicago, Section of Rheumatology, MC 0930, 5841 S. Maryland Avenue, Chicago, IL 60637. Voice: 773-702-0202; fax: 773-702-8702.

mclark@medicine.bsd.uchicago.edu

Ann. N.Y. Acad. Sci. 987: 26–37 (2003). © 2003 New York Academy of Sciences.

efficiently capture antigens. These cells of the innate immune system do not usually respond to specific antigens but rather to the context in which antigenic epitopes are presented. Repetitive epitope arrays and the co-occurrence of Toll-like receptor ligands such as lipopolysaccharide and hypomethylated DNA indicate danger to the host and the need to mount an immune response.[1,2] Productive recognition of antigen activates programs that induce dendritic cells to migrate to lymphoid organs where they mature into efficient APCs for $CD4^+$ naïve T cells. Dendritic cells provide an important bridge between the innate and adaptive immune responses.[3]

In contrast to dendritic cells, which reside and capture antigens in the periphery, B lymphocytes usually transit through peripheral tissues to reside primarily in secondary lymphoid organs. These naïve B cells are poor APCs that cannot efficiently initiate immune responses. They first require activation by dendritic and cognate T cells. However, antigen presentation by B lymphocytes is required to mount high-affinity humoral responses and for coordinating antigen-specific cytotoxicity.[4]

B lymphocytes differ from other APCs in several fundamental ways. The most important difference is that B lymphocytes are clonotypic and usually only capture and process antigens recognized by the B cell antigen receptor (BCR). The capture of antigen by the BCR is extremely efficient and comparable to preferred mechanisms of uptake utilized by other APCs. In B lymphocytes, other mechanisms by which antigen might gain access to the endocytic pathway function poorly. Pinocytosis by B cells is much less efficient than BCR-mediated uptake, and lymphocytes are not phagocytic. On many cells of the immune system, capture of antigen-antibody complexes by Fcγ receptors leads to rapid internalization and delivery into the endocytic pathway. However, B cells express a unique FcγR isoform, FcγR II-B1, which contains a cytoplasmic motif that inhibits endocytosis.[5–7] The primacy of the BCR as the portal for entry of antigen ensures that foreign peptides presented in the context of MHC class II are derived from the antigen recognized by the BCR. Immune responses remain focused on antigens originally encountered in contexts indictative of non-self.

It is likely that most relevant antigens encountered by B lymphocytes are polyvalent. Invading organisms display repetitive arrays of limited antigenic epitopes on their outer coats. These antigens co-occur with non-specific B cell activators, such as lipopolysaccharide, that allow humoral responses to be initiated independently of T cell help. Within lymph nodes, B cells often first encounter antigen adsorbed to the surfaces of dendritic cells. The mechanisms by which antigen is adsorbed is still not clear although it involves Fc and complement receptors.[8,9] Recent evidence demonstrates that the BCR can efficiently capture immobilized antigens from the surfaces of these cells.[10,11] The restriction of antigen to a cell surface ensures that even soluble mono- or oligovalent antigens will be presented in polyvalent arrays. Although dendritic cells are important for initiating B lymphocyte responses, subsequent class switching and affinity maturation is dependent upon the surface display of antigens on follicular dendritic cells in the light zone of germinal centers.[4] Like dendritic cells, the adsorption of antigens to FDCs is probably dependent upon Fcγ and complement receptors.

Cross-linking antigens induce a complex program of responses in B lymphocytes that ensure the rapid capture, processing, and productive presentation of antigen.[12–14] These inductive mechanisms enhance the efficiency of antigen processing and allow immune responses to low-affinity antigens.[12,13] In the following sections, we will discuss how BCR cross-linking, and the activation of signaling pathways, enhance the processing of antigens and the loading of MHC class II with immunogenic peptides.

ACCELERATED TRAFFICKING TO THE MIIC

The unligated BCR is constitutively internalized through clathrin-coated pits.[15] Internalization is dependent upon a conserved motif in Igα/Igβ, the immunoreceptor tyrosine-based activation motif (ITAM).[15,16] Phosphorylation of this motif is not required for internalization but is necessary for the recruitment and activation of Syk and the initiation of downstream signaling pathways.[17–19] Internalized unligated receptor complexes can rapidly recycle through early endosomes to be re-expressed on the cell surface.[16]

Cross-linking the BCR rapidly enhances receptor internalization. This may also involve clathrin-coated pits.[84] Receptor cross-linking induces patches and then a predominant cap to form on the cell surface. This cap, containing most of the surface receptor complexes, is internalized within several minutes.[20] In contrast to the constitutive endocytosis of the resting receptor, neither the ITAM motifs in Igα or Igβ, nor any other identified motif, are required for endocytosis of the aggregated receptor[15,20] (unpublished observations). Cross-linking is associated with the rapid translocation of the receptor to cholesterol-rich lipid rafts and the re-organization of the actin cytoskeleton.[21,22] The requirement of the cytoskeleton for receptor endocytosis and trafficking has been demonstrated.[23] The importance of lipid rafts is less clear. Although entry into rafts is not sufficient for receptor internalization and trafficking,[24] one can propose that lipid rafts are important.[25] From studies in both mammalian and yeast systems, localization to lipid rafts could enhance endocytosis in several ways. The cytosolic interface of cholesterol-laden microdomains is rich in phospholipids, such as phosphotidylinositol (4,5)P_2, which have been implicated in endocytosis and receptor trafficking. In both clathrin-dependent and -independent endocytosis, these lipids serve to recruit adaptors containing PH, FYVE, and ENTH domains.[26] Raft localization could also juxtapose the receptor with acylated proteins required for endocytosis.[21] Finally, it has recently been demonstrated that alterations of lipids within the plasma membrane bilayer can directly alter local membrane curvature and favor vesicle formation.[27,28] The applicability of these mechanisms to BCR endocytosis is unknown.

Cross-linking the BCR also accelerates transience through the endocytic pathway to specialized late endosomal compartments rich in MHC class II (MIIC).[29] These Lamp-1$^+$ compartments are acidic and contain cathespins, thiol reductases, and other molecules required for efficient antigen processing.[30–32] Descriptions of this compartment vary suggesting that the MIIC actually consists of multiple specialized compartments with unique properties and specialized functions.[33] It is clear that multiple potential processing compartments can exist in a given cell type and the nature of these compartments can vary with the maturation and activation state of the cell.[34,35] Furthermore, specialized processing and loading compartments derived from earlier in the endocytic pathway have been implicated as potential sites for antigen processing.[36,37] However, in B cells, most studies indicate that compartments derived from late endosomes, collectively referred to as the MIIC, are the primary sites in which antigen-processing and MHC class II peptide loading occurs.[38,39]

Specific targeting mechanisms enrich the MIIC with newly synthesized MHC class II. In the endoplasmic reticulum, MHC class II assembles with trimers of invariant chain (Ii). Part of the aminoterminal domain of Ii (CLIP) occupies the antigen-binding groove while the carboxyterminal domain provides a di-leucine trafficking motif that facilitates targeting to late endosomes.[40–45] Once in the MIIC, Ii is degraded by cathepsin S to the core CLIP peptide protected in the MHC class II groove.[46] The exchange of CLIP for antigenic peptide is catalyzed by the nonclassical molecule H2-M (HLA-DM is the human homologue).[47–51] It is likely that H-2M, by favoring loading of high-affinity peptides, edits the repertoire of peptides presented to $CD4^+$ T cells.[52] Another nonclassical MHC molecule H2-O (HLA-DO), which associates with H2-M and is expressed only in B cells, has been postulated to be a pH-sensitive negative regulator of H2-M and therefore MHC class II peptide loading.[53,54] Such a function is attractive because it would favor the processing and loading of peptides derived from antigens targeted to late endosomes. However, the exact function of H2-O remains controversial.

Ultrastructural and subcellular fraction studies have defined six separate compartments within the B lymphocyte endocytic pathway.[39] From these and other studies, two candidate MIIC compartments have been described. MIIC vesicles can appear as multilaminar organelles reminiscent of lysosomes. More commonly, a complex structure with a limiting membrane studded with Lamp-1 and a lumen containing multivesicular bodies,[55] typical of late endosomes, has been found. The intraluminal vesicles are thought to be derived from transport vesicles that have gained access to the compartment.[56] These vesicles contain the majority of endocytosed aggregated receptor complexes and are rich in MHC class II and tetraspan proteins such as CD82.[57] Newly synthesized MHC class II gain access to the intraluminal bodies by cycling through early endosomes before targeting the MIIC through pathways common with the BCR.[58,59] This trafficking pattern could allow for the placement of MHC class II and endocytosed BCR in close proximity, on the same membrane surface, favoring subsequent loading of receptor-captured peptides (FIG. 1).

The late endosomal compartment, identified by staining with anti-Lamp-1 antibodies, is not homogenous, but consists of discrete MHC class II–bearing compartments that are Ii^+ or Ii^-.[35,39] MHC class II transits first through the Ii^+ vesicles to reside in the Ii^- ones. These latter vesicles are also rich in H2-M and are not accessible to pinocytosed antigen. Therefore, it is likely that the Ii^-/Lamp-1^+ compartment represents the true peptide-loading compartment. The function of the Ii^+ compartment in terms of antigen processing is not known. The accessibility of these vesicles to pinocytosed antigen suggests that they might be part of the normal endocytic route to terminal lysosomes.

Newly formed MHC class II/peptide complexes are rapidly exported to the cell surface and displayed in circumscribed microdomains important for optimal T cell activation. Two distinct types of surface structures have been described. In one, tetraspan proteins, such as those found in the intraluminal vesicles of the MIIC, are thought to form compact membrane microdomains enriched in MHC class II.[60] Classical membrane rafts have also been described.[61] In both cases, it is likely that these microdomains are formed in the MIIC and are then exported to the cell surface. It is not known how these domains are formed or whether the cholesterol rafts provided by the endocytosed BCR are involved.[24] Interestingly, MHC class II/peptide–containing exosomes can be secreted by B cells.[62] These structures, which are capa-

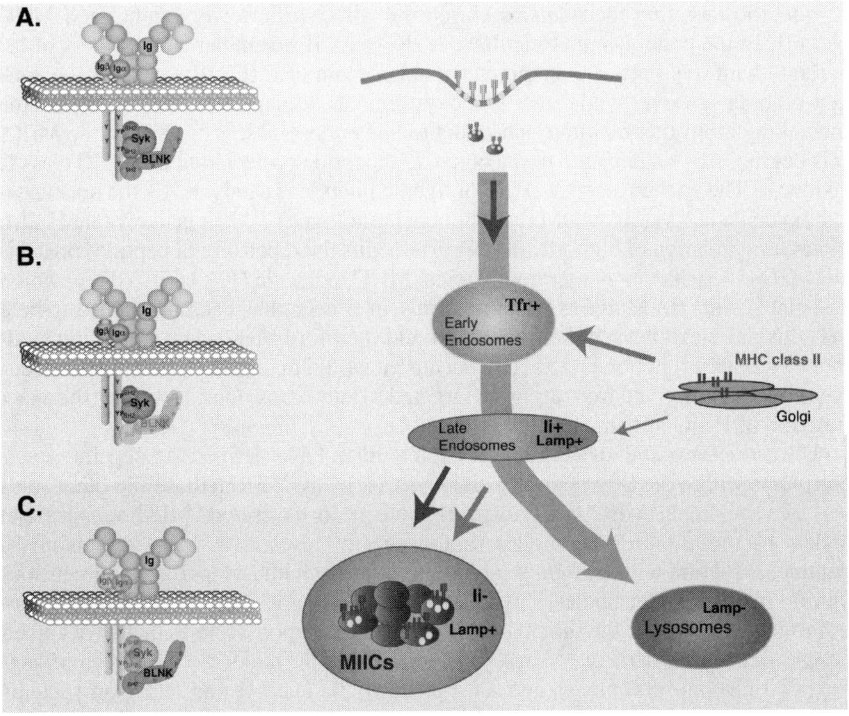

FIGURE 1. Model of B antigen receptor trafficking. (**A**) Cross-linking the BCR induces the tyrosine phosphorylation of Igα ITAM and non-ITAM tyrosines leading to the recruitment of Syk and BLNK and the activation of downstream pathways. Receptor aggregation is also associated with rapid translocation to cholesterol-rich rafts and subsequent endocytosis. The assembly of Syk and BLNK on Igα is required for transit through sequential checkpoints en route to the MIIC. (**B**) Proper sorting through early endosomes requires the cytosolic tails of both Igα and Igβ. The Igα ITAM tyrosines are also required as is the recruitment of Syk. Newly synthesized MHC class II can also traffic through early endosomes. Although this may allow peptide loading in earlier compartments, it is more likely that it facilitates common sorting to late endosomes. (**C**) Entry into MIICs and MIIC remodeling requires the recruitment of BLNK to the BCR complex. Normally, endocytosed BCR complexes accumulate in intraluminal multivesicular vesicles that are also rich in MHC class II. The close apposition of MHC class II and antigen may favor peptide loading. Receptor entry is associated with the translocation and coalescence of the MIIC into a single perinuclear single complex. The common requirement of BLNK for receptor entry into the MIIC and for remodeling of the compartments suggests that common signaling pathways regulate both processes.

ble of activating cognate T cells, have the same structure and bear the same markers as the intraluminal multivesicular bodies of the MIIC. These observations suggest that the mechanisms segregating MHC class II and the BCR within the MIIC dictate how peptides are subsequently presented to cognate T cells.

Recent evidence from our laboratory and others indicate that receptor trafficking to the MIIC is complex, highly regulated and intimately dependent upon the activation of specific signaling molecules. Phosphorylation of tyrosines within the ITAMs of Igα and Igβ is required to recruit and activate Syk.[63] Additionally, the phosphorylation of a non-ITAM tyrosine in Igα recruits the linker molecule BLNK, which serves as a scaffold on which Grb2, PLCγ2, Vav, and Nck are assembled.[64] The formation of this multimeric signalsome directly on the receptor is required for the mobilization of intracellular calcium and the efficient activation of the Erk family of kinases. To study the role of these domains in receptor trafficking, we utilized a chimeric receptor system in which partial receptor complexes containing cytosolic tails of Igα and/or Igβ could be formed and then aggregated.[17] From these studies it was apparent that Igα and Igβ were both necessary and sufficient for rapid trafficking to the MIIC.[20] Chimeric complexes containing only Igα were retained in early endosomes. In contrast to an earlier report, receptor complexes containing only Igβ were rapidly delivered to terminal lysosomes and degraded (Ref. 20; W. Song, personal communication). Within Igα, the ITAM tyrosines were required for normal trafficking. Consistent with the known role of the Igα ITAM in Syk activation, expression of a dominant negative of Syk-inhibited receptor trafficking. These data indicate that proper trafficking of the BCR requires Igα, Igβ, and the activation of Syk.

The recruitment of BLNK to the BCR is required for entry into the MIIC.[65] Receptor complexes, in which the Igα non-ITAM tyrosines were mutated and BLNK recruitment was prevented, sorted normally through early endosomes. Furthermore, they came to reside within close proximity of the Lamp-1^+Ii$^-$ MIIC (Ref. 65; unpublished observation). However, these receptor-laden transport vesicles were excluded from their lumens. Subcellular fractionation experiments indicated that these receptor complexes were physically attached or tethered to the targeted MIIC vesicles (Siemasko, unpublished observation). Interesting, receptor-targeted MIIC vesicles remained peripherally located and did not form the central multivesicular structure normally induced by ligation of the endogenous BCR. Fusing BLNK to the carboxyterminal tail of Igα rescued both entry into the MIIC and receptor-facilitated antigen presentation. In any system that has been studied in detail, transport vesicles undergo sequential sorting, docking, and fusion events regulated by separate but codependent mechanisms.[66,67] Furthermore, in at least one system, neuron secretion, the secretion of neurotransmitters into the synaptic space is triggered by a signaling event, a rise in cytosolic free calcium.[68,69] However, to our knowledge, this is the first demonstration of receptor trafficking being regulated at the point of entry into a targeted compartment.

These observations suggest a model of receptor trafficking in which specific functional domains within the Igα/Igβ cytosolic tails cooperate to mediate receptor trafficking (FIG. 1). The earliest defined checkpoint is at early endosomes. Complexes lacking Igβ are retained in this compartment even though Igα can recruit both Syk and BLNK. We would postulate that the next decision point in receptor trafficking occurs in Lamp-1^+/Ii$^+$ late endosomes. The BCR domains required for diverting receptor complexes off the normal endocytic route towards the MIIC are not known. BLNK could be responsible although it seems unlikely given that the non-ITAM mutant receptor complexes localize exclusively with the Ii$^-$, Lamp-1^+ vesicles. Finally, entry into the MIIC compartment requires BLNK. The observation that receptor complexes defective in BLNK recruitment are attached to the MIIC vesicles suggest that there is a defect in terminal fusion.[65] The mechanism(s) by which BLNK mediates receptor entry is not known. The signaling capacities of BLNK are undoubtedly important because treatment with PMA and ionomycin restores normal trafficking.

However, receptor-facilitated antigen presentation is still defective, suggesting that BLNK may have several functions.

REGULATION OF ANTIGEN PROCESSING AND PEPTIDE LOADING WITHIN THE MIIC

Stimulation of the BCR induces a series of changes within the MIIC predicted to enhance antigen processing and MHC class II peptide loading. The best-characterized of these changes is the increase in newly synthesized MHC class II trafficking through the endocytic pathway following receptor ligation.[29,70] To date, two mechanisms have been defined. First, cross-linking the BCR induces the PKC-dependent phosphorylation of Ii which enhances MHC class II trafficking from the golgi to the MIIC.[71,72] Second, receptor cross-linking also transiently inhibits cathepsin S activity, thereby preventing Ii degradation and allowing MHC class II to accumulate within the MIIC.[73] The relative importance of these two mechanisms is not known. Although total MHC class II is increased, this might not be as important as the localization of these molecules within the MIIC, the endocytic path taken to arrive there, or the efficient export of MHC class II/peptide complexes to the cell surface.

Stimulation of the BCR also radically alters the biophysical properties of the MIIC. In resting cells, late endosomal vesicles are usually .3 μm in diameter. However, following receptor ligation, these vesicles coalesce into a single perinuclear multivesicular structure.[35] Fusion between vesicles generates large structures of greater than 1 μm in diameter. In both resting and activated cells, Lamp-1^+/Ii^+ and Lamp-1^+/Ii^- vesicles are easily discerned, indicating that there is specificity in how the vesicles fuse which preserves discrete heterogeneous environments within the MIIC complex. This aggregation of late endosomal vesicles becomes the primary target of the endocytosed receptor complex. The purpose of activating or remodeling the MIIC compartments is unclear. It is possible that centralizing the late endosomal compartments within the cell allows for the efficient processing of the large pulse of antigen-laden complexes that are endocytosed. This might be especially important if efficient MHC class II peptide loading requires coordinated or sequential processes performed in different microenvironments. Alternatively, fusion within the MIIC compartments may reflect the activation of processes required for entry of BCR-laden transport vesicles into the MIIC.

REGULATION OF ANTIGEN PROCESSING BY SIGNALING PATHWAYS

Understanding the molecular processes that control receptor trafficking and MIIC regulation is difficult because the BCR both initiates signaling cascades and traffics through regulatory checkpoints to the MIIC. In other words, the BCR comes in intimate contact with the structures and processes that it regulates. Furthermore, many of the signaling molecules implicated in receptor trafficking, such as Syk and BLNK, are recruited directly to the BCR complex. Their recruitment to the receptor could either serve to initiate signaling pathways or provide chaperones for transit through the endocytic pathway. Experiments utilizing pharmacological inhibitors or, in one report, a dominant negative of Syk,[74] suggest that the proximal molecules recruited to the BCR facilitate trafficking by activating permissive signaling pathways.

Such a conclusion is consistent with evolving theories of BCR activation. Following BCR ligation, the phosphorylation of the Y204 non-ITAM tyrosine is rapid and transient as is the assembly of the BLNK nucleated signaling complex on the BCR (Ref. 75; unpublished observation). However, the transient assembly of this complex is sufficient for entry into the MIIC. Usually, receptor cross-linking induces the delivery of an intact BCR complex containing Igα and Igβ to the MIIC.[76] However, under some circumstances, BCR cross-linking can induce a dissociation of Igα/Igβ from membrane-bound immunoglobulin.[77] The Igα/Igβ is then free to associate with surface-expressed MHC class II to form a positive feedback loop for B cell activation.[78] mIgG, but not IgM, has a 25-amino-acid cytosolic domain that allows sorting of dissociated mIgG to the MIIC.[79] The relative importance of the trafficking motif in mIgG is not known. However, it may allow the Igα/Igβ complex to play multiple roles in mediating peripheral lymphocyte activation.

IMPLICATIONS FOR TOLERANCE AND AUTOIMMUNITY

Acquired immune responses are usually initiated by peripheral dendritic cells and other professional APCs, which respond to antigens encountered in inflammatory contexts. In contrast, B lymphocytes are usually recruited to immune response after these initiating APCs have recognized antigen and have imported it to secondary lymphoid organs. However, subsequent generation of high-affinity humoral responses requires B lymphocytes to efficiently present antigenic peptides to T cells. The contribution of B lymphocyte APC function to autoimmune diseases characterized by high-affinity autoantibodies, such as systemic lupus erythematosus, is obvious. It is now clear from analysis of NOD mice that the antigen-presenting capabilities of B lymphocytes are required for some T cell–mediated diseases as well.[80] In both examples, B lymphocytes are necessary for propagation of disease. Therefore, mechanisms controlling B lymphocyte antigen processing might provide attractive targets for the treatment of both B and T cell–mediated autoimmune diseases.

Ideally, these therapeutics would modulate pathways that regulate MHC class II loading, but not B cell maturation or peripheral survival. Signaling pathways downstream of BLNK, which are required for both entry into the MIIC and for MIIC modeling, are particularly attractive. PLCγ2 was initially considered because increases in free intracellular calcium are involved in vesicle transport in other systems. However, inhibiting intracellular calcium mobilization in B lymphocytes had no effect on receptor trafficking (unpublished observation). Vav, which is also recruited to BLNK,[64] serves as a GEF for Rho family members which are important mediators of cytoskeleton remodeling.[81] In preliminary experiments, expression of C3-transferase, which inhibits Rho or expression of a Rac dominant negative, had no effect on BCR trafficking to the MIIC (unpublished observations). These data suggest that other BLNK-dependent pathways are critical for receptor trafficking.

It is possible that the very proximal mechanisms of signal initiation will also provide important future therapeutic targets in the treatment of autoimmune disease. The main determiners of BCR receptor trafficking and MIIC modeling are proximal signaling molecules recruited to the BCR following productive receptor engagement. Normal activation of signaling pathways is associated with the phosphorylation of both ITAM and non-ITAM tyrosines within the Igα/Igβ heterodimer and the coordinated recruitment of both Syk and BLNK to the BCR complex. Each signaling effector facilitates the transition of a different checkpoint in the endocytic pathway. In contrast, paucivalent and/or low-affinity ligands induce a characteristic altered activation of signaling pathways that commit the cell to anergy.[82] It is not known whether this partial signaling is associated with altered receptor phosphorylation as has been reported for the TCR.[83] However, the BCR on anergic B cells is excluded from lipid rafts indicating that very proximal mechanisms of receptor activation are altered.[22] Coupling BCR signaling directly to receptor-facilitated antigen presentation would guarantee that the quality and quantity of B cell antigen receptor engagement determine cell fate decisions.

REFERENCES

1. AKIRA, S., K. TAKEDA & T. KAISHO. 2001. Toll-like receptors: critical proteins linking innate and acquired immunity. Nat. Immunol. **2:** 675–680.
2. ADEREM, A. & R. ULEVITCH. 2000. Toll-like receptors in the induction of the innate immune response. Nature **406:** 782–787.
3. LUDEWIG, B. *et al.* 2001. Dendritic cells in autoimmune diseases. Curr. Opin. Immunol. **13:** 657–662.
4. KELSOE, G. 1996. Life and death in germinal center (redux). Immunity **4:** 107–111.
5. AMIGORENA, S. *et al.* 1992. Cytoplasmic domain heterogeneity and functions of IgG Fc receptors in B lymphocytes. Science **256:** 1808–1812.
6. MINSKOFF, S., K. MATTER & I. MELLMAN. 1998. Fc gamma RII-B1 regulates the presentation of B cell receptor-bound antigens. J. Immunol. **162:** 2732–2740.
7. WAGLE, N.M. *et al.* 1999. Regulation of B cell receptor-mediated MHC class II antigen processing by FcgammaRRIIB1. J. Immunol. **162:** 2732–2740.
8. TEW, J.G. *et al.* 1997. Follicular dendritic cells and presentation of antigen and costimulatory signals to B cells. Immunol. Rev. **156:** 39–52.
9. FU, Y.X. *et al.* 1997. Independent signals regulate development of primary and secondary follicle structure in spleen and mesenteric lymph node. Proc. Natl. Acad. Sci. USA **94:** 5739–5743.
10. BATISTA, F.D. & M.S. NEUBERGER. 2000. B cells extract and present immobilized antigen: implications for affinity discrimination. EMBO J. **19:** 513–520.
11. BATISTA, F.D., D. IBER & M.S. NEUBERGER. 2001. B cells acquire antigen from target cells after synapse formation. Nature **411:** 489–494.
12. ALUVIHARE, V.R. *et al.* 1997. Acceleration of intracellular targeting of antigen by the B-cell antigen receptor: importance depends on the nature of the antigen-antibody interaction. EMBO J. **16:** 3553–3562.
13. BATISTA, F.D. & M.S. NEUBERGER. 1998. Affinity dependence of the B cell response to antigen: a threshold, a ceiling and the importance of off-rate. Immunity **8:** 751–759.
14. SIEMASKO, K. & M.R. CLARK. 2001. The control and facilitation of MHC class II antigen processing by the BCR. Curr. Opin. Immunol. **13:** 32–36.
15. PATEL, K.J. & M.S. NEUBERGER. 1993. Antigen presentation by the B cell antigen receptor is driven by the alpha/beta sheath and occurs independently of its cytoplasmic tyrosines. Cell **74:** 939–946.
16. CASSARD, S. *et al.* 1998. A tyrosine-based signal present in Ig alpha mediates B cell receptor constitutive internalization. J. Immunol. **160:** 1767–1773.
17. LUISIRI, P. *et al.* 1996. Cooperativity and segregation of function within the Igα/β heterodimer of the B cell antigen receptor complex. J. Biol. Chem. **271:** 5158–5163.
18. SANCHEZ, M. *et al.* 1993. Signal transduction by immunoglobulin is mediated through Ig-alpha and Ig-beta. J. Exp. Med. **178:** 1049–1055.

19. KIM, K.M. *et al.* 1993. Differential signaling through the Ig-alpha and Ig-beta components of the B-cell antigen receptor. Eur. J. Immunol. **23:** 911–916.
20. SIEMASKO, K. *et al.* 1999. Igα and Igβ are required for efficient trafficking to late endosomes and to enhance antigen presentation. J. Immunol. **162:** 6518–6525.
21. CHERUKURI, A., M.L. DYKSTRA & S.K. PIERCE. 2001. Floating the raft hypothesis: lipid rafts play a role in immune cell activation. Immunity **14:** 657–660.
22. WEINTRAUB, B.C., J.E. JUN, A.C. BISHOP, *et al.* 2000. Entry of B cell receptor into signaling domains is inibited in tolerant B cells. J. Exp. Med. **191:** 1443–1448.
23. BROWN, B.K. & W. SONG. 2001. The actin cytoskeleton is required for the trafficking of the B cell antigen receptor to the late endosomes. Traffic **2:** 414–427.
24. CHENG, P.C. *et al.* 1999. A role for lipid rafts in B cell antigen receptor signaling and antigen targeting. J. Exp. Med. **190:** 1549–1560.
25. IKONEN, E. 2001. Roles of lipid rafts in membrane transport. Curr. Opin. Cell Biol. **13:** 470–477.
26. SIMONSEN, A. *et al.* 2001. The role of phosphoinositides in membrane transport. Curr. Opin. Cell Biol. **13:** 485–492.
27. SCHMIDT, A. *et al.* 1999. Endophilin I mediates synaptic vesicle formation by transfer of arachidonate to lysophosphatidic acid. Nature **401:** 133–141.
28. HUTTNER, W.B. & J. ZIMMERBERG. 2001. Implications of lipid microdomains for membrane curvature, budding and fission. Curr. Opin. Cell Biol. **13:** 478–484.
29. CHENG, P.C. *et al.* 1999. MHC class II antigen processing in B cells: accelerated intracellular targeting of antigens. J. Immunol. **162:** 7171–7180.
30. WATTS, C. 2001. Antigen processing in the endocytic compartment. Curr. Opin. Immunol. **13:** 26–31.
31. DRAKE, J.R. *et al.* 1999. Involvement of MIIC-like late endosomes in B cell receptor-mediated antigen processing in murine B cells. J. Immunol. **162:** 1150–1155.
32. KLEIJMEER, M.J. *et al.* 1997. Major histocompatibility complex class II compartments in human and mouse B lymphoblasts represent conventional endocytic compartments. J. Cell Biol. **139:** 639–649.
33. NEEFJES, J. 1999. CIIV, MIIC and other compartments for MHC class II loading. Eur. J. Immunol. **29:** 1421–1425.
34. CASTELLINO, F. & R.N. GERMAIN. 1995. Extensive trafficking of MHC-class II-invariant chain complexes in the enodcytic pathway and appearance of peptide-loaded class II in multiple compartments. Immunity **2:** 73–88.
35. SIEMASKO, K. *et al.* 1998. Cutting edge: signals from the B lymphocyte antigen receptor regulate MHC class II containing late endosomes. J. Immunol. **160:** 5203–5208.
36. PIERRE, P. *et al.* 1996. HLA-DM is localized to conventional and unconventional MHC class II-containing endocytic compartments. Immunity **4:** 229–239.
37. AMIGORENA, S. *et al.* 1994. Transient accumulation of new class II MHC molecules in a novel endocytic compartment in B lymphocytes. Nature **369:** 113–120.
38. QIU, Y. *et al.* 1994. Separation of subcellular compartments containing distinct functional forms of MHC class II. J. Cell Biol. **125:** 595–605.
39. FERRARI, G. *et al.* 1997. Distinct intracellular compartments involved in invariant chain degradation and antigenic peptide loading of major histocompatibility complex (MHC) class II molecules. J. Cell Biol. **139:** 1433–1446.
40. LAMB, C.A. *et al.* 1991. Invariant chain targets HLA class II molecules to acidic endosomes containing internalized influenza virus. Proc. Natl. Acad. Sci. USA **88:** 5998–6002.
41. MILLER, J. 1994. Endosomal localization of MHC class II-invariant chain complexes. Immunol. Res. **13:** 244–252.
42. VIVILLE, S. *et al.* 1993. Mice lacking the MHC class II-associated invariant chain. Cell **72:** 635–648.
43. BAKKE, O. & B. DOBBERSTEIN. 1990. MHC class II-associated invariant chain contains a sorting signal for endosomal compartments. Cell **64:** 707–716.
44. LOTTEAU, V. *et al.* 1990. Intracellular transport of class II MHC molecules directed by invariant chain. Nature **348:** 600–605.
45. ZHONG, G., P. ROMAGNOLI & R.N. GERMAIN. 1997. Related leucine-based cytoplasmic targeting signals in invariant chain and major histocompatibility complex class II

molecules control endocytic presentation of distinct determinants in a single protein. J. Exp. Med. **185:** 429–438.
46. SHI, G. *et al.* 1999. Cathepsin S required for normal MHC class II peptide loading and germinal center development. Immunity **10:** 197–206.
47. DENZIN, L.K. & P. CRESSWELL. 1995. HLA-DM iInduces CLIP dissociation from MHC class II ab dimers and facilitates peptide loading. Cell **82:** 155–165.
48. WEBER, D.A., B.D. EVAVOLD & P.E. JENSEN. 1996. Enhanced dissociation of HLA-DR-bound peptides in the presence of HLA-DM. Science **274:** 618–620.
49. SANDERSON, F. *et al.* 1994. Accumulation of HLA-DM, a regulator of antigen presentation, in MHC class II compartments. Science **266:** 1566–1569.
50. SLOAN, V. *et al.* 1995. Mediation by HLA-DM of dissociation of peptides from HLA-DR. Nature **375:** 802–806.
51. SHERMAN, M.A., D.A. WEBER & P.E. JENSEN. 1995. DM enhances peptide binding to class II MHC by release of invariant chain-derived peptide. Immunity **3:** 197–205.
52. NANDA, N.K. & A.J. SANT. 2000. DM determines the cryptic and immunodominant fate of T cell epitopes. J. Exp. Med. **192:** 781–788.
53. DENZIN, L.K. *et al.* 1997. Negative regulation by HLA-DO of MHC class II-restricted antigen processing. Science **278:** 106–109.
54. LILJEDAHL, M. *et al.* 1998. Altered antigen presentation in mice lacking H2-O. Immunity **8:** 233–243.
55. WEST, M.A., J.M. LUCOCQ & C. WATTS. 1994. Antigen processing and class II MHC peptide-loading in human B-lymphoblastoid cells. Nature **369:** 147–151.
56. RIBERDY, J.M. *et al.* 1994. Transport and intracellular distribution of MHC class II molecules associated with invariant chain in normal and antigen processing mutant cell lines. J. Cell Biol. **125:** 1215–1237.
57. HAMMOND, C. *et al.* 1998. The tetraspan protein CD82 is a resident of MHC class II compartments where it associates with HLA-DR, -DM, and -DO molecules. J. Immunol. **161:** 3282–3291.
58. BRACHET, V. *et al.* 1999. Early endosomes are required for major histocompatiblity complex class II transport to peptide-loading compartments. Mol. Biol. Cell **10:** 2891–2904.
59. POND, L. & C. WATTS. 1999. Functional early endosomes are required for maturation of major histocompatibility complex class II molecules in human B lymphoblastoid cells. J. Biol. Chem. **274:** 18049–18054.
60. KROPSHOFER, H. *et al.* 2001. Tetraspan microdomains distinct from lipid rafts enrich select peptide-MHC class II complexes. Nat. Immunol. **3:** 61–68.
61. ANDERSON, H.A., E.M. HILTBOLD & P.A. ROCHE. 2000. Concentration of MHC class II molecules in lipid rafts facilitates antigen presentation. Nat. Immunol. **1:** 156–162.
62. RAPOSO, G. *et al.* 1996. B lymphocytes secrete antigen-presenting vesicles. J. Exp.. Med. **183:** 1161–1172.
63. KUROSAKI, T. *et al.* 1995. Role of the Syk autophosphorylation site and SH2 domains in B cell antigen receptor signaling. J. Exp. Med. **182:** 1815–1823.
64. KABAK, S. *et al.* 2002. The direct recruitment of BLNK to immunoglobulin alpha couples the B-cell antigen receptor to distal signaling pathways. Mol. Cell Biol. **22:** 2524–2535.
65. SIEMASKO, K. *et al.* 2002. Receptor-facilitated antigen presentation requires the recruitment of B cell linker protein to Igalpha. J. Immunol. **168:** 2127–2138.
66. HARBURY, P.A. 1998. Springs and and zippers: coiled coils in SNARE-mediated membrane fusion. Curr. Biol. **6:** 1487–1491.
67. WICKNER, W. & A. HAAS. 2000. Yeast vacuole fusion: a window on organelle trafficking mechanisms. Annu. Rev. Biochem. **69:** 247–275.
68. ROBINSON, L.J. & T.F.J. MARTIN. 1998. Docking and fusion in neurosecretion. Curr. Opin. Cell Biol. **10:** 483–492.
69. CHEN, Y.A. *et al.* 1999. SNARE complex formation is triggered by Ca^{2+} and drives membrane fusion. Cell **97:** 165–174.
70. ZIMMERMAN, V.S. *et al.* 1999. Engagement of B cell receptor regulates the invariant chain-dependent MHC class II presentation pathway. J. Immunol. **162:** 2495–2502.

71. ANDERSON, H.A. & P.A. ROCHE. 1999. Phosphorylation regulates the delivery of MHC class II invariant chain complexes to antigen processing compartments. J. Immunol. **160:** 4850–4858.
72. ANDERSON, H.A. *et al.* 1999. Phosphorylation of the invariant chain by protein kinase C regulates MHC class II trafficking to antigen-processing compartments. J. Immun. **163:** 5435–5443.
73. LANKAR, D. *et al.* 2002. Dynamics of major histocompatibility complex class II compartments during B cell receptor-mediated cell activation. J. Exp. Med. **195:** 461–472.
74. LANKAR, D. *et al.* 1998. Syk tyrosine kinase and B cell antigen receptor (BCR) immunoglobulin-alpha subunit determine BCR-mediated major histocompatibility complex class II-restricted antigen presentation. J. Exp. Med. 188: 819–831.
75. KABAK, S. *et al.* 2000. The activation of PLCγ2 by direct recruitment of BLNK/Btk/Vav to the B cell antigen receptor. In preparation.
76. BROWN, B.K. *et al.* 1999. Trafficking of the Igα/Igβ heterodimer with membrane Ig and bound antigen to the major histocompatibility complex class II peptide-loading compartment. J. Biochem. **274:** 11439–11446.
77. VILEN, B.J., T. NAKAMURA & J.C. CAMBIER. 1999. Antigen-stimulated dissociation of BCR mlg from Ig-α/Ig-β: implications for receptor desensitizations. Immunity **10:** 239–248.
78. LANG, P. *et al.* 2001. TCR-induced transmembrane signal by peptide/MHC class II via associated Igα/Igβ dimers. Science **291:** 1537–1540.
79. KNIGHT, A.M. *et al.* 1997. Antigen endocytosis and presentation mediated by human membrane IgG1 in the absence of the Ig-alpha/Ig-beta dimer. EMBO J. **13:** 3842–3850.
80. NOORCHASHM, H. *et al.* 1999. I-Ag7-mediated antigen presentation by B lymphocytes is critical in overcoming a checkpoint in T cell tolerance to islet beta cells of nonobese diabetic mice. J. Immunol. **163:** 743–750.
81. FISCHER, K.-D., K. TEDFORD & J.M. PENNINGER. 1998. Vav links antigen-receptor signaling to the actin cytoskeleton. Semin. Immunol. 10.
82. HEALY, J.I. *et al.* 1997. Different nuclear signals are activated by the B cell receptor during positive versus negative signaling. Immunity **6:** 419–428.
83. KERSH, E., A. SHAW & P. ALLEN. 1998. Fidelity of T cell activation through multistep T cell receptor zeta phosphorylation. Science **281:** 572–575.
84. STODDART, A. *et al.* 2002. Lipid rafts unite signaling cascades with clathrin to regulate BCR internalization. Immunity **17:** 451–462.

Activation of Rheumatoid Factor (RF) B Cells and Somatic Hypermutation Outside of Germinal Centers in Autoimmune-Prone MRL/*lpr* Mice

MARK J. SHLOMCHIK,[a,b] CHAD W. EULER,[a] SEAN C. CHRISTENSEN,[a,b] AND JACQUELINE WILLIAM[a,b]

[a]*Department of Laboratory Medicine, Yale University School of Medicine, New Haven Connecticut 06520, USA*

[b]*Section of Immunobiology, Yale University School of Medicine, New Haven, Connecticut 06520, USA*

ABSTRACT: Two critical questions need to be answered concerning the origins of autoreactive B cells in autoimmunity. First, how are autoreactive B cells regulated in normal situations? Second, how do such B cells escape tolerance mechanisms during autoimmunity? To address these questions, an Ig transgenic (Tg) mouse system based on the rheumatoid factor (RF) specificity has been developed. Tg mice express either the H or both H and L chains from AM14, an MRL/*lpr*-derived RF. Using this system, it was first shown that RF B cells are neither tolerized nor activated in a normal mouse. New insights into the timing and sites of initial RF B cell activation in MRL/*lpr* mice have been gained recently. RF B cells are activated. It was found, unexpectedly, that RF B cell activation, somatic hypermutation, and selection take place outside of the germinal center. We discuss the implications of this for the regulation of autoreactive B cells as well as for the regulation of hypermutation.

KEYWORDS: rheumatoid factor; somatic hypermutation; immunoglobulin V gene; systemic lupus erythematosus; autoantibody; autoimmunity

INTRODUCTION

Autoreactive B cells play central roles in systemic autoimmune disease.[1,2] They secrete autoantibodies that can be pathogenic, and they also promote T cell autoimmunity. Two critical questions need to be answered concerning the origins of autoreactive B cells in autoimmunity. First, how are autoreactive B cells regulated in normal situations? Second, how do such B cells escape tolerance mechanisms during autoimmunity?

Address for correspondencer: Mark Shlomchik, MD, PhD, Department of Laboratory Medicine, Yale University School of Medicine, 333 Cedar Street, P.O. Box 208035, New Haven, Connecticut 06520-8035. Voice: 203-688-2089; fax: 203-688-2748.
mark.shlomchik@yale.edu

Ann. N.Y. Acad. Sci. 987: 38–50 (2003). © 2003 New York Academy of Sciences.

Substantially more is known about the answer to the first question than to the second. This is largely due to the pioneering studies of Nemazee and Goodnow, who used immunoglobulin (Ig) transgenic (Tg) mice to study the fates of developing B cells that recognized "self." Nemazee found that B cells specific for membrane-bound self-MHC I were deleted if they could not edit their receptors away from the self-reactive specificity.[3–5] In the mice studied by Goodnow, the "autoantigen" was actually hen egg lysozyme (HEL), itself introduced as a transgene. When HEL was expressed as a soluble molecule, HEL-specific B cells developed, but were anergic and also had shortened lifespans in the periphery.[6]

When these results were obtained, it was suggested that breakdowns in either deletion/editing or anergy would be responsible for autoimmunity.[7,8] However, when this idea was tested by crossing the MHC and HEL models onto the strongly autoimmune-prone Fas-deficient genetic background, the most significant finding was that self-tolerance remained intact.[9,10] This in turn implicated other stages of B cell development as potential tolerance checkpoints. Indeed, it has been speculated for some time that autoreactive B cells of relatively low affinity that bypass deletion and anergy might need to be controlled in the periphery.[11,12] In particular, it has been suggested that B cells in germinal centers (GCs) undergoing somatic hypermutation may (need to) be filtered to eliminate newly generated or higher affinity autoreactive B cells.[13–15]

Why was loss of tolerance not observed in the MHC or HEL models? One possible reason is that these systems modeled very high-affinity B cells. These cells can be eliminated efficiently by central tolerance mechanisms, and cannot be the precursors for autoantibodies in autoimmune disease. Another possibility is that the target antigens in the Nemazee and Goodnow systems are different from those that are dominant in systemic autoimmunity. It is well known that nuclear autoantigens, particularly those containing nucleic acids, are preferred targets for autoantibodies.[16] Anti-IgG or rheumatoid factor (RF) autoantibodies are also common in certain systemic autoimmune diseases.[17] It has not been clear why these autoantigens are preferred targets, although it has been suggested that many of these autoantigens contain repeating determinants, are highly charged, or are released and processed in a particular way during cellular apoptosis.[18,19]

To gain insight into how autoantibodies that express typical disease-associated specificities are regulated, we and others created Tg mouse models using the Ig V regions isolated from autoantibody-secreting hybridomas generated from spleens of autoimmune-prone mice. Anti-DNA,[20–24] anti-Sm,[25] and RF[26–30] have been the major specificities studied. Here we will focus on our work with the RF model called AM14. The AM14 RF model has some advantages over these other authentic autoantibody Tg models. First, the RF autoantigen, IgG2a, is a polymorphic protein; AM14 binds only to the IgG2a^a allele and not to the IgG2a^b allele. Thus, we have access to mice that express (IgH^a) or lack (IgH^b) the autoantigen, as there are mouse strains that are congenic for these IgH alleles. Of course, it is impossible to have a mouse that lacks the other major autoAgs, such as DNA. Furthermore, we have already found that AM14 B cells in normal mice develop and populate the periphery even in the presence of the autoantigen and that such cells appear immunocompetent, and not anergized.[26,27] This has allowed us to study the later stages of B cell tolerance and how they break down in autoimmunity, providing novel insights into the way B cell autoimmunity initiates and how it evolves. The results of some of these ongoing studies are described and discussed below.

RESULTS

Since we had already shown that AM14 B cells could develop normally and were immuncompetent in the BALB/c (IgH^a) strain, we turned our attention to whether these cells would undergo spontaneous, autoAg-specific activation in an autoimmune-prone genetic background. Although the original AM14 RF B cell was activated in an autoimmune mouse, and activated anti-IgG2a B cells would be rare in a normal mouse, there had been no direct evidence that AM14 was in fact the result of autoAg-specific activation. We first studied the AM14 Tg model on the B6.Fas^{lpr} (B6/*lpr*) model.[30] Such mice are Fas-deficient and are known to have a systemic autoimmune syndrome characterized by antichromatin and RF autoantibodies.[31–34] We found that older B6/*lpr*.IgH^a AM14 Tg mice often had large numbers of RF antibody-forming cells (AFCs).[30] However, AM14 Tg B6/*lpr* mice (IgH^b) mice that lacked the autoAg did not show significant numbers of RF AFC. Surprisingly, this was true of mice that carried only the H chain of the HL pair that constitutes the complete AM14 antibody. Another interesting feature of these data was the very wide scatter among individual mice in the number of AFCs. This scatter mirrored that seen among inbred MRL/*lpr* mice in the titers of autoantibodies. In the case of AM14 Tg mice, it was even more remarkable in the sense that the B cell repertoire was highly restricted already, yet there was stochastic onset of autoimmunity. Altogether, these results indicated that autoantigen was specifically inducing autoantibody production. However, these studies did not define why, where, when, or how this was occurring.

To better define these parameters, we began to study the H-only Tg mice, this time on the MRL/*lpr* background. MRL/*lpr* mice have a more severe form of autoimmune disease than B6/*lpr* mice, with a spectrum of autoantibodies and clinical features that more resembles systemic lupus erythematosus. The H Tg mice had the advantage that there was a very low background of RF AFC in the autoAg-negative strain, and a very substantial increase in RF AFC was seen in many of the mice as they aged. We used an anti-idiotype antibody (4-44) to the HL pair of AM14 to identify RF B cells in tissue sections. In H Tg mice, this antibody identifies RF B cells that have expressed the same Vκ8 gene found in the original AM14 antibody—or a very similar gene—in combination with Jκ4 or Jκ5; such B cells have RF specificity.

We found that in most older MRL/*lpr* mice there were numerous darkly staining Id^+ cells in the spleens (data not shown[35]). These were located at the edge of the T cell zone, adjacent to the red pulp. This location corresponds to the marginal sinus-bridging channels, a site where the initial plasma cell focus reaction occurs during the immune response to various exogenous antigens. These regions did not contain follicular dendritic cells, as determined by staining with anti-CD21/35 (CR1/2). Perhaps most remarkably, rarely did we see any GCs with detectable idiotype positive cells in AM14 H Tg MRL/*lpr* spleens, even when we conducted extensive searches through multiple serial sections.[35] This was most surprising, due to the extensive evidence of somatic hypermutation in V genes of autoantibodies. From these earlier studies, it was strongly expected to find RF B cells proliferating in GCs.[36–45]

We demonstrated that RF B cells at the T zone–red pulp border were proliferating in two ways. First, we injected mice with BrdU and sacrificed them 2 h later. Only cells actively in the S phase during the course of 2 h would take up BrdU and then be detectable with anti-BrdU. Indeed, we saw many RF B cells in these areas that had taken up BrdU.[35] FACS analysis of spleens prepared in this fashion demonstrated that 10–15% of RF B cells took up BrdU in just 2 h, indicative of a high proliferative rate. This was corroborated by staining for Ki67, which is positive in nuclei of proliferating cells (data not shown). Thus, the T zone–red pulp border is a site of proliferation for RF B cells, and does not simply contain terminally differentiated nondividing plasma cells.

The fact that cells at this site were proliferating led us to ask whether these cells were undergoing active somatic hypermutation. To test this, we decided to microdissect ("pick") Id$^+$ RF B cells from these sites and to sequence their Vκ regions. We reasoned that most if not all of the B cells would express a Vκ8 gene identical to or similar to that in the original AM14 hybridoma, and designed amplification primers accordingly. Indeed, more than three-quarters of all microdissections of 10 or more cells (10–30 was a typical size) could be amplified to yield a band of the correct size. From each pick, we isolated the PCR product, then cloned it to make a bacterial library. Each clone in the library could in principle represent a V region from an individual cell, though undoubtedly there will be repeat representations in the library. Nonetheless, as long as the number of bacterial clones that were sequenced was somewhat less than the number of cells picked, the repeats shouldn't be too numerous. Moreover, sequences that are different are manifestly from different cells, since the PCR error rate is less than 1 bp per 10 sequences. Sample data from this type of analysis are shown in FIGURE 1. In this case, and in many others,[35] we found substantial heterogeneity among the recovered sequences. Moreover, the pattern of shared and unique mutations was typically hierarchical, allowing us to reconstruct genealogies that were consistent with ongoing mutation and division. Because these picks were so small and because most if not all of the sequences within a single pick were related (i.e., shared VJ junctions as well as mutations), we concluded that the cells that were microdissected were undergoing active mutation. This conclusion is also consistent with the ongoing RF B cell proliferation at this site. Had mutation been occurring elsewhere followed by migration, then we would have expected independent B cell clones to be randomized among the microdissected area. Extensive searches for locations in which mutation could have been occurring outside of the T zone–red pulp area were negative. GCs would have been the most likely other site of mutation. However, as mentioned, we rarely saw Id$^+$ GCs. In some spleens, extensive sectioning and staining (one spleen with ~150 sections) failed to reveal any GCs, let alone Id$^+$ GCs; yet ongoing hypermutation was readily detected at extrafollicular sites. We also directly evaluated some of the rare Id$^-$ GCs that were seen; of seven that were microdissected in at least two places, only one gave a PCR product (TABLE 1). Control Id$^+$ GCs induced by immunization with immune complexes were PCR-positive in every case. Most importantly, when we sequenced clones from the positive PCR product, they were unmutated and genetically unrelated to the mutated clones that were isolated from nearby T zone–red pulp border areas (not shown). These latter clones had Vκ8–Jκ4 joins, whereas the joins in the GC were Vκ8–Jκ5. Thus, we could demonstrate that ongoing hypermutation was occurring at the extrafollicular sites and could exclude that the cells were coming from nearby GCs.

Mutation at these sites was occurring at a high rate. To estimate the rate, we used both analytical and simulation-based computer modeling to analyze the branching patterns of the genealogies. When this information was coupled with the knowledge of how many nearby (and presumably clonally related) cells were analyzed, this al-

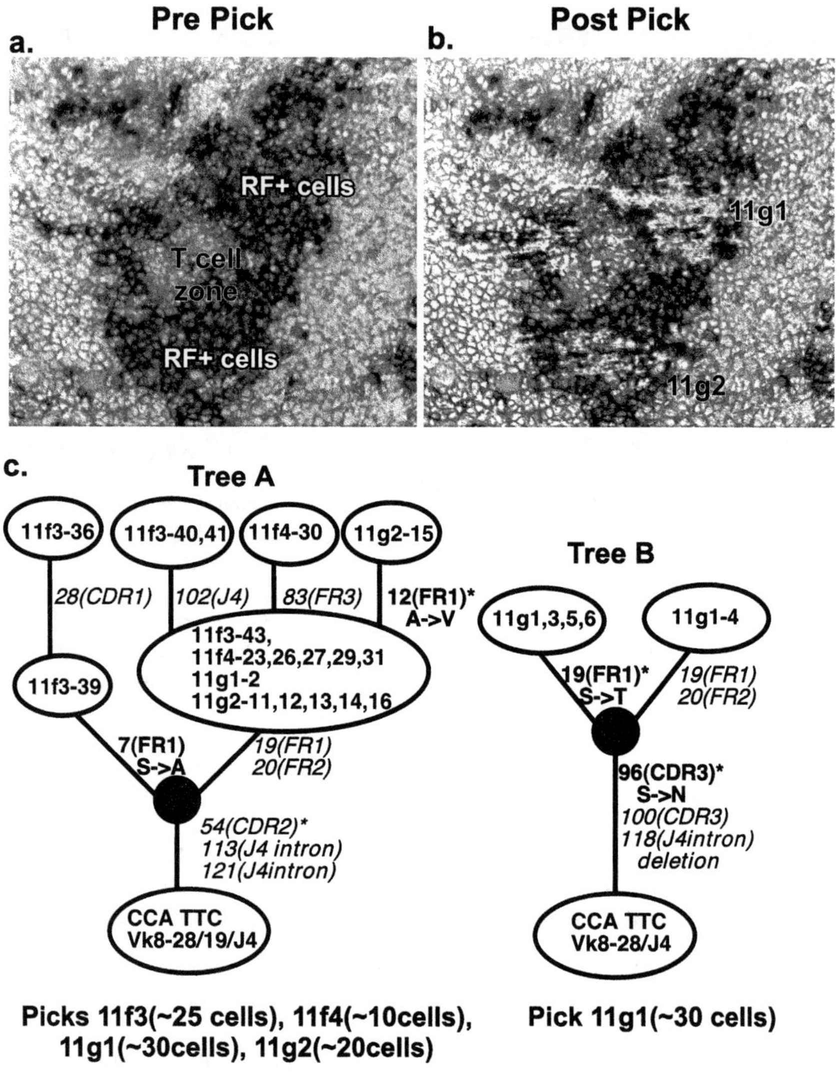

FIGURE 1. Genealogical trees generated from sequences of picks from mouse 2540 demonstrating ongoing somatic V region diversification. (**a**) Spleen section of mouse 2540 prior to microdissection. AM14 Id$^+$ cells are *darkly stained.* The T cell zone (*light gray*) is labeled on the figure. (**b**) The same spleen section following microdissection is demonstrated. Picks 11g1 and 11g2, which had approximately 30 and 20 Id$^+$ cells, respectively, are indicated. (**c**) Individual clones from the picks were assigned to a particular genealogical tree on the basis of shared and unique mutations. Each *open circle* represents one or more clones with the same sequence. The numbers of mutated codons are written at the sides of branches. Regions affected (i.e., FR, CDR, J region) are in *parentheses*. Amino acid changes are written below the codon. Silent mutations are in *italics*. Mutations that have been seen repeatedly in multiple mice, considered hotspots, are indicated by *asterisks*. Inferred intermediates in clonal evolution are represented by *dark circles*. Picks 11f3 and 11f4 (c) were taken from the same cluster as 11g1 and 11g2, in an adjacent section. *Circles* at the base of trees represent germline sequences. (Adapted from William *et al.*[35])

TABLE 1. Comparison of Id⁻ GC picks to Id⁺ GC picks

	Picked	
	Id^- GCs	Id^+ GCs[a]
Number of PCR Positive GCs: Total Picked	1:7	10:10
% +ve	14.29%	100.00%
Number of PCR Positive Picks: Total Picked	2:17[b]	21:27
% +ve	11.76%	77.78%

[a] Id^+ GCs isolated from H Tg Balb/c and $MRL^{+/+}$ mice.
[b] Both PCR-positive picks in the Id-negative GC are Jκ5. In contrast, all studied T zone–red pulp spleen foci are Jκ4, including ones adjacent to the GC.

lowed an excellent estimate of the rate (S. Kleinstein *et al.*, manuscript in preparation). RF B cells proliferating in the extrafollicular regions of the spleen are mutating at an average rate of ~1 mutation per cell per division (~10^{-3}/bp/gen). This rate is similar to, if not higher than, previous estimates of the GC rate. To provide a fairer comparison, we applied our methods to estimate the rate of mutation in Vh regions isolated from GCs formed during the response to nitrophenyl-chicken gamma globulin, using a database provided by G. Kelsoe. Again, the rate was similar or perhaps slightly lower in the GC than in the RF B cells (manuscript in preparation).

It is not yet clear whether these mutations are selected in a process that leads to affinity maturation of RF B cells. However, indirect evidence suggests that there is antigen-based selection, and that affinity maturation does occur. Analysis of R/S patterns among the mutations (TABLE 2) demonstrates strong selection against R mutations in the framework regions (FRs), consistent with the requirement for mutated cells having a functional BCR that can continue to bind antigen. Selection against R mutations was also evident in complementarity-determining region (CDR) one and CDR2. It is likely that these regions are well-suited to binding IgG2a, and there may be many fewer R mutations that improve affinity here than destroy it. Although it is typical to associate selection in favor of R mutations with positive selection (and when significant enrichment of R mutations does occur, it is highly indicative of positive selection), negative selection in CDRs must also occur in the context of selection for Ag binding. In this regard, it seems that only a few Vκ8 genes are capable of reconstituting RF activity in the context of the AM14 H chain, suggesting that there must be certain key features in some of the CDRs that endow ability to bind IgG2a (M.J.S., unpublished observations). Nonetheless, we did see a marked enrichment of R mutations in CDR3 (ratio of 12.5 when the random expected ratio is 4.85; $P = 0.094$; TABLE 2). The *P*-value of this ratio did not quite reach significance in our current data set, most likely owing to the very high ratio expected at random.

Another approach to determining whether there is affinity maturation is to measure the affinity of the RF secreted by the mutant cells. Although it is difficult to directly measure RF affinities, which are typically low, we found it was possible to distinguish low- and higher-affinity RFs by decreasing the density of the target IgG2a on an ELISA plate.[28] As the density of Ag coating is reduced, low-affinity RFs lose binding, whereas there is little effect on higher-affinity RFs. We used this

TABLE 2. Replacement and silent mutation distributions in Vκ8-19 sequences

Region	Observed Replace-ments	Observed Silents	Observed R:S[a]	Theoretically Expected R:S[b]	P^c
CDR1	76	28	2.71	4.11	0.017
CDR2	6	10	0.60	3.00	1.4×10^{-3}
CDR3	25	2	12.50	4.85	0.094
CDRs	107	40	2.68	3.98	8.5×10^{-3}
FR1	43	11	3.91	2.74	0.074
FR2	53	45	1.18	3.56	$<1 \times 10^{-6}$
FR3	73	36	2.03	2.99	0.019
FRs	169	92	1.84	3.04	$4.2\ 10^{-5}$
Jκ4	15	5	3.00	3.13	0.2

[a]R = Replacement; S = silent.

[b]Quotient of total possible R to total possible S, based on inherent replacement frequency of the germline Vκ8-19 gene (Ref. 1).

[c]Probability that excess or scarcity of R mutations resulted in a region only from chance (Ref. 1).

assay preliminarily to assess the affinity of several RF B-cell hybridomas isolated from unstimulated B cells of AM14 H Tg MRL/*lpr* mice. In each case (data not shown), the hybridomas had a higher apparent affinity than the original AM14 antibody (expressed as an IgM isolated from a transfectoma). We are not certain which amino acids in the hybridomas control higher-affinity RFs. We also applied the same assay to supernatants of overnight cultures of splenocytes of AM14 H Tg MRL/*lpr* mice. RF AFCs secrete measurable amounts of RF into such supernatants. Again, these polyclonal supernatants showed binding characteristics consistent with higher affinity than the RF protein encoded by the germline sequence (J.W., C.E., M.S., manuscript in preparation).

DISCUSSION

RFs are common autoantibodies. High titer IgG antibodies are found in RA,[17,46,47] and to a lesser degree in Sjogren's syndrome and systemic lupus erythematosus.[16,48–52] In addition, IgM RFs are frequently found, at least transiently, in normal individuals undergoing immune responses.[53–55] Finally, RFs can accompany some chronic infections, such as tuberculosis.[56] These features make the study of RFs particularly interesting. We have used RFs as a model system to study how they are induced in spontaneous autoimmune diseases and how they are regulated in normal mice.

We originally discovered that RF B cells in AM14 Tg mice are not obviously tolerized in the presence of autoantigen. Instead, they remain immunocompetent, though they are not activated to spontaneously produce autoantibodies.[26,27] This set the stage for studies of RF regulation in autoimmune-prone mice, since it had been shown already that earlier stages of B cell tolerance, like anergy and deletion, did not break down during systemic autoimmunity.[9,10] Our studies in B6/*lpr* mice, and later in MRL/*lpr* mice, have led to a number of surprising findings. These have implications for the control of somatic hypermutation, as well as how autoantibodies are regulated, as will be discussed.

Initially we found that, only in the presence of the autoantigen IgG2a[a] and only in autoimmune-prone (Fas-deficient) strains, RF B cells indeed became spontaneously activated to secrete autoantibody. This was an age-dependent phenomenon, in that it was substantially more prevalent in older mice. Even among an age-matched cohort, however, there was a large degree of mouse-to-mouse variability, indicating a stochastic component to the onset of autoimmunity.[30] Immunohistologic studies revealed that the site of RF B cell proliferation and differentiation was at the T zone–red pulp border.[35] This site appears to be the same as that where B cells initially responding to hapten-carrier immunization proliferate and develop AFCs.[57,58]

This finding created a puzzle, in that we knew from many other studies that RFs and other autoantibodies were highly mutated in MRL/*lpr* mice, yet we saw them proliferating outside of GCs. Microdissection and sequencing of the endogenous (i.e., non-Tg) Vκ genes expressed in B cells of H Tg mice demonstrated clearly that high rate, extensive somatic hypermutation was ongoing in these extrafollicular B cells. Mutation at the T zone–red pulp border demonstrates that mutation does not require a GC *per se*. There have been a few other instances in immunodeficient individuals in which mutation outside of the GC has been claimed. However, mutation in RF Tg Fas-deficient mice differs fundamentally from cases in which mutation was observed in animals with severe immunological defects. In LT-α-deficient mice, mutated Ig V region mRNAs were recovered from spleens following repeated immunization.[59] Since these mice lack follicular dendritic cells and GCs in their spleens, these data were interpreted to mean that mutation could occur outside of GCs. Importantly, the site of mutation in these mice was never determined; and LT-deficient mice do have peanut agglutinin-positive clusters of B cells, and do in some cases have detectable lymph nodes with GCs, thus raising the possibility that mutation occurred in these GCs or GC-like structures.[60,61] Furthermore, only a very low frequency of mutations and of mutated cells was found even after repeated high-dose immunization.[59,62] In contrast, AM14 Tg B cells are undergoing high rate somatic hypermutation. Similarly, some mutation has recently been observed in a subset of IgM^+ B cells in humans deficient in CD40L and in whom GCs are rarely observed.[63] Again, the histologic site of mutation was not defined. Mutation in this subset of B cells in CD40L-deficient humans is thought to represent antigen-independent development of a preimmune repertoire, as occurs in sheep and rabbits, and thus would be quite different from autoantigen-driven hypermutation.

In contrast to the LT-α and CD40L-deficient cases, mutation at the T zone–red pulp border in Fas-deficient mice could not result simply from an inability to form GCs. Fas-deficient animals have normal GC formation and follicular dendritic cell development; and Fas-deficiency in young (8–1-week-old) mice does not affect the kinetics or size of GCs after immunization,[64,65] although GC formation can be reduced in older mice. Indeed, we do see Id^- GCs in some mice in which we also see ongoing mutation at extrafollicular sites (data not shown). Thus, it is extremely unlikely that mutation at these sites is due to inability to form GCs as a direct result of the Fas defect.

We therefore propose that somatic hypermutation is induced whenever B cells are stimulated to proliferate for a substantial number of cell cycles under conditions where antigen and T-cell-derived signals required for mutation are present.[66] Normally, proliferative foci at the T zone–red pulp border begin to involute by 8 days postimmunization and disappear by the peak of the GC reaction.[57] The absence of evident mutation at this site on day ten[67] could be explained by lack of sufficient numbers of cell divisions and/or insufficient T cell help, which may wane as the focus begins to decay. This hypothesis would explain mutation in GCs, *in vitro* under appropriate conditions,[66] and at extrafollicular sites in chronic autoimmune responses. The apparent restriction of mutation to GCs, which was heretofore presumed to be due to special signals present only in the GC, instead is more likely due to the fact that under most circumstances studied to date, B cell responses that last beyond several days are restricted to GCs.

Since Fas-deficient mice can form GCs that support normal somatic hypermutation,[64,65] why then does mutation occur in extrafollicular sites in the case of spontaneous RF autoimmune responses in MRL/*lpr* mice? It is not because RF B cells themselves are incapable of entering GCs or that we cannot detect them there, as we have observed RF B cells in GCs in lymph nodes (and occasionally in spleen) of some of these same autoimmune mice, as well as in normal and autoimmune RF Tg mice that had been immunized with IgG2a-containing immune complexes (manuscript in preparation). Instead we believe mutation occurs at this site because of the unusually long duration of RF B cell proliferation at the T zone–red pulp border in these mice. Such foci are probably sites of sustained proliferation for many weeks and perhaps longer in H Tg MRL/*lpr* mice.

However, this explanation then begs the question of why there is a sustained autoimmune reaction at the T zone–red pulp border. A number of factors may play a role. One is that Ag may preferentially locate at this site. Indeed, we do have evidence for the presence there of cell-associated IgG2a in our autoimmune RF Tg mice (unpublished data). This IgG2a, most likely in the form of immune complexes, could be captured by RF B cells binding via sIg or by FcR^+ cells, including other B cells as well as DCs, which are also abundant at this site.[35,68] In addition, it could be presented by IgG2a-secreting plasma cells that would also reside there. We are investigating the roles of Ag and its location in this process by several means.

The nature of signals received by the RF B cell might also dictate its subsequent migration and differentiation patterns. Leadbetter *et al.*[69] have recently shown that RF B cells that bind chromatin–antichromatin immune complexes are stimulated via Toll-like Receptor 9 (TLR9), a receptor thought to detect the presence of bacterial DNA enriched in CpG dinucleotides.[70] Binding of chromatin–antichromatin immune complexes by RF B cells results in markedly enhanced proliferation *in vitro* compared to control immune complexes and provides a potential link between the innate immune system and autoimmunity. Evidently, chromatin contains sufficient CpG-like DNA to stimulate B cells, and/or B cells that take it up through their BCR are especially efficient at delivering the TLR9 ligand to its receptor. In any case, immune complexes that contain chromatin can provide a signal via TLR9 to cells that take up the complexes efficiently, just as foreign antigens do by providing ligands for a variety of TLRs. We therefore proposed that autoantigens that could provide

TLR signals may be prime targets in systemic autoimmunity and that this might explain the spectrum of common autoantibodies.[69] Less well understood is how TLR signaling that is coincident with BCR signaling will affect the differentiation and migration of B cells *in vivo*. It is tempting to speculate that the unique features, such as the site of proliferation of the AM14 autoimmune response, may relate in part to the presence of a TLR9 ligand in the autoantigen.

Since we have identified proliferating and mutating cells outside GCs and away from follicular dendritic cells, it is interesting to ask how these cells differ from GC cells. This analysis is ongoing, though we have evidence for a population of cells that resembles plasmablasts and another that resembles an activated B cell (manuscript in preparation). Neither expresses the receptor for PNA, as expected from the immunostaining and their location. We are actively pursuing how these cells might differ from similar cells in normal mice, if at all. We are also working on which of the two populations (or both) are undergoing mutation as they proliferate; we already know that both populations contain mutations in their V regions (manuscript in preparation).

Future studies will further characterize the nature of the autoimmune response and address the limiting factors for the onset of autoimmunity. Why does autoimmunity take so long to develop? This latency and the stochastic timing of onset are common features of many autoimmune diseases in humans and in genetically predisposed strains of mice. We believe that the nature and availability of autoantigen, as well as the availability of T cell help, will be important limiting factors. The RF Tg system should be very useful in the near future in testing these hypotheses and developing a better understanding of the initiation and evolution of B cell autoimmunity.

REFERENCES

1. CHAN, O.T.M., M.P. MADAIO & M.J. SHLOMCHIK. 1999. The central and multiple roles of B cells in lupus pathogenesis. Immunol. Rev. **169:** 107–121.
2. SHLOMCHIK, M.J., J. CRAFT & M.J. MAMULA. 2001. From T to B and back again: positive feedback in systemic autoimmune disease. Nature Rev. Immunol. **1:** 147–153.
3. RUSSELL, D.M., Z. DEMBIC, G. MORAHAN, *et al.* 1991. Peripheral deletion of self-reactive B cells. Nature **354:** 308–311.
4. TIEGS, S.L., D.M. RUSSELL & D. NEMAZEE. 1993. Receptor editing in self-reactive bone marrow B cells. J. Exp. Med. **177:** 1009–1020.
5. NEMAZEE, D.A. & K. BURKI. 1989. Clonal deletion of B lymphocytes in a transgenic mouse bearing anti-MHC class-I antibody genes. Nature **337:** 562–566.
6. GOODNOW, C.C., J. CROSBIE, S. ADELSTEIN, *et al.* 1988. Altered immunoglobulin expression and functional silencing of self-reactive B lymphocytes in transgenic mice. Nature **334:** 676–682.
7. GOODNOW, C.C., R.A. BRINK & A.E. ADAMS. 1991. Breakdown of self-tolerance in anergic B lymphocytes. Nature **352:** 532–536.
8. BASTEN, A., R. BRINK, E. ADAMS, *et al.* 1991. Self tolerance in the B-cell repertoire. Immunol. Rev. **122:** 5–19.
9. RATHMELL, J.C., M.P. COOKE, W.Y. HO, *et al.* 1995. CD95 (Fas)-dependent elimination of self-reactive B cells upon interaction with CD4+ T cells. Nature **376:** 181–184.
10. RUBIO, C.F., J. KENCH, D.M. RUSSELL, *et al.* 1996. Analysis of central B cell tolerance in autoimmune-prone MRL/*lpr* mice bearing autoantibody transgenes. J. Immunol. **157:** 65–71.
11. NOSSAL, G.J.V. 1988. Somatic mutations in B lymphocytes: new perspectives in tolerance research. Immunol. Cell Biol. **66:** 105–110.

12. GOODNOW, C.C., S. ADELSTEIN & A. BASTEN. 1990. The need for central and peripheral tolerance in the B cell repertoire. Science **248:** 1373–1379.
13. NOSSAL, G.J.V. 1994. Negative selection of lymphocytes. Cell **76:** 229–239.
14. PULENDRAN, B., G. KANNOURAKIS, S. NOURI, *et al.* 1995. Soluble antigen can cause enhanced apoptosis of germinal-centre B cells. Nature **375:** 331–334.
15. SHOKAT, K.M. & C.C. GOODNOW. 1995. Antigen-induced B-cell death and elimination during germinal-centre immune responses. Nature **375:** 334–338.
16. TAN, E.M. 1989. Antinuclear antibodies: diagnostic markers for autoimmune diseases and probes for cell biology. Adv. Immunol. **44:** 93–151.
17. KARSH, J., S.P. HALBERT, M. ANKEN, *et al.* 1982. Anti-DNA, anti-deoxyribonucleoprotein and rheumatoid factor measured by ELISA in patients with systemic lupus erythematosus, Sjogren's syndrome and rheumatoid arthritis. Int. Arch. Allergy Appl. Immunol. **68:** 60–69.
18. ROSEN, A., R.L. CASCIOLA & J. AHEARN. 1995. Novel packages of viral and self-antigens are generated during apoptosis. J. Exp. Med. **181:** 1557–1561.
19. CASCIOLA-ROSEN, L., F. ANDRADE, D. ULANET, *et al.* 1999. Cleavage by granzyme B is strongly predictive of autoantigen status: implications for initiation of autoimmunity. J. Exp. Med. **190:** 815–826.
20. ERIKSON, J., M.Z. RADIC, S.A. CAMPER, *et al.* 1991. Expression of anti-DNA immunoglobulin transgenes in non-autoimmune mice. Nature **349:** 331–334.
21. GAY, D., T. SAUNDERS, S. CAMPER & M. WEIGERT. 1993. Receptor editing: an approach by autoreactive B cells to escape tolerance. J. Exp. Med. **177:** 999–1008.
22. CHEN, C., M.Z. RADIC, J. ERIKSON, *et al.* 1994. Deletion and editing of B cells that express antibodies to DNA. J. Immunol. **152:** 1970–1982.
23. CHEN, C., Z. NAGY, M.Z. RADIC, *et al.* 1995. The site and stage of anti-DNA B-cell deletion. Nature **373:** 252–255.
24. PEWZNER-JUNG, Y., D. FRIEDMANN, E. SONODA, *et al.* 1998. B cell deletion, anergy, and receptor editing in "knock in" mice targeted with a germline-encoded or somatically mutated anti-DNA heavy chain. J. Immunol. **161:** 4634–4645.
25. SANTULLI-MAROTTO, S., M.W. RETTER, R. GEE, *et al.* 1998. Autoreactive B cell regulation: peripheral induction of developmental arrest by lupus-associated autoantigens. Immunity **8:** 209–219.
26. SHLOMCHIK, M.J., D. ZHARHARY, S. CAMPER, *et al.* 1993. A rheumatoid factor transgenic mouse model of autoantibody regulation. Int. Immunol. **5:** 1329–1341.
27. HANNUM, L.G., D. NI, A.M. HABERMAN, *et al.* 1996. A disease-related RF autoantibody is not tolerized in a normal mouse: implications for the origins of autoantibodies in autoimmune disease. J. Exp. Med. **184:** 1269–1278.
28. WANG, H. & M.J. SHLOMCHIK. 1997. High affinity rheumatoid factor transgenic B cells are eliminated in normal mice. J. Immunol. **159:** 1125–1134.
29. WANG, H.W. & M.J. SHLOMCHIK. 1998. Maternal Ig mediates neonatal tolerance in rheumatoid factor transgenic mice but tolerance breaks down in adult mice. J. Immunol. **160:** 2263–2271.
30. WANG, H. & M.J. SHLOMCHIK. 1999. Autoantigen-specific B cell activation in Fas-deficient rheumatoid factor immunoglobulin transgenic mice. J. Exp. Med. **190:** 639–649.
31. SOBEL, E.S., T. KATAGIRI, K. KATAGIRI, *et al.* 1991. An intrinsic B cell defect is required for the production of autoantibodies in the *lpr* model of murine systemic autoimmunity. J. Exp. Med. **173:** 1441–1449.
32. HALPERN, M.D., C.L. FISHER, P.L. COHEN & R.A. EISENBERG. 1992. Influence of the Ig H chain locus on autoantibody production in autoimmune mice. J. Immunol. **149:** 3735–3740.
33. WARREN, R.W., D.M. SAILSTAD & D.S. PISETSKY. 1984. Monoclonal rheumatoid factors from B6-lpr/lpr mice. Clin. Exp. Immunol. **58:** 731–736.
34. WARREN, R.W., D.M. SAILSTAD, S.A. CASTER & D.S. PISETSKY. 1984. Specificity analysis of monoclonal anti-DNA antibodies from B6-lpr/lpr mice. Arthritis Rheum. **27:** 545–551.
35. WILLIAM, J., C. EULER, S. CHRISTENSEN & M.J. SHLOMCHIK. 2002. Somatic hypermutation outside of GCs in autoantibody responses. Science **297:** 2066–2070.

36. SHLOMCHIK, M.J., A. MARSHAK-ROTHSTEIN, C.B. WOLFOWICZ, *et al.* 1987. The role of clonal selection and somatic mutation in autoimmunity. Nature **328:** 805.
37. SHLOMCHIK, M.J., A.H. AUCOIN, D.S. PISETSKY & M.G. WEIGERT. 1987. Structure and function of anti-DNA antibodies derived from a single autoimmune mouse. Proc. Natl. Acad. Sci. USA **84:** 9150–9154.
38. SHLOMCHIK, M.J., M.A. MASCELLI, H. SHAN, *et al.* 1990. Anti-DNA antibodies from autoimmune mice arise by clonal expansion and somatic mutation. J. Exp. Med. **171:** 265.
39. RADIC, M.Z., M.A. MASCELLI, J. ERIKSON, *et al.* 1989. Structural patterns in anti-DNA antibodies from MRL/lpr mice. Cold Spring Harbor Symp. Quant. Biol. **54:** 933–946.
40. TILLMAN, D.M., N.T. JOU, R.J. HILL & T.N. MARION. 1992. Both IgM and IgG anti-DNA antibodies are the products of clonally selective B cell stimulation in (NZB × NZW)F1 mice. J. Exp. Med. **176:** 761–779.
41. LOSMAN, M.J., T.M. FASY, K.E. NOVICK & M. MONESTIER. 1992. Monoclonal autoantibodies to subnucleosomes from a MRL/Mp-+/+ mouse: oligoclonality of the antibody response and recognition of a determinant composed for histones H2A, H2B, and DNA. J. Immunol. **148:** 1561–1569.
42. MONESTIER, M. 1991. Variable region genes of anti-histone autoantibodies from a MRL/Mp-*lpr/lpr* mouse. Eur. J. Immunol. **21:** 1725–1731.
43. WINKLER, T.H., H. FEHR & J.R. KALDEN. 1992. Analysis of immunoglobulin variable region genes from human IgG anti-DNA hybridomas. Eur. J. Immunol. **22:** 1719–1728.
44. RANDEN, I., D. BROWN, K.M. THOMPSON, *et al.* 1992. Clonally related IgM rheumatoid factors undergo affinity maturation in the rheumatoid synovial tissue. J. Immunol. **148:** 3296–3301.
45. THOMPSON, K.M., M. BORRETZEN, I. RANDEN, *et al.* 1996. V-gene repertoire and hypermutation of rheumatoid factors produced in rheumatoid synovial inflammation and immunized healthy donors. Ann. N.Y. Acad. Sci. **764:** 440–449.
46. WITHRINGTON, R.H., I. TEITSSON, H. VALDIMARSSON & M.H. SEIFERT. 1984. Prospective study of early rheumatoid arthritis. II. Association of rheumatoid factor isotypes with fluctuations in disease activity. Ann. Rheum. Dis. **43:** 679–685.
47. WERNICK, R., J.J. LOSPALLUTO, C.W. FINK & M. ZIFF. 1981. Serum IgG and IgM rheumatoid factors by solid phase radioimmunoassay. Arthritis Rheum. **24:** 1501–1511.
48. IZUI, S., M. ABDELMOULA, Y. GYOTOKU, *et al.* 1984. IgG rheumatoid factors and cryoglobulins in mice bearing the mutant gene *lpr* (lymphroliferation). Rheumatol. Int. **4**(Suppl): 45–48.
49. THEOFILOPOULOS, A.N. & F.J. DIXON. 1985. Murine models of systemic lupus erthematosus. Adv. Immunol. **37:** 269–390.
50. HANG, L., A.N. THEOFILOPOULOS & F.J. DIXON. 1982. A spontaneous rheumatoid arthritis-like disease in MRL/l mice. J. Exp. Med. **155:** 1690–1701.
51. YOUINOU, P.Y., J.W. MORROW, A.W.F. LETTIN, *et al.* 1984. Specificity of plasma cells in the rheumatoid synovium. Scand. J. Immunol. **20:** 307–315.
52. HOWARD, T.W., M.J. IANNINI, J.J. BURGE & J.T. DAVIS. 1991. Rheumatoid factor, cryoglobulinemia, anti-DNA, and renal disease in patients with systemic lupus erythematosus. J. Rheumatol. **18:** 826–830.
53. VAN SNICK, J. & P. COULIE. 1983. Rheumatoid factors and secondary immune responses in the mouse. I. Frequent occurrence of hybridomas secreting IgM anti-IgG1 autoantibodies after immunization with protein antigens. Eur. J. Immunol. **13:** 890–894.
54. NEMAZEE, D. & V. SATO. 1983. Induction of rheumatoid factors in the mouse. Regulated production of autoantibody in secondary humoral response. J. Exp. Med. **158:** 529–545.
55. WELCH, M.J., S. FONG, J. VAUGHAN & D. CARSON. 1983. Increased frequency of rheumatoid factor precursor B lymphocytes after immunization of normal adults with tetanus toxoid. Clin. Exp. Immunol. **51:** 299–304.
56. DJAVAD, N., S. BAS, X. SHI, *et al.* 1996. Comparison of rheumatoid factors of rheumatoid arthritis patients, of individuals with mycobacterial infections and of normal

controls: evidence for maturation in the absence of an autoimmune response. Eur. J. Immunol. **26:** 2480–2486.

57. Jacob, J., R. Kassir & G. Kelsoe. 1991. In situ studies of the primary immune response to (4-hydroxy-3-nitrophenyl)acetyl. I. The architecture and dynamics of responding cell population. J. Exp. Med. **173:** 1165–1175.
58. Liu, Y.J., J. Zhang, P.J. Lane, *et al.* 1991. Sites of specific B cell activation in primary and secondary responses to T cell-dependent and T cell-independent antigens. Eur. J. Immunol. **21:** 2951–2962.
59. Matsumoto, M., S.F. Lo, C.J.L. Carruthers, *et al.* 1996. Affinity maturation without germinal centres in lymphotoxin-α-deficient mice. Nature **382:** 462–466.
60. Fu, Y.-X., G. Huang, M. Matsumoto, *et al.* 1997. Independent signals regulate development of primary and secondary follicle structure in spleen and mesenteric lymph node. Proc. Natl. Acad. Sci. USA **94:** 5739–5743.
61. Koni, P.A., R. Sacca, P. Lawton, *et al.* 1997. Distinct roles in lymphoid organogenesis for lymphotoxins alpha and beta revealed in lymphotoxin beta-deficient mice. Immunity **6:** 491–500.
62. Wang, Y., G. Huang, J. Wang, *et al.* 2000. Antigen persistence is required for somatic mutation and affinity maturation of immunoglobulin. Eur. J. Immunol. **30:** 2226–2234.
63. Weller, S., A. Faili, C. Garcia, *et al.* 2001. CD40-CD40L independent Ig gene hypermutation suggests a second B cell diversification pathway in humans. Proc. Natl. Acad. Sci. USA **98:** 1166–1170.
64. Smith, K.G., G.J. Nossal & D.M. Tarlinton. 1995. Fas is highly expressed in the germinal center but is not required for regulation of the B-cell response to antigen. Proc. Natl. Acad. Sci. USA **92:** 11628–11632.
65. Takahashi, Y., H. Ohta & T. Takemori. 2001. Fas is required for clonal selection in germinal centers and the subsequent establishment of the memory B cell repertoire. Immunity **14:** 181–192.
66. Kallberg, E., S. Jainandunsing, D. Gray & T. Leanderson. 1996. Somatic mutation of immunoglobulin V genes in vitro. Science **271:** 1285–1289.
67. Jacob, J. & G. Kelsoe. 1992. In situ studies of the primary immune response to (4-hydroxy-3-nitrophenyl)acetyl. II. A common clonal origin for periarteriolar lymphoid sheath-associated foci and germinal centers. J. Exp. Med. **176:** 679–687.
68. Garcia De Vinuesa, C., A. Gulbranson-Judge, M. Khan, *et al.* 1999. Dendritic cells associated with plasmablast survival. Eur. J. Immunol. **29:** 3712–3721.
69. Leadbetter, E.A., A.M. Hohlbaum, I.R. Rifkin, *et al.* 2002. Chromatin–IgG complexes activate B cells by dual engagement of IgM and Toll-like receptors. Nature **416:** 603–607.
70. Hemmi, H., O. Takeuchi, T. Kawai, *et al.* 2000. A Toll-like receptor recognizes bacterial DNA. Nature **408:** 740–745.

Coordination of T Cell Activation and Migration through Formation of the Immunological Synapse

MICHAEL L. DUSTIN

Program in Molecular Pathogenesis, Skirball Institute of Biomolecular Medicine and the Department of Pathology, New York University School of Medicine, New York, New York 10016, USA

ABSTRACT: T cell activation is based on interactions of T cell antigen receptors with MHC-peptide complexes in a specialized cell–cell junction between the T cell and antigen-presenting cell—the immunological synapse. The immunological synapse coordinates naïve T cell activation and migration by stopping T cell migration with antigen-presenting cells bearing appropriate major histocompatibility complex (MHC) peptide complexes. At the same time, the immunological synapse allows full T cell activation through sustained signaling over a period of several hours. The immunological synapse supports activation in the absence of continued T cell migration, which is required for T cell activation through serial encounters. Src and Syk family kinases are activated early in immunological synapse formation, but this signaling process returns to the basal level after 30 min; at the same time, the interactions between T cell receptors (TCRs) and MHC peptides are stabilized within the immunological synapse. The molecular pattern of the mature synapse in helper T cells is a self-stabilized structure that is correlated with cytokine production and proliferation. I propose that this molecular pattern and its specific biochemical constituents are necessary to amplify signals from the partially desensitized TCR.

KEYWORDS: adhesion; migration; activation; antigen; receptors; kinases; phosphorylation; signaling

INTRODUCTION

Naïve T cells recirculate between the blood and secondary lymphoid tissues completing an average of two traversals per day.[1,2] Naïve T cell activation *in vivo* results in the removal of antigen-reactive cells from the recirculating pool for approximately two days.[3,4] During this time, the antigen-specific cells are concentrated in draining lymph nodes or the spleen, depending upon the route of antigen entry. After this two-day period, the daughter cells from proliferation of the primary naïve cells em-

Address for correspondence: Michael L. Dustin, Program in Molecular Pathogenesis, Skirball Institute of Biomolecular Medicine and the Department of Pathology, New York University School of Medicine, 540 First Avenue, New York, NY 10016.Voice: 212-263-3207; fax 212-263-5711.

dustin@saturn.med.nyu.edu

Ann. N.Y. Acad. Sci. 987: 51–59 (2003).

igrate from the secondary lymphoid tissues and migrate either to secondary lymphoid tissues or to distinct peripheral tissues. This retention of antigen-specific cells is selective and may be explained by the tendency of antigen-specific T cells to stop migrating following T cell receptor engagement.[5] The stop signal is based on both the Ca^{2+}-dependent suspension of motility in the short term (minutes) and the maintenance of T cell polarity towards the antigen-presenting cells (APCs) in the long term (hours).[5,6] The stop signal was further demonstrated to correlate with a specific molecular pattern in the interface between T cells and APCs.[7] This pattern has been revealed through studies of fixed cell–cell conjugates, live cell–cell conjugates, and interactions between T cells and supported planar bilayers containing major histocompatibility complex (MHC) peptides and adhesion molecules (FIG. 1).[7–12] The initial molecular pattern can be described as a nascent or immature immunological synapse in which T cell receptors (TCRs) are engaged in the periphery of the extensive contact area.[7] This pattern is followed by formation of a mature immunological synapse in which the TCRs are clustered in the center of the contact area.[7,8] The pattern in which the TCRs and MHC peptides are centrally clustered and stabilized over a period of hours is correlated with full T cell activation when both the dose and quality of MHC-peptide complexes is varied. In the mature immunological synapse, the central cluster is also referred to as a central supramolecular activation cluster (cSMAC), and the LFA-1 ring, as a peripheral supramolecular activation cluster (pSMAC).[8] T cells deficient in specific signaling molecules in this pathway, such as WASP and Vav, which are involved in TCR movement on the surface following activation, have defects in T cell activation, suggesting that active T cell movements are important for activation and immunological synapse formation.[13–15]

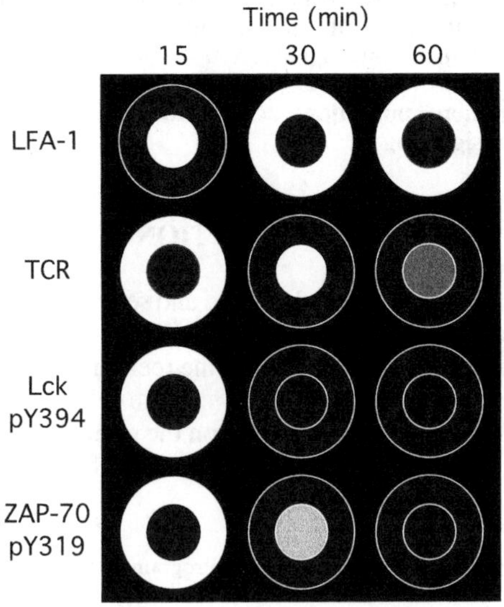

FIGURE 1. Location of signaling in the immunological synapse. Two domains are shown: central and peripheral. Abundance is represented as gray scale from dark (= low) to light (= high). Based on data in Lee *et al.*[12]

While mature immunological synapse formation is a common mode of interaction during T cell signaling *in vitro*, certain environmental conditions, or APCs trigger a different mode of T cell interaction characterized by continued T cell migration.[16,17] This mode of activation is more similar to the classical concept of serial triggering in which a few MHC-peptide complexes engaged a large number of TCRs to sustain signaling over a prolonged period as the T cells continually encounter new and recycled MHC-peptide complexes.[18,19] The concept that environmental signals can be dominant over TCR-mediated polarization signals has been described for certain chemokines,[20] but is likely to extend to other microenvironmental signals. Given that the T cell can respond in different ways to similar stimuli, it's important to understand the function of the immunological synapse and the conditions under which it forms as the primary mode of T cell activation.

MOLECULAR PATTERN FORMATION IN THE IMMUNOLOGICAL SYNAPSE RELATIONSHIP TO THE CALCIUM SIGNAL

Supported planar bilayers containing MHC-peptide complexes and the adhesion molecule ICAM-1 have provided a surprisingly effective surrogate APC allowing dynamic imaging of molecular patterns during T cell activation.[7] ICAM-1 is a ligand for the lymphocyte integrin LFA-1, perhaps the most ubiquitously employed adhesion molecule on T cells. The supported planar bilayers containing only an agonist MHC-peptide complex and ICAM-1 can trigger proliferation of previously activated T cells, and bilayers containing agonist MHC-peptide complexes,[7] ICAM-1, and CD80 can trigger proliferation of naïve T cells.[21] The advantage of the bilayer system is that the ligands can be fluorescently labeled, and fluorescence imaging can then be used to identify sites of interaction between receptors on the T cell and the fluorescent ligands in the bilayer. Since the ligands in the bilayer are initially uniformly distributed and can only interact with the receptors or each other, the interpretation of the images is greatly simplified compared to the situation with cell–cell interaction in which cytoskeletal interactions and lateral heterogeneity of membranes greatly complicates interpretation.[22] For example, MHC-peptide accumulation in the bilayer can be interpreted in terms of TCR–MHC-peptide interactions with the only complexity arising from co-receptors and oligomerization. On the other hand, MHC-peptide accumulation in a cell–cell interface could be due to association with liquid ordered membrane domains (rafts), the cytoskeleton, and vesicular stores of MHC-peptide complexes near the interface in addition to interactions with the T cell receptor and co-receptors. The clarity of interpretation provided by the bilayer system comes at the expense of loss of the potentially important topology, cytoskeletal associations, and membrane heterogeneity present in the APC. Therefore, it is important to use both approaches in a complementary fashion, since neither system alone can address all the important questions. In fact, an important technology for future experiments will be to take these studies into the *in vivo* environment through intravital imaging.

The planar bilayer system and cell–cell systems both concur on the following molecular choreography for immunological synapse formation.[7,23] The LFA-1–ICAM-1 interaction is initiated in the central region of the nascent immunological synapse. In the first moments of contact (<30 s), it is likely that membrane projections bearing

TCRs at the periphery of the LFA-1-dependent contact come close enough to the APC surface to interact with MHC-peptide complexes. This peripheral TCR engagement initiates signals that trigger the tyrosine kinase cascade. The initial evidence for this early signaling came from studies on the Ca^{2+} signal that show that this signal is initiated very early[6,7,10] and in specific situations where peripheral TCR engagement was imaged in parallel.[7,23] The influx of cytoplasmic Ca^{2+} is downstream of activation of phopholipase Cγ through the tyrosine kinase cascade requiring Lck, ZAP-70, and ITK.[24] Thus, early (<30 s) Ca^{2+} signaling strongly implied the Lck is activated within seconds of contact, well before formation of the mature immunological synapse with its focused, central TCR engagement.

The mature immunological synapse is characterized by stabilization of the TCR–MHC-peptide interaction in the central cluster. This was deduced by fluorescence photobleaching recovery experiments.[7,23] In this experiment, the fluorescence of MHC-peptide complexes interacting in the central region of the mature synapse is photobleached with a brief flash from an appropriately focused laser beam, rendering these molecules non-fluorescent, but functionally intact. The fluorescence is then monitored over time to determine if the bleached molecules accumulated in the central cluster will exchange with the fluorescent molecules that diffuse in from outside the small bleached area. While fluorescence in the bilayer did recover, the accumulated fluorescence in the central cluster did not, suggesting that the interaction in this region had become stable. It is possible that this stabilization is based on oligomerization of the TCR-MHC-peptide complex as detected in solution for the 2B4 TCR-E^k-MCC88-103 system.[25] This stabilization process does not appear to be reversible, since migrating T cells were observed to shed TCRs from their surface during disengagement from the central cluster. This stabilization of TCRs may provide a focal point for polarization of T cells in stable synapses and may represent a form of self-catalyzed protein refolding to form these oligomers.

EARLY T CELL SIGNALING THROUGH LCK AND ZAP-70

Classical biochemical studies using antibody cross-linking or APCs suggested that Lck activation, which includes the tyrosine kinase cascade, is transient, but thinking about the newly described immunological synapse and the stable central cluster raised the possibility that *in situ* phosphorylation could be more sustained (FIG. 2).[26,27] Recently, this question was directly addressed using phosphotyrosine specific antibodies to the activation loop of Lck and a key autophosphorylation site of ZAP-70, two key kinases whose activities are central to the classical definition of TCR signaling, to determine the timing and location of tyrosine kinase activation in T cell–splenic APC immunological synapses.[12] Consistent with the classical paradigm, it was found that both Lck and ZAP-70 were activated early in the nascent synapse and were no longer active by the time the mature immunological synapse had formed. The mechanisms that return tyrosine phosphorylation to the basal level are not known, but it was noted that the splenic APC induced dramatic TCR down-regulation just following the formation of the central TCR cluster. This down-regulation may be related to the role of ubiquitin ligases of the cbl family. Interestingly, cbl may be recruited to the central TCR cluster by the CIN85-related protein CD2AP.[28,29] CD2AP binds the cytoplasmic domain of CD2 in a T cell activation dependent man-

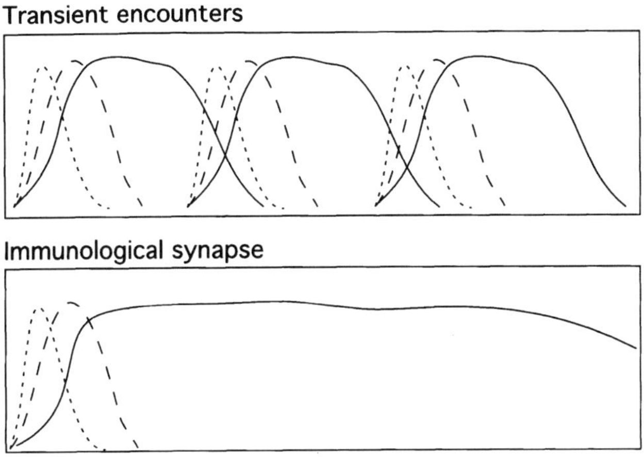

FIGURE 2. Model for signaling kinetics in serial encounters versus the stable synapse. The lines in the top and bottom panels reflect tyrosine kinase signaling (*dashed lines*) and a hypothetical signaling process required for cytokine transcription (*solid line*). This signaling to the nucleus is downstream of activation and of the tyrosine kinase cascade and has a half-life that is determined by the structure of the interface—short in an unstructured interface and longer in a structured interface (immunological synapse).

ner and brings CD2 to the center of the immunological synapse.[9] In addition, CD2AP, like CIN85, may recruit cbl and facilitate TCR degradation. This process may contribute to the rapid return of the TCR signal to basal levels.

The results on tyrosine phosphorylation stimulated an experiment to determine how long the T cell–APC conjugate had to persist to program T cells for full activation.[12] The result was that the immunological synapse had to persist for 2–3 h to commit the T cell to one division and for 3–6 h to commit the T cell to multiple cell divisions. This creates a paradox that the tyrosine phosphorylation of the TCR returns to basal levels by 1 h, but signal integration continues for up to 6 h and likely longer.[30] The events that constitute signaling in this time frame are of great interest. One possibility is that the efficacy of tyrosine kinases for activation of downstream transcription systems is greatly enhanced in the organized immunological synapse. Support for this comes from the observation that Ca^{2+} mobilization is sustained in the immunological synapse over a period of hours.[7,10] This suggests an ongoing process that stimulates Ca^{2+} mobilization, which would be consistent with some level of effective phospholipase Cγ activation through a tyrosine phosphorylation dependent process. Alternatively, the maintenance of transcription factor translocation may be tyrosine kinase independent in the mature synapse and may rely on distinct signaling mechanisms that are set up in the early stages of immunological synapse formation.

Why form stable synapses when repeated early signaling can trigger T cells? As mentioned above, T cells can signal without forming a stable immunological synapse. This alternative mode of activation requires crawling of the T cell over the APC and between APCs, and the signaling is immediately blocked by addition of cytochalasin D.[16,18] Naïve murine T cells interacting with dendritic cells in a collagen gel are stimulated to proliferate through a series of transient interactions with APCs that were each on the order of 12 min, but were repeated many times as the T cells migrated through the three-dimensional collagen gel.[17] With every new interaction between T cells and antigen-positive APCs, the T cells underwent a robust Ca^{2+} flux, suggesting repeated rounds of early TCR signaling that broke off just as the Lck and ZAP-70 activation would be expected to return to baseline, although tyrosine kinase activity was not directly examined. In fact, it may be important that these T cells break contact with the APCs to reset the tyrosine kinases, so that they can undergo repeated rounds of signaling. *In vivo*, it is most likely that stable immunological synapse formation would be used to coordinate T cell activation with migration so that the antigen-specific T cells can be focused in one compartment to contribute to the collective process of making an appropriate response to the pathogen.[31,32] However, this serial encounter mode may be important in different *in vivo* settings, such as the movement of effector cells in the skin and other collagen-rich peripheral tissues. Relevant to this argument is the observation of Anderson and Shaw that collagen in lymph nodes is sequestered in reticular fibers; thus, T cells in lymph nodes are not exposed to collagen.[33,34] I propose that the process of effective activation through formation of a stable immunological synapse evolved to focus antigen-specific T cells on the appropriate secondary lymphoid tissue and further on the appropriate APCs.

FORMATION OF THE IMMUNOLOGICAL SYNAPSE: SELF-ASSEMBLY VERSUS CAPPING

Self-assembly processes have captured the imagination of many biologists because such processes posit the ability to build large-scale biological devices from less complex building blocks. The prevailing view of immunological synapse formation has been based on the paradigm of topological segregation combined with antibody-mediated capping.[35,36] In the capping process, the signaling triggered by cross-linking surface proteins initiates a process of actin polymerization and myosin activation leading to contraction of the actin gel and collection of the actin-associated, cross-linked receptors into a cap at one pole of the cell.[37] Chakraborty and colleagues have employed partial differential equations describing molecular segregation by size, reaction-diffusion processes, and membrane fluctuations to model immunological synapse formation as a self-assembly process that does not require actin-myosin transport.[38] Remarkably, the model can be run with measured parameters and predicts stable immunolologcal synapse formation through the same series of intermediates that are observed experimentally. While the model is in the early stages of development, it raises the important possibility that synapse formation or stability may be based on a self-assembly process that does not require transport or may operate in parallel with cytoskeletal transport mechanisms. While the synapse self-assembly model does not require transport, an intact actin cytoskeleton is required to set the appropriate membrane-bending properties to generate synaptic patterns. Therefore, we can consider two alternatives for synapse formation that should be distinguishable at the single-molecule level. The capping model predicted that molecules are transported to the center of the synapse along straight-line paths, whereas the self-assembly model predicts a biased random walk. If there is a significant element of membrane-driven self-assembly, then some of the genetic experiments may require reinterpretation. It is possible that rather than setting up a capping-like phenomenon, molecules such as Vav and WASP may have a role in adjusting the physical properties of the membrane to favor self-assembly of the synapse.

CONCLUSION

There is still much to learn about the function of the immunological synapse. For helper T cells, both naïve and effector, the synapse formation process is related to antigen dose and strength. Formation of helper T cell immunological synapses appears to be very sensitive to the kinetics of the TCR–MHC-peptide interaction. This may be related to the self-assembly process identified by Qi *et al.*[38] Despite the intrinsic tendency of T cells to form stable synapses when triggered by agonist MHC-peptide complexes and adhesion molecules, the environment can have a strong impact on signaling. If the environment is dominant for control of T cell polarity, the T cell does not stop migration on the APC, but can still be activated through a process of serial encounters with multiple APCs involving a resetting of the tyrosine kinase cascade to allow repeated firing. When the TCR signal is dominant for control the T cell polarity, the T cell stops and uses one round of transient tyrosine kinase signaling to set up a stable synapse. The stable immunological synapse is then required to generate signals that maintain transcription factor activation. The specific nature of these signals is not known and represents an important area for future research.

ACKNOWLEDGMENTS

This work was supported by grants from the National Institutes of Health and the Irene Diamond Professorships in Immunology.

REFERENCES

1. Springer, T.A. 1995. Traffic signals on endothelium for lymphocyte recirculation and leukocyte emigration. Annu. Rev. Physiol. **57:** 827.
2. Jenkins, M.K., A. Khoruts, E. Ingulli, *et al.* 2001. In vivo activation of antigen-specific CD4 T cells. Annu. Rev. Immunol. **19:** 23.
3. Sprent, J., J.F.A.P. Miller & G.F. Mitchell. 1971. Antigen-induced selective recruitment of circulating lymphocytes. Cell. Immunol. **2:** 171.
4. Sprent, J. & J.F.A.P. Miller. 1972. Interaction of thymus lymphocytes with histoincompatible cells. II. Recirculating lymphocytes derived from antigen-activated thymus cells. Cell. Immunol. **3:** 385.
5. Dustin, M.L., S.K. Bromely, Z. Kan, *et al.* 1997. Antigen receptor engagement delivers a stop signal to migrating T lymphocytes. Proc. Natl. Acad. Sci. USA **94:** 3909.
6. Negulescu, P.A., T. Kraieva, A. Khan, *et al.* 1996. Polarity of T cell shape, motility, and sensitivity to antigen. Immunity **4:** 421.
7. Grakoui, A., S.K. Bromley, C. Sumen, *et al.* 1999. The immunological synapse: a molecular machine controlling T cell activation. Science **285:** 221.
8. Monks, C.R., B.A. Freiberg, H. Kupfer, *et al.* 1998. Three-dimensional segregation of supramolecular activation clusters in T cells. Nature **395:** 82.

9. DUSTIN, M.L., M.W. OLSZOWY, A.D. HOLDORF, *et al.* 1998. A novel adapter protein orchestrates receptor patterning and cytoskeletal polarity in T cell contacts. Cell **94:** 667.
10. WÜLFING, C., M.D. SJAASTAD & M.M. DAVIS. 1998. Visualizing the dynamics of T cell activation: intracellular adhesion molecule 1 migrates rapidly to the T cell/B cell interface and acts to sustain calcium levels. Proc. Natl. Acad. Sci. USA **95:** 6302.
11. KRUMMEL, M.F., M.D. SJAASTAD, C. WULFING & M.M. DAVIS. 2000. Differential clustering of CD4 and CD3zeta during T cell recognition. Science **289:** 1349.
12. LEE, K.-H., A.D. HOLDORF, M.L. DUSTIN, *et al.* 2002. Tyrosine kinase signaling precedes formation of the immunological synapse. Science **295:** 1539.
13. FISCHER, K.D., Y.Y. KONG, H. NISHINA, *et al.* 1998. Vav is a regulator of cytoskeletal reorganization mediated by the T- cell receptor. Curr. Biol. **8:** 554.
14. SNAPPER, S.B., F.S. ROSEN, E. MIZOGUCHI, *et al.* 1998. Wiskott-Aldrich syndrome protein-deficient mice reveal a role for WASP in T but not B cell activation. Immunity **9:** 81.
15. WÜLFING, C., A. BAUCH, G.R. CRABTREE & M.M. DAVIS. 2000. The Vav exchange factor is an essential regulator in actin-dependent receptor translocation to the lymphocyte-antigen-presenting cell interface. Proc. Natl. Acad. Sci. USA **97:** 10150.
16. UNDERHILL, D.M., M. BASSETTI, A. RUDENSKY & A. ADEREM. 1999. Dynamic interactions of macrophages with T cells during antigen presentation. J. Exp. Med. **190:** 1909.
17. GUNZER, M., A. SCHAFER, S. BORGMANN, *et al.* 2000. Antigen presentation in extracellular matrix: interactions of T cells with dendritic cells are dynamic, short lived, and sequential. Immunity **13:** 323.
18. VALITUTTI, S., M. DESSING, K. AKTORIES, *et al.* 1995. Sustained signalling leading to T cell activation results from prolonged T cell receptor occupancy: role of T cell actin cytoskeleton. J. Exp. Med. **181:** 577.
19. VALITUTTI, S., S. MÜLLER, M. CELLA, *et al.* 1995. Serial triggering of many T-cell receptors by a few peptide-MHC complexes. Nature **375:** 148.
20. BROMLEY, S.K., D.A. PETERSON, M.D. GUNN & M.L. DUSTIN. 2000. Cutting edge: hierarchy of chemokine receptor and TCR signals regulating T cell migration and proliferation. J. Immunol. **165:** 15.
21. BROMLEY, S.K., A. IABONI, S.J. DAVIS, *et al.* 2001. The immunological synapse and CD28-CD80 interactions. Nat. Immunol. **2:** 1159.
22. DUSTIN, M.L., L.M. FERGUSON, P.Y. CHAN, *et al.* 1996. Visualization of CD2 interaction with LFA-3 and determination of the two-dimensional dissociation constant for adhesion receptors in a contact area. J. Cell Biol. **132:** 465.
23. JOHNSON, K.G., S.K. BROMLEY, M.L. DUSTIN & M.L. THOMAS. 2000. A supramolecular basis for CD45 regulation during T cell activation. Proc. Natl. Acad. Sci. USA **97:** 10138.
24. SCHAEFFER, E.M., J. DEBNATH, G. YAP, *et al.* 1999. Requirement for Tec kinases Rlk and Itk in T cell receptor signaling and immunity. Science **284:** 638.
25. REICH, Z., J.J. BONIFACE, D.S. LYONS, *et al.* 1997. Ligand-specific oligomerization of T cell-receptor moleucles. Nature **387:** 617.
26. DUSTIN, M.L. & A.C. CHAN. 2000. Signaling takes shape in the immune system. Cell **103:** 283.
27. BROMLEY, S.K., W.R. BURACK, K.G. JOHNSON, *et al.* 2001. The immunological synapse. Annu. Rev. Immunol. **19:** 375.
28. PETRELLI, A., G.F. GILESTRO, S. LANZARDO, *et al.* 2002. The endophilin-CIN85-Cbl complex mediates ligand-dependent downregulation of c-Met. Nature **416:** 187.
29. SOUBEYRAN, P., K. KOWANETZ, I. SZYMKIEWICZ, *et al.* 2002. Cbl-CIN85-endophilin complex mediates ligand-induced downregulation of EGF receptors. Nature **416:** 183.
30. IEZZI, G., K. KARJALAINEN & A. LANZAVECCHIA. 1998. The duration of antigenic stimulation determines the fate of naive and effector T cells. Immunity **8:** 89.
31. DUSTIN, M.L. & A.R. DE FOUGEROLLES. 2001. Reprograming T cells: the role of extracellular matrix in coordination of T cell activation and migration. Curr. Opin. Immunol. **13:** 286.

32. DUSTIN, M.L., P.M. ALLEN & A.S. SHAW. 2001. Environmental control of immunological synapse formation and duration. Trends Immunol. **22:** 192.
33. EBNET, K., E.P. KALDJIAN, A.O. ANDERSON & S. SHAW. 1996. Orchestrated information transfer underlying leukocyte endothelial interactions. Annu. Rev. Immunol. **14:** 155.
34. GRETZ, J.E., C.C. NORBURY, A.O. ANDERSON, *et al.* 2000. Lymph-borne chemokines and other low molecular weight molecules reach high endothelial venules via specialized conduits while a functional barrier limits access to the lymphocyte microenvironments in lymph node cortex. J. Exp. Med. **192:** 1425.
35. HOLSINGER, L.J., I.A. GRAEF, W. SWAT, *et al.* 1998. Defects in actin-cap formation in Vav-deficient mice implicate an actin requirement for lymphocyte signal transduction. Curr. Biol. **8:** 563.
36. DUSTIN, M.L. & J.A. COOPER. 2000. The immunological synapse and the actin cytoskeleton: molecular hardware for T cell signaling. Nat. Immunol. **1:** 23.
37. BRAUN, J., K. FUJIWARA, T.D. POLLARD & E.R. UNANUE. 1978. Two distinct mechanisms for redistribution of lymphocyte surface macromolecules. I. Relationship to cytoplasmic myosin. J.Cell Biol. **79:** 409.
38. QI, S.Y., J.T. GROVES & A.K. CHAKRABORTY. 2001. Synaptic pattern formation during cellular recognition. Proc. Natl. Acad. Sci. USA **98:** 6548.

Intrinsic T Cell Defects in Systemic Autoimmunity

PHILIP L. KONG,[a,b] JARED M. ODEGARD,[a,b] FARIDA BOUZAHZAH,[a] JIN-YOUNG CHOI,[a] LEAH D. EARDLEY,[a,b] CHRISTINA E. ZIELINSKI,[a] AND JOSEPH E. CRAFT[a,b]

[a]*Section of Rheumatology, Department of Internal Medicine, and* [b]*Section of Immunobiology, Yale School of Medicine, New Haven, Connecticut 06520, USA*

ABSTRACT: Systemic lupus erythematosus (SLE) is an autoimmune disease characterized by loss of T cell tolerance to nuclear antigens. Studies in mice and humans have demonstrated that T cells from individuals with lupus are abnormal. Here, we review the known T cell defects in lupus and their possible biochemical nature, genetic causes, and significance for lupus pathogenesis.

KEYWORDS: autoimmunity; systemic lupus erythematosus (SLE); CD4 T cells; anergy; immunological tolerance

INTRODUCTION

The ability to distinguish between self and non-self is the *sine qua non* of the adaptive immune system. T cells are taught to be self-reverent in the thymus, but there are instances in which they violate these instructions and produce an anti-self response. Our fundamental understanding of how self-tolerance is maintained may be most powerfully elucidated if we study a case in which it fails: systemic lupus erythematosus (SLE).

SLE is characterized by high-titer autoantibodies against ubiquitous nuclear self-antigens. Pathogenesis depends on CD4 T cells, primarily by providing B cell help.[1–4] The T cell requirement was originally demonstrated with neonatal thymectomy,[5] and subsequently confirmed by experiments with αCD4- and αThy1-depleting antibodies[6,7] and with knockout animals.[8–11] Furthermore, it has been shown that activation of autoantigen-specific T cells is required for full disease penetrance, as replacement of potentially autoreactive T cells in MRL/Fas^{lpr} mice with T cells of a single specificity leads to the abrogation of disease, despite transgenic T cell activation.[12] The autoantigens recognized by T cells are related to those targeted by lupus autoantibodies, including chromatin and ribonucleoproteins.[13,14] Interestingly, T cells can promote disease in mice that lack the ability to generate secreted antibodies, implicating their direct role in tissue injury.[15]

Address for correspondence: Joseph E. Craft, Box 208031, LCI 610, 333 Cedar Street, New Haven, Connecticut 06520-8031. Voice: 203-785-7063; fax: 203-785-5415.
joseph.craft@yale.edu

Ann. N.Y. Acad. Sci. 987: 60–67 (2003). © 2003 New York Academy of Sciences.

Given these critical but rudimentary insights into lupus development, it is natural to ask what, if any, roles do T cells have in lupus susceptibility. Are T cells innocent—that is, are they co-opted or a part of the crime by association—or are they intrinsically guilty in contributing to disease susceptibility? Our own work, consistent with the work of others, suggests that CD4 T cells from individuals with lupus carry genetic defects that contribute to disease susceptibility. Understanding these abnormalities will give us theoretical insights into the workings of the immune system, as well as potential targets for therapeutic interventions.

INTRINSIC T CELL ABNORMALITIES IN LUPUS

Activation of autoreactive T cells in lupus is thought to be a consequence of peripheral tolerance abrogation, since central tolerance appears intact.[16] The etiology of this loss in tolerance is currently unknown. We believe that lupus T cells have intrinsic defects that render them more susceptible to activation through their T cell receptor (TCR)-CD3 complex after contact with self-peptides. This hypothesis stems in part from several observations. First, T cells from lupus-prone mice have increased numbers of activated T cells *in vivo*.[17–19] Second, a genetic locus on chromosome 7 from lupus-prone New Zealand mice (NZM; New Zealand mixed) contributes to a lower threshold of T cell activation and a higher threshold for apoptotic death.[20] Third, T cells from SLE patients appear to have abnormalities in TCR signaling[21–27] and apoptosis,[28,29] as well as in expression of effector molecules, including CD40 ligand (CD154).[29–31]

The interpretation of the studies of T cells isolated from humans and mice with lupus is complicated by the heterogeneity and activation history of such cells. Moreover, these studies, and the genetic studies of lupus mice, have not conclusively shown that the identified abnormalities are intrinsic to the T cells, rather than the results of extrinsic effects such as antigen presentation. Indeed, macrophages from lupus-prone mice have been shown to be abnormal in their cytokine profiles[32–34]; B cells are also genetically hyper-responsive in lupus animals.[35] Such abnormalities extrinsic to T cells could account for the observed T cell phenotype.

The use of TCR transgenic animals allows us to examine lupus T cell biology in more controlled conditions—highly purified cells with known history, well-characterized antigens, genetic and environmental homogeneity—making the results more interpretable and conclusive. Our laboratory, for example, uses the AND TCR transgenic system that recognizes $PCC_{88–104}$ peptide presented by $I\text{-}E^k$ with high affinity.[36] Related altered peptide ligands are available that have single amino acid changes in the TCR binding positions, giving rise to reduced affinities.[37] Naïve AND T cells derived from the lupus-prone $MRL/+^{Fas\text{-}lpr}$ animals were compared to equivalent cells from H-2 matched non-autoimmune strains (B10.BR, CBA/CaJ, AKR/J). It was found that, upon stimulation with various peptides under identical conditions, MRL.AND cells proliferated faster and made more IL-2 than their non-autoimmune counterparts.[38] They also developed a stronger effector phenotype, as suggested by CD40L expression. The differences were particularly notable at lower doses of full agonist stimulation, and with altered peptide ligand (APL) stimulation. This leads to the conclusion that MRL T cells have a lower threshold of activation. This finding has since been extended to other T cell stimuli (unpublished results), suggesting that the defect is not peculiar to the AND system, but rather linked to TCR-CD3 signaling in general.

An intrinsic defect that may alter the behavior of T cells in the face of a low-avidity TCR interaction has important consequences. Given that central tolerance appears to be largely intact in lupus models, it is likely that most, if not all, T cells that react with ubiquitous self-antigens with high avidity would be deleted in the thymus. We therefore propose a model in which the major culprits in lupus are peripheral T cells that are reactive to these antigens with low avidities. T cells, in their normal course of recirculation throughout the body, will encounter MHC:self-peptide complexes and interact weakly on a regular basis. In normal animals, this "tickling" induces transient signaling[39] that may promote their survival in the periphery,[40] but is of insufficient strength for activation. In lupus animals, where an intrinsic T cell defect exists, this low avidity interaction may be sufficient to trigger an activation program. Particularly, this activation event may occur in the setting of abundant costimulation, such as during an infection, while chronic inflammation could promote further autoreactive T cell priming. Therefore, T cell hyper-responsiveness may cooperate with other immune system defects to promote a cycle of tissue damage and further pathological priming against self.[41]

A separate but related observation made recently using the same transgenic system suggested that MRL T cells are less susceptible to anergy induction in the periphery than non-autoimmune T cells.[42] In these experiments, AND T cells from various strains were adoptively transferred into identical F_1 hosts that expressed the cognate antigen PCC as a self-antigen. As expected, the transferred T cells encountered the cognate antigen and became unresponsive to secondary challenges *ex vivo*. This state of anergy was much less apparent in MRL T cells, which retained much of their ability to proliferate and make interleukin-2 (IL-2) when rechallenged. The implication of this is clear—avoidance of anergy induction by ubiquitous self-antigens in self-reactive T cells can directly lead to a breach in peripheral tolerance.

By utilizing a transgenic system and excluding extrinsic factors that may confound the results, these studies clearly demonstrate that T cells from lupus-prone mice have intrinsic defects that may alter their fate in the lifetime of an animal. This observation needs to be confirmed in other lupus-prone strains and in SLE patients, and some of these studies are underway. We shall now consider the origin of these intrinsic defects.

THE GENETICS OF LUPUS

While lupus is a relatively uncommon disease in natural populations, numerous transgenic and single-gene knockout animals develop lupus-like conditions or have some features of lupus disease. The genes involved are diverse—they include both activating and inhibitory genes, and affect T cells, B cells, and the innate immune system.[43] That all these models develop lupus-like syndromes suggests that lupus is a disease caused by a generalized immune dysregulation. This is unlike other organ-specific autoimmune diseases where the nature of the autoantigen (its expression level or distribution, for example) has a large influence on pathogenesis.

As valuable as these transgenic and knockout models are in enhancing our understanding of immune function and regulation, it is clear that most cases of natural lupus (in humans or mice) are not caused by single-gene defects. With the rare exception of early complement deficiencies,[44,45] which presumably affect immune complex clearance, lupus is a complex, multigenic trait. Various genetic screens in human lupus have identified over a dozen separate loci that contribute to lupus susceptibility.[46–49] Only a handful of these loci from different screens overlap, highlighting the complexity of the human disease and the difficulties in studying it. The causes for the apparent discrepancies are unclear, but may include population heterogeneity, use of different disease criteria, and the choice of the traits screened.

The genetics of spontaneous lupus in inbred mice models is also complex and multigenic. However, our ability to manipulate the genes and control the environment of the animals makes this complexity more tractable. Several genome-wide screens have been carried out in murine lupus,[50–52] and it is noted that many of the loci overlap significantly.[43] While it is unclear whether these loci from independent screens represent the same genes or not, they become the obvious focus of in-depth analysis. Several groups use the congenic animal approach to separate out the effects of an individual locus, avoiding the confounding effect of genetic interactions. Significant progress has been made recently in understanding the contributions of these loci to pathology,[53,54] their interactions,[55] and most importantly, the precise molecular nature of these genetic lesions.[56,57]

GENETICS OF INTRINSIC T CELL DEFECTS

Intrinsic T cell defects that contribute to lupus susceptibility probably have a genetic root. Genetic ablation of the p53-effector gene Gadd45,[58] as well as the transcriptional modulator E2F2,[59] give rise to lowered activation threshold in T cells in these animals, reminiscent of the phenotype observed in MRL animals. Both knockout animals develop severe systemic autoimmunity. There is also evidence that anergy induction defects have a genetic basis—mice lacking the adaptor molecule Cbl-b develop spontaneous autoimmunity characterized by autoantibody production and infiltration of activated T and B cells in multiple organs.[60] Interestingly, diminished phosphorylation of Cbl/Cbl-b molecules correlates with prolonged CD40L expression in lupus individuals.[61] This defect was linked to an abnormal extracellular-regulated kinase (ERK) signaling pathway, as blockade of this pathway with a pharmacological agent abolishes CD40L hyper-expression.[61]

While it is unlikely that natural, multigenic lupus involves any null mutations in any of these related genes, it is likely that some of the loci associated with lupus susceptibility influence T cells in a similar but more subtle way. Perhaps the strongest evidence that lupus T cell defects are genetic in origin comes from the NZM.Sle3 congenic animals.[20] These mice have changes in T cell activation *in vivo* and *in vitro*, as well as defects in death and survival. At this point the molecular nature of these differences is not clear, and defining it will be a challenge.

An illustration of the difficulty that is ahead of us in this endeavor comes from the studies of the H-2 congenic animals. The H-2 locus has been identified as a strong lupus-susceptibility locus in multiple screens.[50–52] For example, H-2^k carried by MRL animals confers susceptibility to lupus when compared to H-2^b carried by B6 animals.[51] Despite this strong and persistent link, the effect of the H-2 locus is relatively weak. When the susceptible H-2^k haplotype was replaced by the resistant

H-2^b haplotype in MRL/+$^{Fas\text{-}lpr}$ animals, some differences in autoantibody production were found. Yet no differences were seen in kidney pathology, T cell activation status, or mortality.[62]

INTRINSIC T CELL DEFECTS: WHO REALLY CARES?

We have reviewed so far the evidence that T cells in both human and murine lupus are intrinsically abnormal, and an elaboration of what some abnormalities in the MRL mouse model are. We propose that these intrinsic T cell defects contribute to the generalized dysregulation of the immune system characteristic of lupus. The challenge now is to deepen our understanding of the observations by asking three questions.

Do Intrinsic T Cell Defects Contribute to Lupus Susceptibility?

Up to now most of the evidence that supports the link between intrinsic T cell defects and lupus susceptibility is circumstantial. To test this notion directly, efforts are underway to construct bone marrow chimeras that contain all lupus-prone cell types, with the exception of non-autoimmune-derived T cells. The phenotype of these animals will determine if intrinsic abnormalities in T cells are necessary for full lupus expression. In this or similar experimental approaches, care must be taken so that only the T cell compartment is affected by the manipulations.

Which Molecular Mechanisms Lead to T Cell Defects?

Numerous reports have suggested that signaling in lupus T cells is altered.[21–27] While these observations are suggestive, many of them suffer from the caveats described earlier—poorly defined cell populations with respect to specificity, activation status, and stimulation history. Accordingly, similar biochemical approaches must be applied to more rigorous models to dissect the molecular nature of intrinsic defects such as antigen hyper-responsiveness and anergy avoidance.

As detailed earlier, one of the goals is to understand the genetic heterogeneity of natural lupus susceptibility. Through genetic screens and congenic animals, we are beginning to identify candidate genes that may contribute to disease. While human lupus may not involve the same genes, it is likely that the molecular pathways involved overlap significantly. As noted earlier, however, the contribution of each gene or even each locus (which probably carries a haplotype of susceptible alleles) is only partial. It will take substantial effort to prove any one of them to be causative.

How Do Intrinsic T Cell Defects Contribute to Lupus?

Models of lupus pathogenesis described here or by others are either somewhat incomplete or unsubstantiated. However, an understanding of the role of T cells will be critical for an accurate and comprehensive description of this systemic autoimmune disease. Many additional issues need to be resolved: Is central tolerance really intact in lupus? What is the nature of the autoreactive T cells, with respect to their specificity and avidity? How and where does T cell triggering take place? What is the role of co-stimulation and/or inflammation in the induction and propagation of the autoimmune response? The answers to these questions would yield a solid model for the role of T cells, and the contribution of specific T cell defects, in the etiology of systemic lupus erythematosus. This is important not only for our intellectual curiosity, but to guide the development of an effective and specific therapy.

REFERENCES

1. DAIKH, D., *et al.* 1997. The CD28-B7 costimulatory pathway and its role in autoimmune disease. J. Leukoc. Biol. **62:** 156–162.
2. EARLY, G.S., *et al.* 1996. Anti-CD40 ligand antibody treatment prevents the development of lupus-like nephritis in a subset of New Zealand black x New Zealand white mice: response correlates with the absence of an anti-antibody response. J. Immunol. **157:** 3159–3164.
3. PENG, S.L., *et al.* 1997. αβ T cell regulation and CD40 ligand dependence in murine systemic autoimmunity. J. Immunol. **158:** 2464–2470.
4. VOLL, R.E., *et al.* 1997. Histone-specific Th0 and Th1 clones derived from systemic lupus erythematosus patients induce double-stranded DNA antibody production. Arthritis Rheum. **40:** 2162–2171.
5. STEINBERG, A.D., *et al.* 1980. Effects of thymectomy or androgen administration upon the autoimmune disease of MRL/Mp-lpr/lpr mice. J. Immunol. **125:** 871–873.
6. WOFSY, D., *et al.* 1985. Treatment of murine lupus with monoclonal anti-T cell antibody. J. Immunol. **134:** 852–857.
7. SANTORO, T.J., *et al.* 1988. The contribution of L3T4+ T cells to lymphoproliferation and autoantibody production in MRL-lpr/lpr mice. J. Exp. Med. **167:** 1713–1718.
8. JEVNIKAR, A.M., *et al.* 1994. Prevention of nephritis in major histocompatibility complex class II-deficient MRL-lpr mice. J. Exp. Med. **179:** 1137–1143.
9. KOH, D.R., *et al.* 1995. Murine lupus in MRL/lpr mice lacking CD4 or CD8 T cells. Eur. J. Immunol. **25:** 2558–2562.
10. CHEN, S.Y., *et al.* 1996. The natural history of disease expression in CD4 and CD8 gene-deleted New Zealand black (NZB) mice. J. Immunol. **157:** 2676–2684.
11. PENG, S.L., *et al.* 1996. Murine lupus in the absence of αβ T cells. J. Immunol. **156:** 4041–4049.
12. PENG, S.L., *et al.* 1996. Induction of nonpathologic, humoral autoimmunity in lupus-prone mice by a class II-restricted, transgenic αβ T cell. Separation of autoantigen-specific and -nonspecific help. J. Immunol. **157:** 5225–5230.
13. KALIYAPERUMAL, A., *et al.* 1996. Nucleosomal peptide epitopes for nephritis-inducing T helper cells of murine lupus. J. Exp. Med. **183:** 2459–2469.
14. CROW, M.K., *et al.* 1994. Autoantigen-specific T cell proliferation induced by the ribosomal P2 protein in patients with systemic lupus erythematosus. J. Clin. Invest. **94:** 345–352.
15. CHAN, O.T., *et al.* 1999. A novel mouse with B cells but lacking serum antibody reveals an antibody-independent role for B cells in murine lupus. J. Exp. Med. **189:** 1639–1648.
16. FATENEJAD, S., *et al.* 1998. Central T cell tolerance in lupus-prone mice: influence of autoimmune background and the lpr mutation. J. Immunol. **161:** 6427–6432.
17. SABZEVARI, H., *et al.* 1997. G1 arrest and high expression of cyclin kinase and apoptosis inhibitors in accumulated activated/memory phenotype CD4+ cells of older lupus mice. Eur. J. Immunol. **27:** 1901–1910.
18. ROZZO, S.J., *et al.* 1994. Evidence for polyclonal T cell activation in murine models of systemic lupus erythematosus. J. Immunol. **153:** 1340–1351.
19. ISHIKAWA, S., *et al.* 1998. A subset of CD4+ T cells expressing early activation antigen CD69 in murine lupus: possible abnormal regulatory role for cytokine imbalance. J. Immunol. **161:** 1267–1273.
20. MOHAN, C., *et al.* 1999. Genetic dissection of Sle pathogenesis: Sle3 on murine chromosome 7 impacts T cell activation, differentiation, and cell death. J. Immunol. **162:** 6492–6502.

21. Vassilopoulos, D., *et al.* 1995. TCR/CD3 complex-mediated signal transduction pathway in T cells and T cell lines from patients with systemic lupus erythematosus. J. Immunol. **155:** 2269–2281.
22. Tsokos, G.C. 1992. Lymphocyte abnormalities in human lupus. Clin. Immunol. Immunopathol. **63:** 7–9.
23. Kammer, G.M. 1999. High prevalence of T cell type I protein kinase A deficiency in systemic lupus erythematosus. Arthritis Rheum. **42:** 1458–1465.
24. Kammer, G.M., *et al.* 1996. Deficient type I protein kinase A isozyme activity in systemic lupus erythematosus T lymphocytes. II. Abnormal isozyme kinetics. J. Immunol. **157:** 2690–2698.
25. Liossis, S.N., *et al.* 1998. Altered pattern of TCR/CD3-mediated protein-tyrosyl phosphorylation in T cells from patients with systemic lupus erythematosus: deficient expression of the T cell receptor zeta chain. J. Clin. Invest. **101:** 1448–1457.
26. Blasini, A.M., *et al.* 1998. Protein tyrosine kinase activity in T lymphocytes from patients with systemic lupus erythematosus. J. Autoimmun. **11:** 387–393.
27. Brundula, V., *et al.* 1999. Diminished levels of T cell receptor zeta chains in peripheral blood T lymphocytes from patients with systemic lupus erythematosus. Arthritis Rheum. **42:** 1908–1916.
28. Kovacs, B., *et al.* 1996. Defective CD3-mediated cell death in activated T cells from patients with systemic lupus erythematosus: role of decreased intracellular TNF-α. Clin. Immunol. Immunopathol. **81:** 293–302.
29. Budagyan, V.M., *et al.* 1998. The resistance of activated T-cells from SLE patients to apoptosis induced by human thymic stromal cells. Immunol. Lett. **60:** 1–5.
30. Vakkalanka, R.K., *et al.* 1999. Elevated levels and functional capacity of soluble CD40 ligand in systemic lupus erythematosus sera. Arthritis Rheum. **42:** 871–881.
31. Kato, K., *et al.* 1999. The soluble CD40 ligand sCD154 in systemic lupus erythematosus. J. Clin. Invest. **104:** 947–955.
32. Alleva, D.G., *et al.* 1998. Intrinsic defects in macrophage IL-12 production associated with immune dysfunction in the MRL/++ and New Zealand black/white F1 lupus-prone mice and the Leishmania major-susceptible BALB/c strain. J. Immunol. **161:** 6878–6884.
33. Alleva, D.G., *et al.* 1997. Aberrant cytokine expression and autocrine regulation characterize macrophages from young MRL+/+ and NZB/W F1 lupus-prone mice. J. Immunol. **159:** 5610–5619.
34. Alleva, D.G., *et al.* 2000. Aberrant macrophage cytokine production is a conserved feature among autoimmune-prone mouse strains: elevated interleukin (IL)-12 and an imbalance in tumor necrosis factor-α and IL-10 define a unique cytokine profile in macrophages from young nonobese diabetic mice. Diabetes **49:** 1106–1115.
35. Mohan, C., *et al.* 1997. Genetic dissection of systemic lupus erythematosus pathogenesis: Sle2 on murine chromosome 4 leads to B cell hyperactivity. J. Immunol. **159:** 454–465.
36. Kaye, J., *et al.* 1989. Selective development of CD4+ T cells in transgenic mice expressing a class II MHC-restricted antigen receptor. Nature **341:** 746–749.
37. Lyons, D.S., *et al.* 1996. A TCR binds to antagonist ligands with lower affinities and faster dissociation rates than to agonists. Immunity **5:** 53–61.
38. Vratsanos, G.S., *et al.* 2001. CD4(+) T cells from lupus-prone mice are hyperresponsive to T cell receptor engagement with low and high affinity peptide antigens: a model to explain spontaneous T cell activation in lupus. J. Exp. Med. **193:** 329–337.
39. Dorfman, J.R., *et al.* 2000. CD4+ T cell survival is not directly linked to self-MHC-induced TCR signaling. Nat. Immunol. **1:** 329–335.
40. Viret, C., *et al.* 1999. Designing and maintaining the mature TCR repertoire: the continuum of self-peptide:self-MHC complex recognition. Immunity **10:** 559–568.
41. Craft, J., *et al.* 1999. Autoreactive T cells in murine lupus: origins and roles in autoantibody production. Immunol. Res. **19:** 245–257.
42. Bouzahzah, F., *et al.* 2003. CD4+ T cells from lupus-prone mice avoid antigen specific tolerance induction in vivo. J. Immunol. **170:** 741–748.
43. Theofilopoulos, A.N. & D.H. Kono. 1999. The genes of systemic autoimmunity. Proc. Assoc. Am. Physicians **111:** 228–40.

44. WALPORT, M.J., *et al.* 1997. Complement deficiency and autoimmunity. Ann. N.Y. Acad. Sci. **815:** 267–281.
45. MOULDS, J.M., *et al.* 1992. Genetics of the complement system and rheumatic diseases. Rheum. Dis. Clin. North Am. **18:** 893–914.
46. GAFFNEY, P.M., *et al.* 2000. Genome screening in human systemic lupus erythematosus: results from a second Minnesota cohort and combined analyses of 187 sib-pair families. Am. J. Hum. Genet. **66:** 547–556.
47. GAFFNEY, P.M., *et al.* 1998. A genome-wide search for susceptibility genes in human systemic lupus erythematosus sib-pair families. Proc. Natl. Acad. Sci. USA **95:** 14875–14879.
48. LINDQVIST, A.K. & M.E. ALARCON-RIQUELME. 1999. The genetics of systemic lupus erythematosus. Scand. J. Immunol. **50:** 562–571.
49. SHAI, R., *et al.* 1999. Genome-wide screen for systemic lupus erythematosus susceptibility genes in multiplex families. Hum. Mol. Genet. **8:** 639–644.
50. VIDAL, S., *et al.* 1998. Loci predisposing to autoimmunity in MRL-Fas lpr and C57BL/6-Fas lpr mice. J. Clin. Invest. **101:** 696–702.
51. MOREL, L., *et al.* 1994. Polygenic control of susceptibility to murine systemic lupus erythematosus. Immunity **1:** 219–229.
52. HAYWOOD, M.E., *et al.* 2000. Identification of intervals on chromosomes 1, 3, and 13 linked to the development of lupus in BXSB mice. Arthritis Rheum. **43:** 349–355.
53. MOREL, L., *et al.* 1996. Production of congenic mouse strains carrying genomic intervals containing SLE-susceptibility genes derived from the SLE-prone NZM2410 strain. Mamm. Genome **7:** 335–339.
54. SOBEL, E.S., *et al.* 2002. The major murine systemic lupus erythematosus susceptibility locus Sle1 results in abnormal functions of both B and T cells. J. Immunol. **169:** 2694–2700.
55. MOHAN, C., *et al.* 1999. Genetic dissection of lupus pathogenesis: a recipe for nephrophilic autoantibodies. J. Clin. Invest. **103:** 1685–1695.
56. ROZZO, S.J., *et al.* 2001. Evidence for an interferon-inducible gene, Ifi202, in the susceptibility to systemic lupus. Immunity **15:** 435–443.
57. BOACKLE, S.A., *et al.* 2001. Cr2, a candidate gene in the murine Sle1c lupus susceptibility locus, encodes a dysfunctional protein. Immunity **15:** 775–785.
58. SALVADOR, J.M., *et al.* 2002. Mice lacking the p53-effector gene Gadd45a develop a lupus-like syndrome. Immunity **16:** 499–508.
59. MURGA, M., *et al.* 2001. Mutation of E2F2 in mice causes enhanced T lymphocyte proliferation, leading to the development of autoimmunity. Immunity **15:** 959–970.
60. BACHMAIER, K., *et al.* 2000. Negative regulation of lymphocyte activation and autoimmunity by the molecular adaptor Cbl-b. Nature **403:** 211–216.
61. YI, Y., *et al.* 2000. Regulatory defects in Cbl and mitogen-activated protein kinase (extracellular signal-related kinase) pathways cause persistent hyperexpression of CD40 ligand in human lupus T cells. J. Immunol. **165:** 6627–6634.
62. KONG, P.L. *et al.* 2003. Role of H-2 haplotype in *Fas*-intact lupus-prone MRL mice: association with autoantibodies but not renal disease. Arthritis Rheum. In press.

Opsonization of Apoptotic Cells and Its Effect on Macrophage and T Cell Immune Responses

SUN JUNG KIM,[a] DEBRA GERSHOV,[b] XIAOJING MA,[a]
NATHAN BROT,[a,b] AND KEITH B. ELKON[c]

[a]*Department of Microbiology and Immunology and* [b]*Hospital for Special Surgery, Weill Medical College of Cornell University, New York, New York 10021, USA*

[c]*Division of Rheumatology, University of Washington, Seattle, Washington 98195, USA*

ABSTRACT: Genetic studies in mice indicate that predisposition to lupus-like diseases is caused by at least three mechanisms: (1) alterations in the threshold of activation of lymphocytes or macrophages; (2) defective signaling for activation-induced cell death; and (3) reduced clearance of apoptotic cells. To define the mechanisms whereby lupus develops in mice with deficiencies in either C1q, serum amyloid P component (SAP, the mouse counterpart of C-reactive protein, or CRP), or serum IgM, we studied the efficiency of phagocytosis of apoptotic cells using serum with varying levels of C1q, CRP, or IgM; we also examined the immune response to ingestion of dying cells under these conditions. Deficiency of C1q led to impaired macrophage phagocytosis of apoptotic cells, whereas CRP augmented phagocytosis, largely through recruitment of the early complement components. Like CRP, normal polyclonal IgM bound to apoptotic cells and activated complement on the cell surface. Similarly, direct binding as well as absorption experiments revealed that CRP and IgM antibodies had a similar ligand recognition specificity, namely lysophospholipids containing phosphorylcholine. IL-12 provides a pivotal link between macrophages and the T cell response to ingested material. We observed that necrotic cells induced IL-12 p40 expression, whereas phagocytosis of apoptotic cells profoundly reduced IL-12 production from stimulated macrophages. Furthermore, soluble factors from macrophages that had ingested apoptotic cells suppressed interferon-γ production by activated T cells. These findings suggest that phospholipid exposure on apoptotic cells promotes opsonization by serum proteins leading to activation of complement, macrophage ingestion, and T cell suppression. We discuss how deficient opsonization or processing of dying cells leads to autoimmunity.

KEYWORDS: lupus; autoimmunity; apoptosis; complement; antibody

INTRODUCTION

Prior to 1990, the only known genetic risk factors for systemic lupus erythematosus (SLE) were a weak association with HLA and deficiencies in early complement components. Whereas little else about risk factors in humans has been learned

Address for correspondence: Keith B. Elkon, Division of Rheumatology, Box 356428, University of Washington, Seattle, Washington 98195. Voice: 206-543-3414; fax: 206-685-9397. elkon@u.washington.edu

Ann. N.Y. Acad. Sci. 987: 68–78 (2003).

since, research efforts with mice over the last decade have been of immense value in the identification of genes that cause lupus when over- or under-expressed. Based on recent discoveries in mice and limited information in humans, the "causes" of lupus can broadly be divided into three categories:

(1) defective peripheral tolerance caused by alterations in the expression of genes regulating survival and cell death (such as reduced expression of Fas/Fas ligand, Bim, PTEN, IL-2/IL-2R, overexpression of Bcl-2);

(2) defective clearance of dead and dying cells (deficiencies of early complement components, secreted IgM, pentraxins, DNAse1, and mer fall into this category); and

(3) reduced threshold of lymphocyte activation (such as deficiencies of lyn, CD22, cbl-b, FCRγ2b).

These three categories are expressed in TABLE 1 as follows: (1) apoptosis signals, (2) apoptosis clearance, and (3) activation threshold. The second category, which covers the role of defective clearance of dead and dying cells, will be the focus of this article.

NORMAL PATHWAY OF APOPTOTIC CELL CLEARANCE

Caenorhabditis elegans has served as a model organism to identify genes regulating apoptosis.[1] During development, dying cells are rapidly engulfed by neighboring cells. Approximately seven genes that encode two partially overlapping pathways for uptake and ingestion of apoptotic cells have been identified (reviewed in Ref. 2). One pathway is triggered by the receptor, CED-1, which recruits an adaptor, CED-6. The second pathway utilizes CED-2, 5, 10, and 12 culminating in rac activation, GTP/GDP exchange, and cytoskeletal alterations associated with phagocytosis.

Unlike *C. elegans*, mammals have specialized phagocytic cells that ingest apoptotic cells, and these cells express a much larger number of receptors that have been implicated in the recognition and tethering of apoptotic cells (reviewed in Ref. 3). The receptors include scavenger receptors, a specific phosphitidylserine (PS) receptor, and complement receptors (FIG. 1).

RELATIONSHIP BETWEEN APOPTOTIC CELL CLEARANCE AND IMMUNE RESPONSES

As shown in TABLE 1, deficiency of at least five different types of protein are implicated in provoking susceptibility to lupus through a known or suspected role in the clearance of dying cells or their debris. The involvement of DNAse1 deficiency in mice[4] as well as, possibly, in humans[5] appears to provide an intuitive link between cell death and anti-DNA antibodies. However, the precise role of DNAse1 in degradation of DNA within the dying cell, the extracellular compartment, or the phagocyte remains to be determined.

Perhaps the most compelling case that links defective clearance of apoptotic cells with a lupus-like disease occurs in mice with the targeted deletion of mer. Mer is a receptor belonging to the Axl, Tyro protein tyrosine kinase family that is exclusively

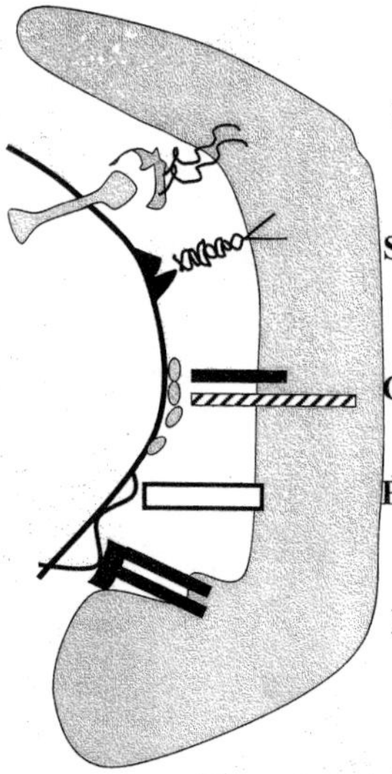

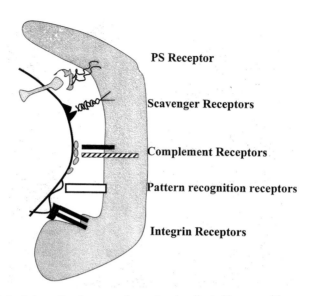

FIGURE 1. Schematic of groups of receptors implicated in recognition or phagocytosis of apoptotic cells. Macrophages recognize alterations on the surface of dying cells. Alterations in phospholipids figure prominently and include the "flip" of phosphatidylserine that is recognized by a unique receptor as well as exposure of lysophosphorylcholine (lyso-PC). Exposure of lyso-PC leads to binding of CRP and natural IgM antibodies followed by activation of complement, as discussed in the text. Other macrophage receptors shown here, such as CD14 or pattern recognition receptors, may bind directly to altered cell ligands (ICAM3) or may also require opsonins, such as thrombospondin, to interact with the class B macrophage scavenger receptor, CD36.[30] Additional receptors and ligands not shown here have also been implicated in this process (see Gregory *et al.*[31] for review).

TABLE 1. Genetic alterations associated with lupus-like disease

Apoptosis Signals	Apoptosis Clearance	Activation Threshold
Fas, Fas ligand	C1q, C4, C2	Lyn
Bcl-2 (overexpression)	SAP/CRP	CD22
Bim	Secreted IgM	Cbl-b
IL-2/IL-2R	DNAse1	FCRγ2b
PTEN (heterozygote)	Mer	PD-1
		BLYS/BAFF (over-expression)
		p21
		CD45 (gain of function)
		Gadd45a

KEY: The lupus phenotype is associated with loss of function except for genes where over-expression, gain of function, or heterozygote is indicated.

expressed in myeloid cells. Deficiency of mer is associated with increased numbers of apoptotic cells in the thymus, reduced phagocytosis of apoptotic cells, and a typical lupus-like illness.[6] Targeted deletion of other members of this family leads to more severe disease.[7]

COMPLEMENT PROMOTES REMOVAL OF DEAD AND DYING CELLS

More controversial are the mechanisms whereby deficiencies of components of the classical complement pathway, serum amyloid P component (SAP)/C-reactive protein (CRP), and serum IgM (TABLE 1) predispose to lupus. Here, we provide evidence that all three of these serum components function to opsonize dying cells and facilitate their "safe clearance"—particularly from sites of inflammation. Serum opsonins are less likely to play a role in closed tissue compartments such as the thymus.

Complement proteins comprise more than 20 serum or cell-associated proteins that function in both innate and adaptive immune responses. Complement may be ac-

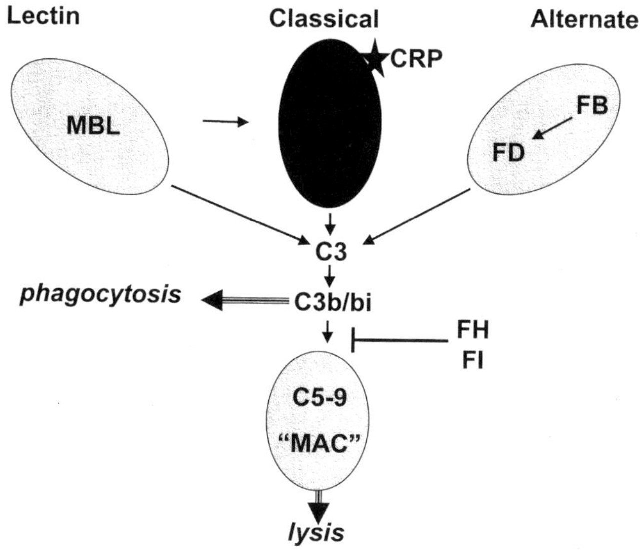

FIGURE 2. Overview of complement activation. The mannose-binding lectin (MBL) pathway is activated by mannose-containing ligands on foreign organisms as well as by some apoptotic cells.[32] The best-known activators of the classical pathway are antibodies and CRP, whereas certain bacterial products and IgA antibodies can initiate alternative pathway engagement. All three pathways of complement activation converge on C3 leading to its cleavage and to the covalent binding of C3b onto the activating surface. Opsonization of the complement activator is critical since this allows the phagocyte expressing complement receptors to "recognize" the activator and respond appropriately (see text and FIG. 6). Complement activation on self-cells is attenuated by soluble regulatory proteins (factor H is a cofactor for factor I) as well as cell-membrane-bound regulatory proteins that prevent the formation of the C3 convertase. If the C3 convertase does assemble, it activates the C5 convertase leading to formation of the membrane attack complex (MAC). Ultimately, this results in cell lysis and necrosis.

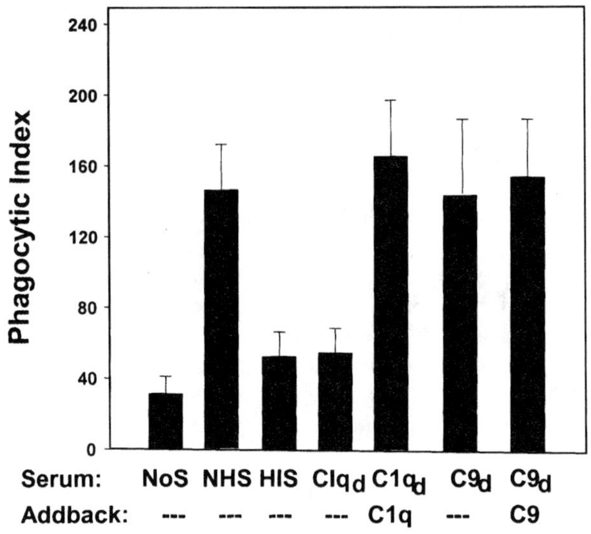

FIGURE 3. C1q deficiency leads to deficient phagocytosis of apoptotic cells. The efficiency of phagocytosis of autologous neutrophils was compared in 15% human serum (NHS) that was selectively deficient in the complement components, C1q, or C9. To verify that a specific complement component was responsible for the effect observed, add-back experiments were performed with the missing factor (C1q, 250 μg/mL or C9, 60 μg/mL). HIS: heat-inactivated serum (56°C for 30 min); NoS: no serum; $_d$: deficient. The mean ± SD of four experiments is shown. (Modified from Mevorach *et al.*[9])

tivated by three different pathways, as shown in FIGURE 2. All three initiating pathways converge to promote the proteolytic cleavage of C3, which results in the covalent binding of C3bi to the activating surface. This is a critical step since it allows the activating particle to interact with the complement receptors, CR3 and CR4, on the phagocytes. In addition, receptors for other complement components as well as a constellation of phagocyte "pattern recognition receptors"[8] expressed on the macrophage, determine whether an immunosuppressive or inflammatory response is generated (see below).

Many explanations have been put forward to account for the well-known tight association between deficiencies of the early complement components (especially C1q) and SLE. This association could be explained by increased infections, alterations in the composition of immune complexes, modulation of T or B cell function, or deficient clearance of self-antigens. In support of the clearance hypothesis, we[9] and others[10] have detected complement on the surface of apoptotic cells. Furthermore, we observed that deficiency of C1q retards the phagocytosis of dying cells *in vitro* and that this defect can be reconstituted with purified C1q (FIG. 3).[9] Taken together with the observations that C1q-deficient mice develop a lupus-like disease associated with increased numbers of apoptotic cells in their kidneys,[11] it is reasonable to propose that these findings are causally linked.

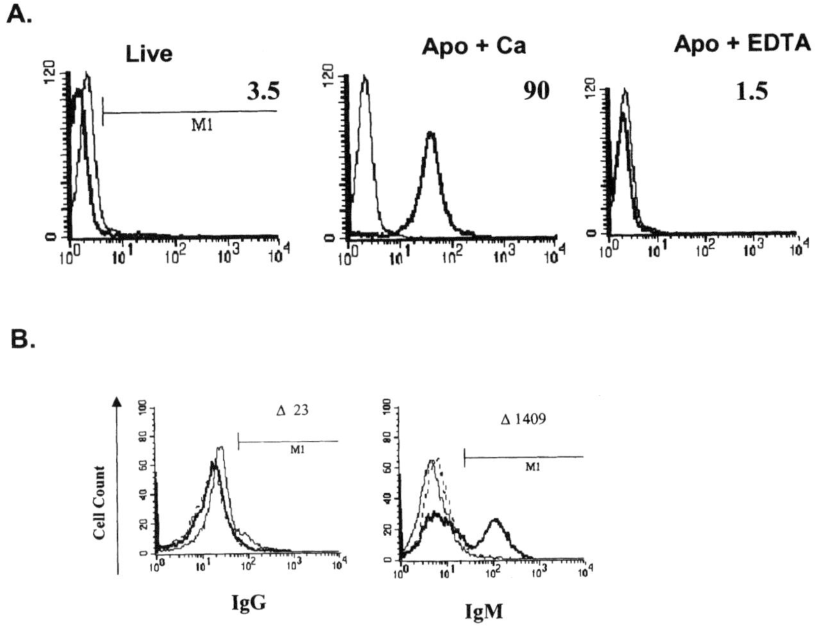

FIGURE 4. Classical pathway activators, CRP, and IgM bind to apoptotic cells. (**A**) Live or apoptotic Jurkat T cells were incubated with purified CRP (50 μg/mL) for 20 min at room temperature in the presence of Ca^{2+} or EDTA (10 mM). CRP binding was assessed by flow cytometry using either isotype control (*thin line*) or anti-CRP antibody (*thick line*). (**B**) Live (*thin line*) or apoptotic (*thick line*) cells were incubated with normal IgG (10 mg/mL) or IgM (1 mg/mL) for 20 min at 37°C. Cells were stained with appropriate secdondary antibody and analyzed by flow cytometry. The apoptotic cells stained with secondary antibody only are shown by the *dotted line*. (Modified from Gershov *et al.*[19])

SAP/CRP BINDS TO APOPTOTIC CELLS AND TO NUCLEAR ANTIGENS

Pentraxins, which constitute a family of highly conserved cyclic pentameric proteins, include SAP and the human homologue of mouse SAP, that is, CRP (reviewed in Ref. 12). Whereas SAP is constitutively expressed at low levels in human serum, release of pro-inflammatory cytokines, such as IL-1, IL-6, and TNF-α, results in up to a 1,000-fold elevation of CRP.[12] CRP was initially characterized by its binding to the C-polysaccharide of Streptococcus pneumonia,[13] suggesting a primary function in clearance of encapsulated bacteria. However, CRP in certain species does not bind to C-polysaccharide.[12] Furthermore, the striking degree of sequence homology between species and the lack of polymorphisms within species[14] suggest that the selection pressures to maintain homeostatic function are more important than pressures from infectious organisms. Since the demonstration of CRP deposition predominantly along the membranes of damaged cells in non-infectious tissue injury,[15] increasing attention has focused on the role of CRP binding to self-molecules.

CRP specifically binds to phosphocholine in membrane bilayers[16] as well as to H1-containing chromatin[17] and to snRNPs.[18]

We observed that CRP bound to the surface of apoptotic cells in a calcium-dependent manner (FIG. 4A) and that CRP amplified classical pathway activation on the surface of the dying cell.[19] Surprisingly, rather than enhancing lysis of the cell, CRP attenuated deposition of the membrane attack complex (MAC) (see FIG. 1). Attenuation of the MAC was caused by the recruitment of factor H, a complement regulatory serum protein that not only acts as a co-factor for factor I, but also accelerates the decay of the C3 and C5 convertases.[20] We have suggested that the contrasting effects of CRP on the early and late complement pathways that act in concert to opsonize and protect apoptotic cells from lysis is the primary function of this acute phase protein in early tissue injury.[19]

NATURAL IgM AUTOANTIBODIES BIND TO PHOSPHOLIPIDS ON DYING CELLS

Although elevated levels of CRP, as occurs in inflammation, promote complement activation on apoptotic cells, complement is deposited on dying cells even when CRP levels are low. Since IgM is a potent activator of the classical complement pathway, and since serum IgM-deficient mice develop lupus,[21,22] we tested whether natural IgM antibodies bind to dying cells and promote complement activation. Normal serum IgM, but not IgG, bound to apoptotic cells (FIG. 4B). Furthermore, mouse serum that was deficient in IgM bound 50–60% less C3 compared to wild-type serum.[23] These observations indicate that natural IgM in serum is, in part, responsible for activating the classical complement pathway on apoptotic cells.

Natural IgM antibodies comprise low-affinity antibodies with reactivity to lipids, sugars, as well as proteins. Since Western blot analysis failed to detect reactivity with membrane protein antigens (results not shown), we examined reactivity with lipid antigens. Despite the translocation of phosphatidylserine to the surface of apoptotic cells and the presence of antibodies to negatively charged phospholipids in the sera of autoimmune patients, absorption studies revealed that the IgM antibodies did not bind to this phospholipid. Instead, both direct binding and absorption studies demonstrated that the natural IgM antibodies bound to lysophosphatidylcholine and phosphorylcholine (PC).[23] This finding is significant since a specific PLA_2 is activated during apoptosis and is implicated in membrane remodeling and the generation of lysophospholipids.[24] In the context of autoimmunity, the phosphorylcholine antigen has long been of interest in association with cross-reactivity to DNA as well as the somatic mutation of anti-PC antibodies to anti-DNA.[25]

T CELL PRIMING AND T CELL SUPPRESSION

One of the most potent cytokines required to prime T cells for a Th1-type immune response is IL-12.[25] To understand the link between the innate and adaptive immune responses to dying cells, we studied the production of IL-12 following ingestion of apoptotic or necrotic cells. When resting macrophages were allowed to ingest apoptotic autologous T cells, little IL-12 was produced (baseline values: p40 = 3,000 and

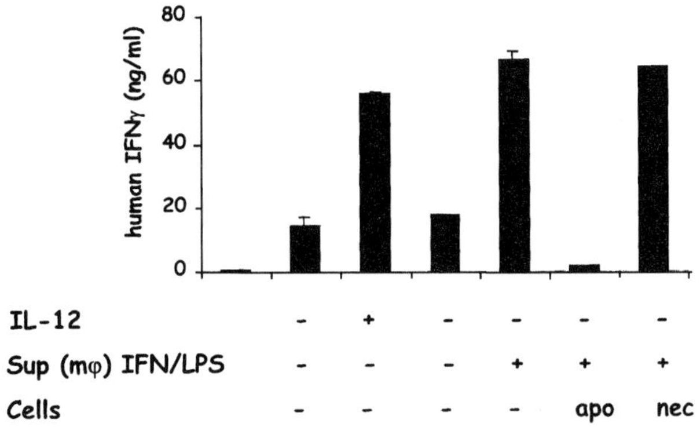

FIGURE 5. Phagocytosis of dying cells modulates T cell priming. Human monocyte-derived macrophages were activated by 100 U/mL of IFNγ and LPS or by IFNγ and LPS together with apoptotic (apo) or necrotic (nec) cells. On the following day, supernatants were harvested and examined for their effect on IFNγ production by ConA-activated T cells. IFNγ production was quantified by ELISA at 24 h. T cells activated by recombinant human IL-12 (2 ng/mL) were used as a positive control.

p70 = 200 pg/mL; values 1 h following ingestion: p40 = 500 and p70 = 5 pg/mL). In contrast, ingestion of cells made necrotic by three freeze-thaw cycles induced high levels of IL-12 p40 and a small amount of bioactive heterodimer, p70 (500 and 50 pg/mL, respectively). When activated macrophages—by LPS alone or LPS plus interferonγ (IFNγ)—ingested apoptotic cells, IL-12 p40 and p70 were strongly inhibited (80–90%), whereas phagocytosis of necrotic cells did not result in significant changes. Surprisingly, antibody neutralization experiments revealed that inhibition was not mediated through anti-inflammatory cytokines known to attenuate IL-12 expression (IL-10, TGFβ1, or PGE2), indicating that novel mechanisms may be involved.

To determine the effect of macrophage ingestion of apoptotic cells on T cell function, we collected the supernatants from macrophages that had ingested apoptotic cells and examined their effect on IFNγ production by concanavalin A (Con A)-treated T cells. As shown in FIGURE 5, supernatants from IFNγ and LPS-activated macrophages, containing high concentrations of IL-12, stimulated IFNγ equivalent to that observed by the addition of recombinant IL-12. When supernatants from macrophages that had ingested apoptotic cells were added to activated T cells, IFNγ production was profoundly suppressed, whereas necrotic cells maintained high levels of this cytokine. Taken together, these findings provide the necessary link between the innate and adaptive immune response to dying cells and suggest that inhibition of IL-12 production by macrophages that have ingested apoptotic cells may be critical to the avoidance of autoimmunity. Processing and presentation of antigens derived from dying cells is a related and important question. This topic is discussed by Steinman *et al.*[26] in this volume.

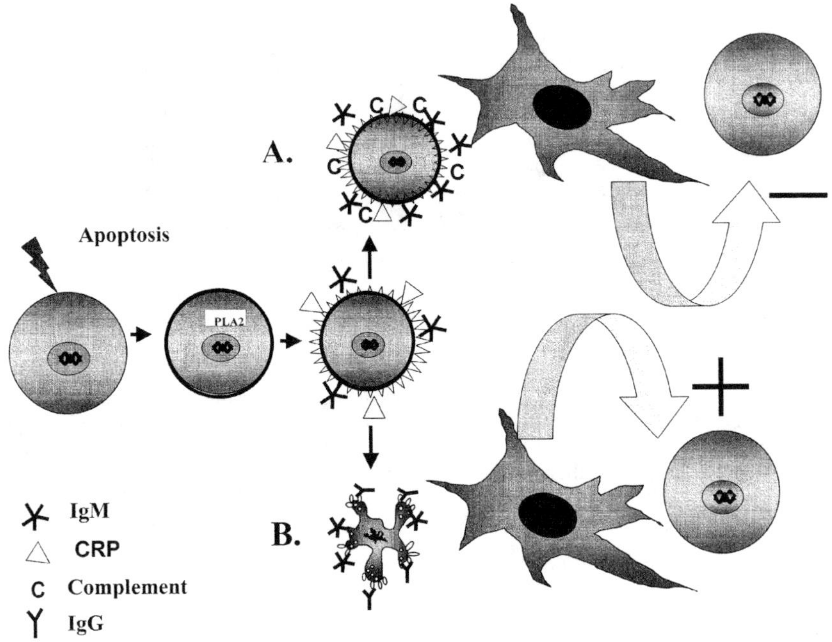

FIGURE 6. Proposed role of serum factors in the clearance of dying cells. Apoptotic cells activate PLA_2, which results in exposure of lysophospholipids, including lyso-PC, on the cell membrane. Lyso-PC is recognized by IgM and/or CRP, and this recognition activates the classical complement pathway (**A**). Macrophages phagocytose complement-coated cells and produce immunosuppressive cytokines such as TGF-β. In contrast, when opsonins are low or absent (**B**), either the cells undergo post-apoptotic necrosis and/or are seen by IgG antibodies. In either case, phagocytosis of this cargo leads to pro-inflammatory cytokine production (see text). (Modified from Kim *et al.*[23])

DOES COMPLEMENT OPSONIZATION OF DYING CELLS PREVENT AUTOIMMUNITY?

Based on the observations discussed above, we propose a model to account for the association of serum opsonins with systemic autoimmunity (FIG. 6). Apoptotic cells not only translocate phosphatidylserine to the outside of the membrane, they activate phospholipases that remodel their membranes to expose altered phospholipids.[23,24] These altered phospholipids include lysophosphatidylcholine and phosphorylcholine, which are ligands for CRP and are also antigenic for natural IgM antibodies. Both of these serum proteins activate the classical complement components leading to efficient removal of the cell. Although C3b/bi binds to CR3 and CR4 on macrophages, it is possible that engagement of other complement receptors[27] as well as non-complement receptors are also involved in recognition and signal transduction by phagocytes. Precise identification of these receptors is important since engulfment of apoptotic cells,[28] including those opsonized with CRP,[19] induce the expression of immunosuppressive cytokines such as TGF-β.

Failure to rapidly remove apoptotic cells on the other hand, allows dying cells to undergo post-apoptotic necrosis. Necrotic cells promote a pro-inflammatory immune response,[28,29] most likely accounting for an immune response to self.

REFERENCES

1. HORVITZ, H., S. SHAHAM & M. HENGARTNER. 1994. The genetics of programmed cell death in the nematode *Caenorhabditis elegans*. Cold Spring Harbor Symp. Quant. Biol. **59:** 377–385.
2. HENGARTNER, M.O. 2001. Apoptosis: corralling the corpses. Cell **104:** 325–328.
3. SAVILL, J. & V. FADOK. 2000. Corpse clearance defines the meaning of cell death. Nature **407:** 784–788.
4. NAPIREI, M., H. KARSUNKY, B. ZEVNIK, *et al.* 2000. Features of systemic lupus erythematosus in DNAse1-deficient mice. Nat. Genet. **25:** 177–181.
5. YASUTOMO, K., T. HORIUCHI, S. KAGAMI, *et al.* 2001. Mutation of DNAse1 in people with systemic lupus erythematosus. Nat. Genet. **28:** 313–314.
6. SCOTT, R.S., E.J. MCMAHON, S.M. POP, *et al.* 2001. Phagocytosis and clearance of apoptotic cells is mediated by mer. Nature **411:** 207–211.
7. LU, Q. & G. LEMKE. 2001. Homeostatic regulation of the immune system by receptor tyrosine kinases of the Tyro 3 family. Science **293:** 306–311.
8. MEDZHITOV, R. & C.A. JANEWAY. 1997. Innate immunity: the virtues of a nonclonal system of recognition. Cell **91:** 295–298.
9. MEVORACH, D., J. MASCARENHAS, D.A. GERSHOV, *et al.* 1998. Complement-dependent clearance of apoptotic cells by human macrophages. J. Exp. Med. **188:** 2313–2320.
10. KORB, L.C. & J.M. AHEARN. 1997. C1q binds directly and specifically to surface blebs of apoptotic human keratinocytes. J. Immunol. **158:** 4525–4528.
11. BOTTO, M., C. DELL'AGNOLA, A.E. BYGRAVE, *et al.* 1998. Homozygous C1q deficiency causes glomerulonephritis associated with multiple apoptotic bodies. Nature Gen. **19:** 56–59.
12. BALTZ, M.L., F.C. DE BEER, A. FEINSTEIN, *et al.* 1982. Phylogenetic aspects of C-reactive protein and related proteins. Ann. N.Y. Acad. Sci. **389:** 49–75.
13. TILLETT, W.S. & T. FRANCIS. 1930. Serological reactions in pneumonia with a non-protein fraction of pneumococcus. J. Exp. Med. **52:** 561–571.
14. THOMPSON, D., M.B. PEPYS & S.P. WOOD. 1999. The physiological structure of human C-reactive protein and its complex with phosphocholine. Structure **7:** 169–177.
15. KUSHNER, I., L. RAKITA & M.H. KAPLAN. 1963. Studies of acute-phase protein. II. Localization of Cx-reactive protein in heart in induced myocardial infarction in rabbits. J. Clin. lnvest. **42:** 286–292.
16. VOLANAKIS, J.E. & K.W.A. WIRTZ. 1979. Interaction of C-reactive protein with artificial phosphatidylcholine bilayers. Nature **281:** 155–157.
17. ROBEY, F.A., K.D. JONES, T. TANAKA, *et al.* 1984. Binding of C-reactive protein to chromatin and nucleosome core particles. J. Biol. Chem. **259:** 7311–7316.
18. DUCLOS, T.W. 1989. C-reactive protein reacts with the U1 small nuclear ribonucleoprotein. J. Immunol. **143:** 2553–2559.
19. GERSHOV, D., S. KIM, N. BROT, *et al.* 2000. C-reactive protein binds to apoptotic cells, protects the cells from assembly of the terminal complement components, and sustains an antiinflammatory innate immune response: implications for systemic autoimmunity. J. Exp. Med. **192:** 1353–1364.
20. VIK, D.P., P. MUNOZ-CANOVES, D.D. CHAPLIN, *et al.* 1990. Factor H. Curr. Top. Microbiol. Immunol. **153:** 147–162.
21. BOES, M. 2000. Role of natural and immune IgM antibodies in immune responses. Mol. Immunol. **37:** 1141–1149.
22. EHRENSTEIN, M.R., H.T. COOK & M.S. NEUBERGER. 2000. Deficiency in serum immunoglobulin (Ig)M predisposes to development of IgG autoantibodies. J. Exp. Med. **191:** 1253–1258.

23. KIM, S., D. GERSHOV, X. MA, *et al.* 2002. I-PLA2 activation during apoptosis promotes the exposure of membrane lysophosphatidylcholine leading to binding by natural IgM antibodies and complement activation. J. Exp. Med. **196:** 655–665.
24. ATSUMI, G., M. TAJIMA, A. HADANO, *et al.* 1998. Fas-induced arachidonic acid release is mediated by Ca^{2+}-independent phospholipase A2 but not cytosolic phospholipase A2, which undergoes proteolytic inactivation. J. Biol. Chem. **273:** 13870–13877.
25. SHARMA, A., D.A. ISENBERG & B. DIAMOND. 2001. Crossreactivity of human anti-ds-DNA antibodies to phosphorylcholine: clues to their origin. J. Autoimmun. **16:** 479–484.
26. STEINMAN *et al.* 2003. Dendritic cell function *in vivo* during the steady state: a role in peripheral tolerance. Ann. N.Y. Acad. Sci. **987:** this volume.
27. EGGLETON, P., K.B. REID, U. KISHORE, *et al.* 1997. Clinical relevance of calreticulin in systemic lupus erythematosus. Lupus **6:** 564–571.
28. FADOK, V.A., D.L. BRATTON, A. KONOWAL, *et al.* 1998. Macrophages that have ingested apoptotic cells in vitro inhibit proinflammatory cytokine production through autocrine/paracrine mechanisms involving TGF-beta, PGE2, and PAF. J. Clin. Invest. **101:** 890–898.
29. STERN, M., J. SAVILL & C. HASLETT. 1996. Human monocyte derived macrophage phagocytosis of senescent eosinophils undergoing apoptosis. Am. J. Pathol. **149:** 911–921.
30. SAVILL, J., N. HOGG, Y. REN, *et al.* 1992. Thrombospondin cooperates with CD36 and the vitronectin receptor in macrophage recognition of neutrophils undergoing apoptosis. J. Clin. Invest. **90:** 1513–1522.
31. GREGORY, C.D. 2000. CD14-dependent clearance of apoptotic cells: relevance to the immune system. Curr. Opin. Immunol. **12:** 27–34.
32. OGDEN, C.A., A. DE CATHELINEAU, P.R. HOFFMAN, *et al.* 2002. C1q and mannose binding lectin engagement of cell surface calreticulin and CD91 initiates macropinocytosis and uptake of apoptotic cells. J. Exp. Med. **194:** 781–795.

Major Peptide Autoepitopes for Nucleosome-Centered T and B Cell Interaction in Human and Murine Lupus

SYAMAL K. DATTA

Rheumatology Division, Northwestern University Medical School, Chicago, Illinois 60611, USA

ABSTRACT: The potential cross-reactivity of normal T and B cells to nuclear antigens is vast, probably due to their "education" by apoptotic cell antigens in generative lymphoid organs. Despite this "nucleocentric repertoire," as we call it, the peripheral immune system normally remains tolerant or ignorant of the products of apoptosis. However, the T helper (Th) cells, and also B cells of lupus, have a regulatory defect in the expression of CD40 ligand (CD40L). A sustained hyper-expression of CD40L by lupus T cells can be triggered by subthreshold stimuli, and is associated with impaired phosphorylation of Cbl-b, a critical downregulatory molecule in T cell signal transduction. This CD40L hyper-expression abnormally prolongs co-stimulatory signals to autoimmune B cells, and it probably instigates APC (dendritic cells, resting anti-DNA B cells, and macrophages) to present apoptotic cell autoantigens in an immunogenic fashion. We have identified the dominant nucleosomal epitopes that are critical for cognate interactions between autoimmune Th cells and anti-DNA B cells in lupus. By scanning of overlapping synthetic peptides, and by mass spectrometry of naturally processed peptides, five major epitopes in nucleosomal histones were localized, namely $H1'_{22-42}$, $H2B_{10-33}$, $H3_{85-105}$, $H4_{16-39}$, and $H4_{71-94}$. The autoimmune T cells as well as B cells of lupus recognize these epitopes, and with age, autoantibodies against the peptide epitopes cross-react with nuclear autoantigens. Moreover, the peptide autoepitopes can be promiscuously presented and recognized by lupus T cells in the context of diverse MHC alleles. This cross-reactivity opens up the possibility of developing "universally" tolerogenic peptides for therapy of lupus in humans despite their MHC diversity. Indeed, tolerogenic therapy with a single histone peptide epitope can halt the progression of established glomerulonephritis in lupus-prone mice by "tolerance spreading" that inactivates a broad spectrum of autoimmune T and B cells in concert.

KEYWORDS: systemic lupus erythematosus; autoimmunity; T cell–B cell interaction; anti-DNA autoantibody; tolerance therapy; peptide epitopes; nucleosomes; apoptosis; CD40 ligand

Address for correspondence: Syamal K. Datta, M.D., Rheumatology Division, Ward 3-315, Northwestern University Medical School, 303 East Chicago Avenue, Chicago, IL 60611. Voice: 312-503-0535; fax: 312-503-0994.
skd257@northwestern.edu

Ann. N.Y. Acad. Sci. 987: 79–90 (2003). © 2003 New York Academy of Sciences.

INTRODUCTION

Genes encoding pathogenic anti-DNA autoantibodies are present in normal mice,[1–3] and indeed, the frequency of peripheral B cells with the potential to produce DNA-binding antibodies is considerably high even in normal mice.[4–6] Nevertheless, these low-affinity autoimmune B cells do not expand or mutate to become pathogenic, because appropriate T helper (Th) cells are normally inoperative.[7] We and others have shown that the production of nephritogenic anti-DNA autoantibodies in SLE is driven by cognate interactions between select populations of autoimmune Th and B cells.[8–14]

COGNATE INTERACTIONS BETWEEN AUTOIMMUNE T AND B CELLS OF LUPUS

To define the primary immunogen that drives the pathogenic autoimmune response, we cloned the autoimmune Th cells from the (SWR × NZB)F_1 (SNF_1) mice with lupus nephritis, as well as from patients with systemic lupus erythematosus (SLE). Only 12–15% of all T cells that are activated *in vivo* in SLE have the ability to induce the production of the pathogenic variety of IgG anti-DNA autoantibodies *ex vivo*.[10,11,15,16] In the SNF_1 model, representative pathogenic autoantibody-inducing Th clones rapidly induce immune-deposit glomerulonephritis when transferred into young pre-autoimmune mice.[16,17] The antigen-binding CDR loops of the T cell receptors (TCRs) expressed by these *special* (functionally characterized and disease-relevant) pathogenic Th clones of lupus contain recurrent motifs of anionic residues, indicating their selection by autoantigens with reciprocally charged residues.[12,16] Indeed, a majority of such pathogenic Th clones of lupus respond to charged determinants in nucleosomal antigens, and nucleosome-specific T cells are detectable in lupus-prone mice long before they produce pathogenic autoantibodies.[18,19] In addition, immunization of pre-autoimmune SNF_1 mice, but not normal mice, with native nucleosome particles precipitates severe lupus nephritis.[18]

PEPTIDE AUTOEPITOPES FOR AUTOIMMUNE T CELLS AND B CELLS AS WELL

In the SNF_1 model, by testing 154 overlapping 15-mer peptides spanning the four core histones of nucleosomes, we localized the critical peptide autoepitopes for lupus nephritis-inducing Th cells, determining their amino acid (aa) positions: 10–33 of H-2B ($H2B_{10–33}$), 16–39 and 71–94 of H4 ($H4_{16–39}$ and $H4_{71–94}$), and 85–102 of H3 ($H3_{85–102}$). These peptides could accelerate lupus nephritis *in vivo* by triggering Th1 cells that induced the production of pathogenic anti-nuclear autoantibodies.[19] To detect other dominant epitopes, we eluted peptides from MHC class II molecules of a nucleosome-specific B cell line that was fed with crude chromatin. The eluted peptides were purified by reverse-phase high-performance liquid chromatography (RP-HPLC) and tested for their ability to stimulate autoimmune Th clones of lupus, and then analyzed by matrix-assisted-laser-desorption time-of-flight mass spectrometry (MALDI-TOF-MS). Amino acid sequences of stimulatory fractions, deduced by electrospray ionization mass spectrometry (ESI-MS/MS), revealed three new autoepitopes.[20] One of the epitope's sequence was homologous to histone $H1'_{22–42}$. The $H1'_{22–42}$ peptide stimulates autoimmune Th cells to augment the production of pathogenic anti-nuclear antibodies and is much more potent than the core histone epitopes in accelerating glomerulonephritis in pre-autoimmune SNF_1 mice. Remarkably, a marked expansion of Th1 cells responsive to the $H1'_{22–42}$ epitope occurs spontaneously in SNF_1 mice very early in life. Moreover, a large proportion of $H1'_{22–42}$-specific T cell clones cross-react with one or more of the core histone epitopes described above, but not with epitopes in other lupus autoantigens.

Surprisingly, the nucleosomal peptides behave as "universal" epitopes and they are promiscuously recognized by the SNF_1 (I-$A^{d/q}$)-derived lupus Th cells even when presented by MHC class II (I-A molecules) of diverse mouse haplotypes, and even human HLA-DR molecules.[21] Due to reciprocal charge interaction, the lupus TCRs probably contact the nucleosomal peptide-complexed with MHC promiscuously and sustain TCR signaling. Competitive binding assays with soluble class II molecules indicated that the peptides bound to the MHC groove.[21] Thus the recognition of nucleosomal peptides by lupus TCRs is MHC-dependent, but unrestricted due to a peptide-dominant interaction. Indeed, susceptibility to lupus nephritis has not been directly linked to genes for any particular MHC class II molecules, but may be attributed to other genes located in the MHC region.[22] The promiscuity of lupus TCRs has profound implications regarding the selection of nucleosome-specific Th cells in the lupus-prone thymus and their defective regulation in the periphery. These results moreover indicate that "universally tolerogenic" epitopes could be designed for antigen-specific therapy of lupus patients with diverse HLA alleles.

Importantly, the nephritogenic epitopes are located in regions of histones that are also targeted by autoantibodies from lupus B cells (B cell epitopes), as well as sites that contact with DNA in the nucleosome particle, indicating how the epitopes could be protected during autoantigen processing and preferentially presented to the Th cells.[19,23,24] Strikingly, with the onset of lupus nephritis, B cells producing autoantibodies specific for the $H1'_{22–42}$ epitope evolve to become highly cross-reactive with nuclear autoantigens. Convergence of T and B cell epitopes in $H1'_{22–42}$ and the ability to elicit a cross-reactive response makes it highly dominant.[20] Interestingly, in an experimentally induced model, a peptide mimic of DNA behaves similarly to the $H1'_{22–42}$ epitope in inducing cross-reactive autoantibodies.[25] In the past, serologic studies had pointed to histone H1 being an important target for pathogenic autoantibodies in SLE,[26,27] but the peptide epitope had not been identified.

HUMAN LUPUS

The pathogenic autoantibody-inducing Th cells of patients with lupus also recognize nucleosomal proteins,[12,28] and they have TCRs homologous to the promiscous TCRs found in murine lupus T cells.[12,21] Despite the heterogeneity of human disease and epitope spreading, we were able to find the autoepitopes in core histones of nucleosomes that are recurrently recognized by the autoimmune T cells of lupus patients. It is noteworthy that nucleosome-specific T cells from lupus patients are extremely difficult to maintain *in vitro*, because nucleosomes are released in culture by dying cells. Unlike mice, human T cells express MHC class II molecules, and can present antigens to other T cells in the cultures, but such T–T cell antigen presentation usually inactivates (anergizes) the T cells. Our methods circumvented such problems.[12,28,29] While deriving the T cell lines, we did not stimulate the lupus T cells with any exogenously added nucleosomal peptides. Repeated stimulation *in vitro* with nucleosomes, or nucleosomes complexed with foreign antigens, has been used to rescue nucleosome-reactive T cells from normal subjects.[30,31] However, deliberate stimulation with autoantigen *in vitro* might actually cause apoptosis of the T cells that were primed spontaneously *in vivo* in patients with active lupus, yielding results that might not be representative of disease.[32] The pre-activated autoimmune T cells need to be rested first and fed with low doses of IL-2 before doing functional studies.[10–12,15,21,28,29] We confirmed and narrowed down the major autoepitopes for lupus further by studying freshly obtained PBMC T cells from patients, using flow cytometry assays to detect rapid intracellular accumulation of cytokines after short-term peptide stimulation.[28] Proliferation assays or limiting dilution analyses are not sensitive enough for such epitope mapping, because they take a few days of culturing during which activation-induced cell death occurs.[32]

Notably, the recurrent autoepitopes that we identified for human lupus T cells, namely $H2B_{10-33}$, $H3_{95-105}$, $H4_{16-39}$, and $H4_{71-94}$, overlap with the major autoepitopes for nephritogenic T cells in lupus-prone SNF_1 mice.[19,28] These peptides, and native nucleosomes, preferentially stimulated interferon-γ (IFNγ), producing T cells in the patient's PBMC, as is the case in murine lupus.[19,28] The nephritogenic autoantibodies in murine lupus that fix complement and bind to Fc receptors in inflammatory cells, belong to the IFNγ-dependent IgG subclasses,[33] and the importance of IFNγ in murine lupus nephritis is well established.[34,35] On the other hand, the importance of IL-10 (a Th2 cytokine?) in the maintenance of autoantibody production with the progression of lupus has also been well documented, but exact mechanisms are not known.[18,36–38]

All the human lupus T cell autoepitopes have multiple HLA-DR binding motifs (TABLE 1), and as in SNF_1 mice, they are located in the histone regions that are also targeted by lupus B cells (autoantibodies) and sites that contact with DNA in the native nucleosome particle. Therefore, these immunodominant epitopes could probably be used as "universal" tolerogens for inhibiting both autoimmune T and B cell populations in lupus patients, despite their diversity of HLA alleles. Other laboratories have confirmed the importance of nucleosome-specific T cells in lupus.[30,39,40]

Remarkably, T cell responses to nucleosomes and its peptide epitopes were detected *in vitro* in all patients with inactive lupus we tested, although their disease was in long-term remission and their serum anti-DNA antibody levels were not elevated. The peptides were probably not being presented *in vivo* in the inactive patients, due to lack of an adequate number of competent APC (such as an activated population of anti-DNA B cells). More important, however, the ability to respond to nucleosomal peptides is not sufficient for full helper competence of lupus T cells. Indeed, T cells from inactive patients do not help in anti-DNA autoantibody production,[10] although they respond to nucleosomes *in vitro*.[28] In lupus-prone mice, T cells responsive to nucleosomes are expanded very early in life, but the ability of the same nucleosome-specific T cells to help in the production of pathogenic anti-DNA autoantibodies appears several months later, along with an appearence of anti-DNA au-

TABLE 1. Multiple HLA-DR binding motifs in histone autoepitopes[a]

Histone Autoepitopes	HLA-DR Binding Motifs		
	HLA-DR1, 4, 7	HLA-DR3	HLA-DR8
$H4_{71\text{-}94}$	TYTEHAKRKT**V**TAMD**V**VYALKRQG	TYTEHAKRKTVTAMDVVYA**L**KR**Q**G	TYTEHAKRKTVTAMDVV**Y**ALK**R**QG
$H4_{16\text{-}39}$	KRHRK**V**LRDN**I**QGITKPAIRRLAR	KRHRK**V**LR**D**NIQGITKPAIRRLAR	
$H3_{94\text{-}108}$	EASEA**Y**LVGL**F**EDTN	EASEAYLVG**L**FE**D**TN	
$H1'_{22\text{-}42}$	STDHPKYSD**M**IVAA**I**QAEKNR		
$H2B_{10\text{-}33}$	PKKGS**K**KAV**T**KAQKKDGKKRKRSR		

[a]The major autoepitopes that are recurrently recognized by human lupus T cells[28] were aligned to various HLA-DR motifs published previously[74–76]:

HLA-DR1, DR4, DR7: {LIVMF} XXXX {STPALIVMC}
{RKHLIY} XXXX {YTNQCDERSW} XXX AMVKWLHYIFP
{RKHLIY} XXX {YTNQCDERSW} XXX AMVKWLHYIFP
HLA-DR3: {FILVY} XX {DNQT}
HLA-DR8: {FIVLY} XXX {HKR}

Any single amino acid inside a pair of brackets { } is one of the most probable anchor residues at that position of the motif. In each of the histone autoepitopes, the putative anchor residues are highlighted by bold and underlined letters. An X indicates that any of the twenty amino acids can be present in that position; also, each X represents an individual amino acid position in the sequence. In addition to the motifs indicated in this table, $H4_{71\text{–}94}$ contains multiple motifs for HLA-DR1,3,4,7,8 and also HLA-DR5 motif. (Updated from Lu *et al.* [28])

toantibodies in serum of the animals.[8,18] Thus, additional signals and maturational events might be required for the specialized function of helping in pathogenic autoantibody production.

The pathogenic "anti-DNA" autoantibodies that are induced by the lupus Th cells form immune complexes with nucleosomes. The positively charged, nucleosomal histone residues in such immune complexes may mediate binding to negatively charged residues in heparan sulfate or collagen of the glomerular basement membrane leading to lupus nephritis, as reviewed in Refs. 41–43. Thus, nucleosomes are one of the primary immunogens that initiate the cognate interaction between pathogenic Th and B cells of lupus. Nucleosomes are routinely released from apoptotic cells, and this event by itself is not unique to lupus.[44,45] However, the spontaneous expansion of nucleosome-specific T cells is a lupus-specific event that occurs very early in life.[18,19] These Th cells are essential for sustaining the pathogenic autoantibody-producing B cells of lupus.[17]

CROSS-REACTIVITY AND TOLERANCE SPREADING

It is amazing how cross-reactive and inter-connected is the immune repertoire against nuclear autoantigens—at the B cell level,[1,46] at the T cell level,[19–21,47] and at the interface of Th and B cells.[19,20,28,47] Furthermore, a single lupus Th clone can help a dsDNA-specific, an ssDNA-specific, a histone-specific, an HMG-specific, or a nucleosome-specific B cell, because each of these B cells by binding to its respective epitope on the whole chromatin, can take it up and process and then present the relevant peptide epitope in the chromatin to the Th clone,[12,18,48] resulting in intermolecular help. Therefore, inhibiting the multipotent Th cells of lupus could block this diversification of response to multiple epitopes.

Indeed, we have been able to delay or even reverse established lupus nephritis in the SNF_1 model by administering the nucleosomal peptide epitopes in tolerogenic regimens (FIG. 1).[47] This therapeutic result established a dominant role of the peptide autoepitopes in disease. Remarkably, tolerogenic therapy with any one of the epitopes could inhibit the production of a spectrum of pathogenic, anti-nuclear autoantibodies, probably due to the promiscuity of lupus T cell receptors and the ability of an individual pathogenic Th cell to help multiple autoimmune B cells. As mentioned above, a single peptide from a histone in the nucleosome can be recognized by multiple autoimmune T cells with diverse TCRs, and conversely, a single autoimmune T cell can recognize multiple nucleosomal peptides that are structurally different.[19,21] Thus, a single epitope may tolerize a spectrum of autoimmune Th cells, and tolerizing one set of Th cells would deprive help for multiple autoimmune B cells (*tolerance spreading*). Moreover, a peptide epitope that is recognized by both autoimmune T and B cells, such as $H4_{16–39}$, is a very effective tolerogen, probably by directly inactivating the B cells.[47]

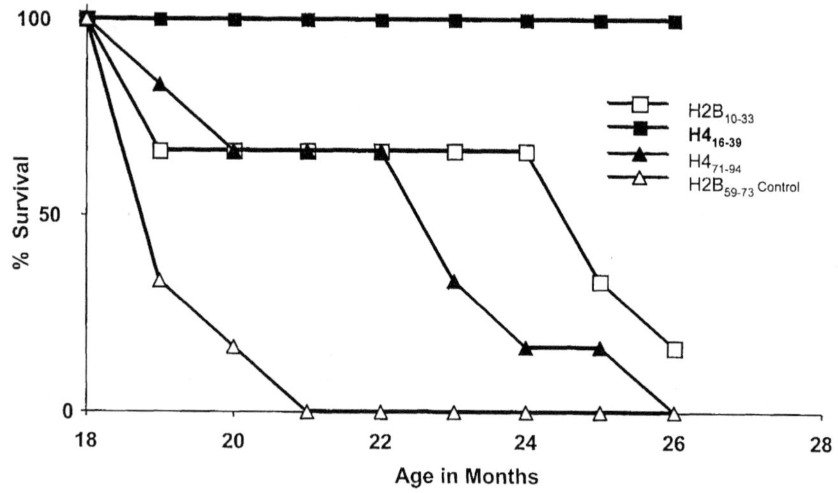

FIGURE 1. Treatment of established lupus nephritis with nucleosomal peptides. SNF_1 mice that were 18 months old and that had established glomerulonephritis were injected with respective peptides at the dose of 300 μg/mouse per month intravenously. The animals were followed by proteinuria measurements, and continued to receive the peptide injections every month until they died. The results are expressed as the percentage of animals that survived. As a negative control, an I-A^d-binding but non-stimulatory peptide, $H2B_{59–73}$, was used.[20] (Modified with permission from Kaliyaperumal *et al.*[47])

CD40L HYPEREXPRESSION AND ABNORMAL REACTIVITY TO APOPTOTIC CELL ANTIGENS IN LUPUS

Blocking the interaction between CD40L (CD40 ligand or CD154) on the autoimmune Th cells and CD40 on autoimmune B cells, even briefly with anti-CD40L antibody therapy in lupus-prone mice, produces unexpected long-term benefits.[17,29] But even more interesting is the prolonged hyper-expression of CD40L in lupus T cells that occurs irrespective of disease status. In normal T cells, CD40L expression is tightly regulated.[49,50] Persistent and increased expression of CD40L could disrupt this regulatory checkpoint and allow autoimmune cells to expand. T cells in lupus-prone SNF_1 mice,[17] and not only T cells,[29,51] but also B cells[29,52] of lupus patients express abnormally high levels of CD40L without any deliberate stimulation.[29]

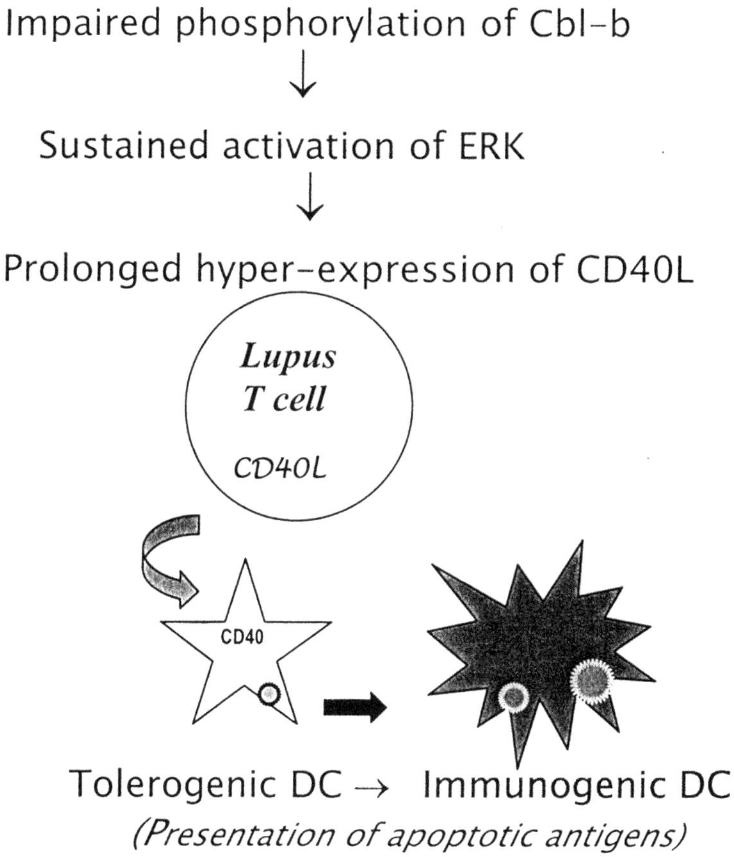

FIGURE 2. A scheme showing how impaired phosphorylation of Cbl-b could lead to a sustained and increased expression of CD40L by lupus T cells on receiving sub-threshold stimuli. The T cells then would signal via CD40 on APC to activate them to present nucleosomal antigens from apoptotic cells. DC: dendritic cells.

Moreover, upon stimulation with *suboptimal* doses of mitogens or anti-receptor antibodies, the upregulation of CD40L in the T and B cells[29] from lupus patients in *long-term remission*, and in T cells from *pre-autoimmune* lupus-prone mice,[17] is markedly greater than normal, indicating an intrinsic regulatory defect associated with a lowered threshold of activation of T cells, as well as B cells of lupus.[29,53–55] Interestingly, the regulatory defect in lupus is specific for CD40L, but not other activation molecules such as CD69.[29,55] Notably, CD40L hyper-expression in B cells might be masked by CD40 on the same B cell, necessitating a pre-blocking step with anti-CD40 antibody to reveal CD40L hyper-expression consistently.[56] In further work, we showed that the sustained hyper-expression of CD40L is associated with an impairment of phosphorylation of Cbl-b in lupus T cells.[55] Cbl-b is an adapter molecule that is important for downregulating critical kinases in the T cell signaling pathway by mediating their ubiquitination.[57,58] Consequently, lupus T cells show persistent activation of the MAP kinase, ERK, which might stabilize CD40L mRNA.[55,59] Activated ERK is also responsible for activation of the transcription factor AP-1 that possibly cooperates with NFAT to increase CD40L transcription.[60,61] Indeed, Cbl-b knockout mice were shown to develop lupus-like autoimmune disease.[62,63] These observations are significant because apoptotic cells, from which nucleosomes are derived, are poorly immunogenic in normal subjects, unless they exceed physiologic levels, or are presented in an inflammatory context.[64–68] The potential repertoire of both T cells and B cells for nuclear autoantigens is vast due to cross-reactive or degenerate recognition and possibly positive selection by nuclear antigens from apoptotic cells in thymus and bone marrow.[4,19,21,46,47] T cells that could recognize nucleosomal peptides are present in normal subjects,[18,30,31,39,69] but they are not spontaneously activated and expanded as in lupus. The possession of promiscuous receptors (TCRs) and abnormally prolonged hyper-expression of CD40L by lupus T cells[7,17,29,55] probably lowers the activation threshold for recognition of apoptotic cell antigens.[7,70,71] A sustained hyper-expression of CD40L by T and B cells in lupus, even with sub-threshold stimuli,[29,55] could instigate tolerogenic dendritic cells[72,73] and resting "anti-DNA" B cells to present nucleosomes in an immunogenic fashion (FIG. 2).[7]

Thus, identification of major nucleosomal peptide autoepitopes for the T cells of human lupus might be important for understanding how such autoimmune T cells arise, for tracking such T cells using peptide-MHC tetramers, and for developing antigen-specific therapy. Importantly, our initial studies in murine lupus indicate that tolerance therapy with nucleosomal peptides would be effective in *established disease*.[47]

ACKNOWLEDGMENTS

This work was supported by Grants R37 AR39157 (MERIT) and RO1 AI41985 from the National Institutes of Health (U.S. Public Health Service).

REFERENCES

1. GAVALCHIN, J., J. NICKLAS, J.W. EASTCOTT, *et al.* 1985. Lupus prone (SWR x NZB)F1 mice produce potentially nephritogenic autoantibodies inherited from the normal SWR parent. J. Immunol. **134:** 885–894.

2. GHATAK, S., T. O'KEEFE, T. IMANISHI-KARI & S.K. DATTA. 1990. Selective strain distribution pattern of a germline VH gene for a pathogenic anti-DNA autoantibody family. Int. Immunol. **2:** 1004–1012.
3. O'KEEFE, T.L., S.K. DATTA & T. IMANISHI-KARI. 1992. Cationic residues in pathogenic anti-DNA autoantibodies arise by mutations of a germline gene that belongs to a large VH gene subfamily. Eur. J. Immunol. **22:** 619–624.
4. DATTA, S.K., B.D. STOLLAR & R.S. SCHWARTZ. 1983. Normal mice express idiotypes related to autoantibody idiotypes of lupus mice. Proc. Natl. Acad. Sci. USA **80:** 2723–2727.
5. CONGER, J.D., B.L. PIKE & G.J. NOSSAL. 1987. Clonal analysis of the anti-DNA repertoire of murine B lymphocytes. Proc. Natl. Acad. Sci. USA **84:** 2931–2935.
6. SOUROUJON, M., M.E. WHITE-SCHARFF, J. ANDRE-SCHWARTZ, *et al.* 1988. Preferential autoantibody reactivity of the preimmune B cell repertoire of normal mice. J. Immunol. **140:** 4173–4179.
7. DATTA, S.K. 2000. Positive selection for autoimmunity. Nature Med. **6:** 259–261.
8. DATTA, S.K., H. PATEL & D. BERRY. 1987. Induction of a cationic shift in IgG anti-DNA autoantibodies: role of T helper cells with classical and novel phenotypes in three murine models of lupus nephritis. J. Exp. Med. **165:** 1252–1268.
9. ANDO, D.G., E.E. SERCARZ & B.H. HAHN. 1987. Mechanisms of T and B cell collaboration in the in vitro production of anti-DNA antibodies in the NZB/NZW F1 murine SLE model. J. Immunol. **138:** 3185–3190.
10. SHIVAKUMAR, S., G.C. TSOKOS & S.K. DATTA. 1989. T cell receptor αβ expressing double negative (CD4–/CD8–) and CD4+ T helper cells in humans augment the production of pathogenic anti-DNA autoantibodies associated with lupus nephritis. J. Immunol. **143:** 103–112.
11. RAJAGOPALAN, S., T. ZORDAN, G.C. TSOKOS & S.K. DATTA. 1990. Pathogenic anti-DNA autoantibody inducing T helper cell lines from patients with active lupus nephritis: isolation of CD4–/CD8– T helper cell lines that express the γδ T-cell receptor. Proc. Natl. Acad. Sci. USA. **87:** 7020–7024.
12. DESAI-MEHTA, A., C. MAO, S. RAJAGOPALAN, *et al.* 1995. Structure and specificity of T-cell receptors expressed by pathogenic anti-DNA autoantibody-inducing T cells in human lupus. J. Clin. Invest. **95:** 531–541.
13. NAIKI, M., B.-L. CHIANG, D. CAWLEY, *et al.* 1992. Generation and characterization of cloned helper T cell lines for anti-DNA responses in NZB.H-2bm12 mice. J. Immunol. **149:** 4109–4115.
14. ZENG, D., M.K. LEE, J. TUNG, *et al.* 2000. Cutting edge: a role for CD1 in the pathogenesis of lupus in NZB/NZW mice. J Immunol **164:** 5000–5004.
15. SAINIS, K. & S.K. DATTA. 1988. CD4+ T cell lines with selective patterns of autoreactivity as well as CD4–/CD8– T helper cell lines augment the production of idiotypes shared by pathogenic anti-DNA autoantibodies in the NZB x SWR model of lupus nephritis. J. Immunol. **140:** 2215–2224.
16. ADAMS, S., P. LEBLANC & S.K. DATTA. 1991. Junctional region sequences of T-cell receptor β chain genes expressed by pathogenic anti-DNA autoantibody-inducing T helper cells from lupus mice: Possible selection by cationic autoantigens. Proc. Natl. Acad. Sci. USA **88:** 11271–11275.
17. MOHAN, C., Y. SHI, J.D. LAMAN & S.K. DATTA. 1995. Interaction between CD40 and its ligand gp39 in the development of murine lupus nephritis. J. Immunol. **154:** 1470–1480.
18. MOHAN, C., S. ADAMS, V. STANIK & S.K. DATTA. 1993. Nucleosome: a major immunogen for the pathogenic autoantibody-inducing T cells of lupus. J. Exp. Med. **177:** 1367–1381.
19. KALIYAPERUMAL, A., C. MOHAN, W. WU & S.K. DATTA. 1996. Nucleosomal peptide epitopes for nephritis-inducing T helper cells of murine lupus. J. Exp. Med. **183:** 2459–2469.
20. KALIYAPERUMAL, A., M.A. MICHAELS & S.K. DATTA. 2002. Naturally processed chromatin peptides reveal a major autoepitope that primes pathogenic T and B cells of lupus. J. Immunol. **168:** 2530–2537.

21. SHI, Y., A. KALIYAPERUMAL, L. LU, *et al.* 1998. Promiscuous presentation and recognition of nucleosomal autoepitopes in lupus: role of autoimmune T cell receptor alpha chain. J. Exp. Med. **187:** 367-378.
22. WAKELAND, E.K., K. LIU, R.R. GRAHAM & T.W. BEHRENS. 2001. Delineating the genetic basis of systemic lupus erythematosus. Immunity **15:** 397–408.
23. STEMMER, C., P. RICHALET-SECORDEL, M.C.J. VAN BRUGGEN, *et al.* 1996. Dual reactivity of several monoclonal anti-nucleosome antibodies for double stranded DNA and a short segment of histone H3. J. Biol. Chem. **271:** 21257–21261.
24. LUGER, K., A.W. MADER, R.K. RICHMOND, *et al.* 1997. Crystal structure of the nucleosome core particle at 2.8 Å resolution. Nature **389:** 251–260.
25. KHALIL, M., K. INABA, R. STEINMAN, *et al.* 2001. T cell studies in a peptide-induced model of systemic lupus erythematosus. J. Immunol. **166:** 1667–1674.
26. HARDIN, J.A. & J.O. THOMAS. 1983. Antibodies to histones in systemic lupus erythematosus: localization of prominent autoantigens on histone H1 and H2B. Proc. Natl. Acad. Sci. USA **80:** 7410–7414.
27. SCHETT, G., R.L. RUBIN, G. STEINER, *et al.* 2000. The lupus erythematosus cell phenomenon: comparative analysis of antichromatin antibody specificity in lupus erythematosus cell-positive and -negative sera. Arthritis Rheum. **43:** 420–428.
28. LU, L., A. KALIYAPERUMAL, D.T. BOUMPAS & S.K. DATTA. 1999. Major peptide autoepitopes for nucleosome-specific T cells of human lupus. J. Clin. Invest. **104:** 345–355.
29. DESAI-MEHTA, A., L. LU, R. RAMSEY-GOLDMAN & S.K. DATTA. 1996. Hyperexpression of CD40 ligand by B and T cells in human lupus and its role in pathogenic autoantibody production. J. Clin. Invest. **97:** 2063–2073.
30. VOLL, R.E., E.A. ROTH, I. GIRKONTAITE, *et al.* 1997. Histone-specific Th0 and Th1 clones derived from systemic lupus erythematosus patients induce double-stranded DNA antibody production. Arthritis Rheum. **40:** 2162–2171.
31. ANDREASSEN, K., U. MOENS, H. NOSSENT, *et al.* 1999. Termination of human T cell tolerance to histones by presentation of histones and polyomavirus T antigen provided that T antigen is complexed with nucleosomes. Arthritis Rheum. **42:** 2449–2460.
32. BIEGANOWSKA, K.D., L.J. AUSUBEL, Y. MODABBER, *et al.* 1997. Direct ex vivo analysis of activated, Fas-sensitive autoreactive T cells in human autoimmune disease. J. Exp. Med. **185:** 1585–1594.
33. GAVALCHIN, J. & S.K. DATTA. 1987. The NZB x SWR model of lupus nephritis. II. Autoantibodies deposited in renal lesions show a restricted idiotypic diversity. J. Immunol. **138:** 138–148.
34. PENG, S.L., J. MOSLEHI & J. CRAFT. 1997. Roles of interferon-gamma and interleukin-4 in murine lupus. J. Clin. Invest. **99:** 1936–1946.
35. BALOMENOS, D., R. RUMOLD & A.N. THEOFILOPOULOS. 1998. Interferon-gamma is required for lupus-like disease and lymphoaccumulation in MRL-lpr mice. J. Clin. Invest. **101:** 364–371.
36. ISHIDA, H., T. MUCHAMUEL, S. SAKAGUCHI, *et al.* 1994. Continuous administration of anti-interleukin 10 antibodies delays onset of autoimmunity in NZB/W F1 mice. J. Exp. Med. **179:** 305–310.
37. LLORENTE, L., W. ZOU, Y. LEVY, *et al.* 1995. Role of interleukin 10 in the B lymphocyte hyperactivity and autoantibody production of systemic lupus erythematosus. J. Exp. Med. **181:** 839–844.
38. NAKAJIMA, A., S. HIROSHE, H. YAGITA & K. OKOMURA. 1997. Roles of IL-4 and IL-12 in the development of lupus in NZB/W F1 mice. J. Immunol. **158:** 1466–1472.
39. KRETZ-ROMMEL, A., S.R. DUNCAN & R.L. RUBIN. 1997. Autoimmunity caused by disruption of central T cell tolerance: a murine model of drug induced lupus. J. Clin. Invest. **99:** 1888–1896.
40. BRUNS, A., S. BLASS, G. HAUSDORF, *et al.* 2000. Nucleosomes are major T and B cell autoantigens in systemic lupus erythematosus. Arthritis Rheum. **43:** 2307–2315.
41. BACH, J.F., S. KOUTOUZOV & P.M. VAN ENDERT. 1998. Are their unique autoantigens triggering autoimmune diseases? Immunol. Rev. **164:** 139–155.
42. BERDEN, J.H. 1997. Lupus nephritis. Kidney Int. **52:** 538–558.

43. LEFKOWITH, J.B. & G.S. GILKESON. 1996. Nephritogenic autoantibodies in lupus: current concepts and continuing controversies. Arthritis Rheum. **39:** 894–903.
44. CASCIOLA-ROSEN, L.A., G. ANHALT & A. ROSEN. 1994. Autoantigens targeted in systemic lupus erythematosus are clustered in two populations of surface structures on apoptotic keratinocytes. J. Exp. Med. **179:** 1317–1330.
45. LORENZ, H.-M., M. GRUNKE, T. HIERONYMUS, *et al.* 1997. In vitro apoptosis and expression of apoptosis-related molecules in lymphocytes from patients with systemic lupus erythematosus and other autoimmune diseases. Arthritis Rheum. **40:** 306–317.
46. GAVALCHIN, J., R.A. SEDER & S.K. DATTA. 1987. The NZB x SWR model of lupus nephritis. I. Cross-reactive idiotypes of monoclonal anti-DNA antibodies in relation to antigenic specificity, charge and allotype: identification of interconnected idiotype families inherited from the normal SWR and the autoimmune NZB parents. J. Immunol. **138:** 128–137.
47. KALIYAPERUMAL, A., M.A. MICHAELS & S.K. DATTA. 1999. Antigen-specific therapy of murine lupus nephritis using nucleosomal peptides: tolerance spreading impairs pathogenic function of autoimmune T and B cells. J. Immunol. **162:** 5775–5783.
48. DATTA, S.K. & A. KALIYAPERUMAL. 1997. Nucleosome-driven autoimmune response in lupus: pathogenic T helper cell epitopes and costimulatory signals. Ann. N.Y. Acad. Sci. **815:** 155–170.
49. ROY, M., T. WALDSCHMIDT, A. ARUFFO, *et al.* 1993. The regulation of the expression of gp39, the CD40 ligand, on normal and cloned CD4+ T cells. J. Immunol. **151:** 2497–2510.
50. FULEIHAN, R., R. NARYANASWAMY, A. HORNER, *et al.* 1994. Cyclosporine A inhibits CD40 ligand expression in T lymphocytes. J. Clin. Invest. **93:** 1315–1320.
51. KOSHY, M., D. BERGER & M.K. CROW. 1996. Increased expression of CD40 ligand on systemic lupus erythematosus lymphocytes. J. Clin. Invest. **98:** 826–837.
52. DEVI, B.S., S. VAN NOORDIN, T. KRAUSZ & K.A. DAVIES. 1998. Peripheral blood lymphocytes in SLE: hyperexpression of CD154 on T and B lymphocytes and increased number of double negative T cells. J. Autoimmunity **11:** 471–475.
53. LIOSSIS, S.-N., B. KOVACS, G. DENNIS, *et al.* 1996. B cells from patients with systemic lupus erythematosus display abnormal antigen receptor-mediated early signal transduction events. J. Clin. Invest. **98:** 2549–2557.
54. BLOSSOM, S. & K.M. GILBERT. 1999. Antibody production in autoimmune BXSB mice: CD40L expressing B cells need fewer signals for polyclonal antibody synthesis. Clin. Exp. Immunol. **118:** 147–153.
55. YI, Y., M. MCNERNEY & S.K. DATTA. 2000. Regulatory defects in Cbl and mitogen-activated protein kinase (extracellular signal-related kinase) pathways cause persistent hyperexpression of CD40 ligand in human lupus T cells. J. Immunol. **165:** 6627-6634.
56. HIGUCHI, T., Y. AIBA, T. NOMURA, *et al.* 2002. Cutting edge: ectopic expression of CD40 ligand on B cells induces lupus-like autoimmune disease. J. Immunol. **168:** 9–12.
57. LUPHER, M.L., N. RAO, M.J. ECK & H. BAND. 1999. The Cbl protooncoprotein: a negative regulator of immune receptor signal transduction. Immunol. Today. **20:** 375–382.
58. JOAZEIRO, C.A., S.S. WING, H. HUANG, *et al.* 1999. The tyrosine kinase negative regulator c-Cbl RING type, E2 dependent, ubiquitin-protein ligase. Science **286:** 309–312.
59. FORD, G.S., B. BARNHART, S. SHONE & L.R. COVEY. 1999. Regulation of CD154 (CD40L) mRNA stability during T cell activation. J. Immunol. **162:** 4037–4044.
60. TSYTSYKOVA, A.V., E.N. TSITSIKOV & R.S. GEHA. 1996. The CD40L promoter contains NF-AT-binding motifs which require AP-1 binding for activation of transcription. J. Biol. Chem. **271:** 3763–3770.
61. SCHUBERT, L.A., G. KING, R. Q. CRON, *et al.* 1995. The human gp39 promoter: two distinct NF-AT-binding elements contribute independently to transcriptional activation. J. Biol. Chem. **270:** 29624–29627.

62. CHIANG, Y.J., H.K. KOLE, K. BROWN, *et al.* 2000. Cbl-b regulates the CD28 dependence of T-cell activation. Nature **403:** 216–220.
63. BACHMAIER, K., C. KRAWCZYK, I. KOZIERADZKI, *et al.* 2000. Negative regulation of lymphocyte activation and autoimmunity by the molecular adapter Cbl-b. Nature **403:** 211-216.
64. VOLL, R.E., M. HERRMANN, E.A. ROTH, *et al.* 1997. Immunosuppressive effects of apoptotic cells. Nature **390:** 350–351.
65. FADOK, V.A., D.L. BRATTON, A. KONOWAL, *et al.* 1998. Macrophages that have ingested apoptotic cells in vitro inhibit proinflammatory cytokine production through autocrine/paracrine mechanisms involving TGF-beta, PGE2, and PAF. J. Clin. Invest. **101:** 890–898.
66. MEVORACH, D., J.L. ZHOU, X. SONG & K.B. ELKON. 1998. Systemic exposure to irradiated apoptotic cells induces autoanitbody production. J. Exp. Med. **188:** 387–392.
67. CHEN, Z., S.B. KORALOV & G. KELSOE. 2000. Complement C4 inhibits systemic autoimmunity through a mechanism independent of complement receptors CR1 and CR2. J. Exp. Med. **192:** 1339–1352.
68. ROVERE, P., M.G. SABBADINI, C. VALLINOTO, *et al.* 1999. Dendritic cell presentation of antigens from apoptotic cells in a proinflammatory context: role of opsonizing anti-beta2-glycoprotein I antibodies. Arthritis Rheum. **42:** 1412–1420.
69. DECKER, P., A. LE MOAL, J.P. BRIAND & S. MULLER. 2000. Identification of a minimal T cell epitope recognized by antinucleosome Th cells in the C-terminal region of histone H4. J. Immunol. **165:** 654–662.
70. KRETZ-ROMMEL, A. & R.L. RUBIN. 2000. Disruption of positive selection of thymocytes causes autoimmunity. Nature Med. **6:** 298–305.
71. VRATSANOS, G.S., S. JUNG, Y.M. PARK & J. CRAFT. 2001. CD4(+) T cells from lupus-prone mice are hyperresponsive to T cell receptor engagement with low and high affinity peptide antigens: a model to explain spontaneous T cell activation in lupus. J. Exp. Med. **193:** 329–337.
72. STEINMAN, R.M., S. TURLEY, I. MELLMAN & K. INABA. 2000. The induction of tolerance by dendritic cells that have captured apoptotic cells. J. Exp. Med. **191:** 411–416.
73. GROHMANN, U., F. FALLARINO, S. SILLA, *et al.* 2001. CD40 ligation ablates the tolerogenic potential of lymphoid dendritic cells. J. Immunol. **166:** 277–283.
74. SOUTHWOOD, S., J. SIDNEY, A. KONDO, *et al.* 1998. Several common HLA-DR types share largely overlapping peptide binding repertoires. J. Immunol. **160:** 3363–3373.
75. CHICZ, R.M., R.G. URBAN, J.C. GORGA, *et al.* 1993. Specificity and promiscuity among naturally processed peptides bound to HLA-DR alleles. J. Exp. Med. **178:** 27–47.
76. CHICZ, R.M., R.G. URBAN, W.S. LANE, *et al.* 1992. Predominant naturally processed peptides bound to HLA-DR1 are derived from MHC-related molecules and are heterogeneous in size. Nature **358:** 764–768.

Mechanisms of Autoantibody Diversification to SLE-Related Autoantigens

UMESH S. DESHMUKH,[a] FELICIA GASKIN,[a,b] JANET E. LEWIS,[a] CAROL C. KANNAPELL,[a] AND SHU MAN FU[a]

Division of Rheumatology and Immunology, Department of Internal Medicine, [a]the Specialized Center of Research on Systemic Lupus Erythematosus, and [b]Departments of Psychiatric Medicine and Neurology, University of Virginia, Charlottesville, Virginia 22908, USA

ABSTRACT: Systemic lupus erythematosus is a prototype of systemic autoimmunity with autoantibodies (autoAbs) to ribonucleoproteins such as Ro/La, snRNP, dsDNA, and other cellular constituents. A/J mice were used to explore the mechanism of autoAb diversification with recombinant proteins and synthetic peptides. Previous studies showed that $Ro60_{316-335}$ induced Abs to Ro60, La, and snRNP proteins. Specific Abs to determinants outside $Ro60_{316-335}$ were detected. Absorption experiments showed that Abs to La and snRNP proteins were due to the induction of anti-Ro60 Abs cross-reactive with these peptides. With snRNP proteins, SmD, SmB, and A-RNP as immunogens, specific patterns of intermolecular spreading were obtained in addition to Abs to the immunogens. With SmD-immunized mice, specific Abs to A-RNP and SmB were detected. With SmB as the immunogen, specific Abs to A-RNP were detected in the majority of the mice. Only in a rare incident, specific Abs to SmD were induced. In A-RNP-immunized mice, only Abs to the 70-kD U1-RNP were seen. In all cases, Abs capable of precipitating snRNP particles were detected. Thus, the intermolecular epitope spreading is immunogen-dependent. Evidence for the presence of cross-reactive T cells to more than one autoAg was obtained. The Ag-dependent unique patterns of Ab diversification will facilitate analyses of patients' sera. These results have implications regarding the nature of the Ag-driven autoimmune process.

KEYWORDS: autoantibodies; epitope spreading; cross-reactivity; systemic lupus erythematosus; snRNP; T cell; B cell

Systemic lupus erythematosus (SLE) is a prototypic autoimmune disorder affecting skin, joints, muscles, kidneys, heart, lungs, blood, and the central nervous system. It is characterized by diverse autoantibodies (autoAbs) to dsDNA; histones; ribonucleoproteins such as Ro60, La, Ro52, Sm, and other components of the snRNP particle; ribosomal proteins; Ku; Su; and other cellular proteins. Recently, it has been established that T cells play important roles in the pathogenesis of SLE, including direct

Address for correspondence: Shu Man Fu, Division of Rheumatology and Immunology, Department of Internal Medicine, University of Virginia School of Medicine, Box 800412, Charlottesville, VA 22908. Voice: 434-924-9627; fax: 434-924-9578.
sf2e@virginia.edu

Ann. N.Y. Acad. Sci. 987: 91–98 (2003).

end organ damage by cytotoyic T cells and the induction of autoAb formation. Many of the autoAbs form immune complexes leading to complement fixation and end organ damage. Emerging evidence suggests that interplay among genetic and environmental factors is important in the pathogenesis of SLE.

During the past several years, our laboratory has been interested in the mechanisms of autoAb diversification to SLE-related ribonucleoprotein autoantigens. Initial studies used recombinant Ro60 and its peptides as immunogens.[1,2] Recently we have extended our analysis to the snRNP system.[3] These data will be reviewed and their implications will be discussed.

IMMUNIZATION WITH Ro60 AND ITS PEPTIDES: THE GENERATION OF CROSS-REACTIVE ANTIBODIES TO La, SmD, AND U1-RNP AND INTERMOLECULAR EPITOPE SPREADING

In the initial study, inhibition experiments showed that some of the Abs to Ro60 were cross-reactive with other ribonucleoproteins, such as La, SmD, and U1-RNP, when recombinant mouse or human Ro60 ribonucleoproteins were used as immunogens.[1] The fact that Ro60 removed all the Ab activity to La, SmD, and U1-RNP indicates the absence of specific Ab to these ribonucleoproteins. Thus, it was concluded that there was no intermolecular epitope spreading.

The mapping of T and B epitopes was done on both mouse and human Ro60 with overlapping Ro60 peptides. Mouse $Ro60_{316–335}$ (KARIHPFHILIALETYRAGH) was found to have dominant T and B cell epitopes. Human $Ro60_{316–335}$ KARIHPF-HILVALETYKTGH, which differs from its mouse homologue by three amino acids, was used as a source of immunogens. It induced a vigorous T cell proliferative response and Ab production. This T cell response could be recalled by $mRo60_{316–335}$ *in vitro*. It induced Ab to the immunogens and Ro60, La, SmD, and U1-RNP as well as antibodies to the nuclei and the Golgi complex. Intramolecular B cell epitope spreading was shown to be induced by the demonstration that specific Abs to epitopes on other parts of Ro60 were in the immune sera.[2] These epitopes were shown to be a conformational epitope present in native Ro60 and an epitope present on a Ro60 fragment, $Ro60_{128–285}$. Abs to La, SmD, and U1-RNP were also produced. However, these Ab activities were absorbed completely by Ro60 although the immunizing peptide was not able to do so. It was surprising that similar results were obtained with the mouse homologue as the immunogen, with the exception that anti-Golgi complex Abs were not induced. Attempts to demonstrate T cell epitope spreading to other parts of the Ro60 or to other ribonucleoproteins of interest were not successful. The induced T and B cell responses persisted more than 150 days after the initial immunizations. It is of interest to note that IgM Ab response also persisted over this long period of time. Hybridomas secreting IgM monoclonal Abs specific for Ro60 had been generated from splenic cells late in the course of the immune response. The coding sequences of these IgM Abs showed significant somatic mutations, suggesting that these IgM-secreting B cells were from the memory B cell pool.[4] As a control for these experiments, $hRo60_{441–465}$, which differs from its mouse homologue by four amino acids, was used as immunogen. This peptide induced a vigorous T cell response and Ab production to the peptide, Ro60, and Ro52. However, all these Ab activities were completely removed by the immunizing peptide. It was also shown that the mouse homologue was not able to recall the induced T cell proliferative response *in vitro* and that the immune response was a short-lived one.

From these observations, it is concluded that in order to induce intramolecular B epitope spreading with a molecular mimic such as human Ro60 peptides, the immunogen must contain both T and B cell epitopes recognized by the recipient, and a similar T cell epitope must be present on the endogenous autoantigen. These data also support the thesis that endogenous autoantigen plays a role in sustaining the autoimmune response.

IMMUNE RESPONSES TO snRNP: ANTIGEN-DEPENDENT DISTINCT B CELL EPITOPE SPREADING IN MICE IMMUNIZED WITH INDIVIDUAL PROTEINS OF THE snRNP PARTICLE

The clinical observation that autoAb specificity studies of lupus sera have shown that the immune responses are often to sets of physically linked proteins such as Ro60/La and snRNP led Hardin to postulate that the autoimmune response in lupus is driven by the whole particles.[5] This is the "particle hypothesis." This hypothesis is the basis for the prevalent view of mechanism for intermolecular epitope spreading,[6–8] which is depicted in FIGURE 1. This interstructural T cell help mechanism is based on the observation that a single T cell can provide help to multiple B cells with specificities to the monomers of a hetero-polymeric protein complex. The hypothesis places a greater role on B cells as antigen presenting cells in the Ig responses to polymeric autoantigens. The immune responses to the snRNP have been studied considerably both in man and mouse. The polymeric nature of the snRNP particle, the conservation of many of the coding sequences of individual proteins between man and mouse, and the specificity of some of these autoAbs for SLE are the main features for the intense investigative efforts on this autoantigen in lupus research. We

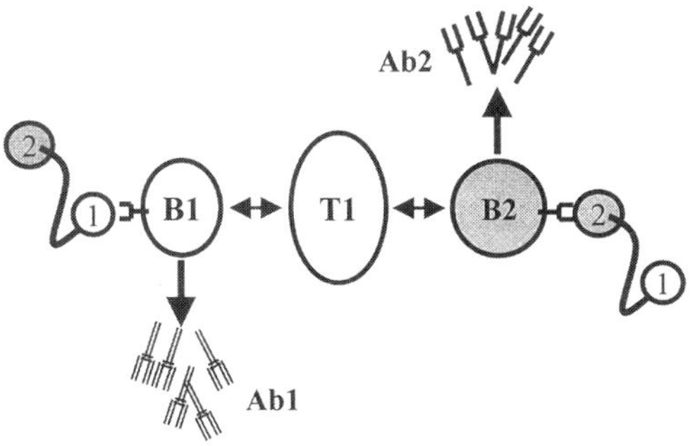

FIGURE 1. Prevalent view on the mechanisms of intermolecular epitope spreading. B cells (B1) reactive with antigen 1 receive help from T cells (T1) reactive with peptides from antigen 1 and generate antibody 1 (Ab1). B cells (B2) reactive with antigen 2 capture the entire antigenic particle and process and present peptides from antigen 1 to T1 cells. The ensuing T–B interaction results in the generation of antibody specificity 2 (Ab2).

have also taken advantage of these unique features of the snRNP particle. Some of our studies on the snRNP proteins have been summarized in a recent publication.[3]

We have generated recombinant proteins for SmD, SmB, and A-RNP as immunogens. Because of the presence of 6-His tag in these recombinant proteins, Western blot analysis with mouse lymphoma cell lysate as the substrate and immunoprecipitation with *in vitro* translated products of these proteins were the assays used to avoid the complication of the 6-His tag's antigen contribution to the system. In these experiments, alum instead of complete Freund's adjuvant was used as the adjuvant. The use of alum as adjuvant avoids some of the complications in the interpretation of the results. Serial sera from eight A/J mice immunized with SmD were analyzed. By two months after the initial immunization, all eight mice made IgG Abs to SmD, A-RNP, and SmB/B'. In addition, several mice also generated Abs to proteins with molecular weights between 16 and 21 kDa. None of the eight mice injected with alum alone made these Abs. These reactivities were confirmed by immunoprecipitation experiments. Absorption experiments with recombinant proteins detected the generation of specific Abs to A-RNP and SmB/B'. The Western blot analysis of the pooled immune sera, both untreated and absorbed, is shown in FIGURE 2. SmD, A-RNP, and SmB/B′ completely removed all the Abs specific for these proteins respectively with substantial reactivities remaining for the other two proteins. It is of interest to note that all three proteins removed the Ab reactivities to other proteins,

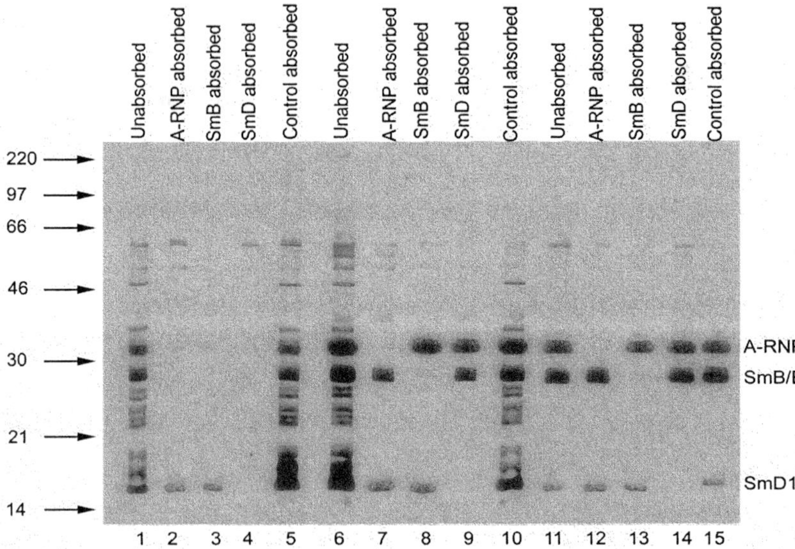

FIGURE 2. Intermolecular epitope spreading to A-RNP and SmB/B′ occurs in A/J mice immunized with SmD. Pooled sera from immunized mice at multiple time points were absorbed with different antigens coupled to sepharose beads. The reactivities of unabsorbed and absorbed sera were analyzed in Western blots, employing WEHI 7.1 cell extracts. Lanes 1–5 are sera obtained 1 month post immunization. Lanes 6–10 represent sera at 2 months post-immunization, and lanes 11–15 represent sera at 3 months post-immunization.

TABLE 1. Immunogen-dependent intermolecular B cell epitope spreading within the snRNPs

SmD	→	A-RNP, SmB/B′
SmB	→	A-RNP, SmD
A-RNP	→	70-kDa protein

suggesting that there are considerable amounts of polyreactive Abs in the immune sera. Similar results were obtained in immunoprecipitation experiments. Thus, it was concluded that with SmD as the immunogen, there was intermolecular epitope spreading to A-RNP and SmB/B'. Similar analyses were done with A-RNP and SmB/B′ as immunogen, leading to the conclusion that the patterns of intermolecular epitope spreading are dependent on the immunogen. In all immunized mice, Abs capable of precipitating native snRNP particles were induced with all three immunogens. These results are summarized in TABLE 1.

It is also of note that Abs capable of precipitating the native snRNP particle often appear later than those for the immunogens. This observation supports the hypothesis of interstructural T cell help proposed by Craft and colleagues[6–8] and the "particle hypothesis" proposed by Hardin.[5] However, these hypotheses are not congruent with the observation that distinct patterns of epitope spreading are detected with each immunogen. This incongruence compels us to postulate the additional hypothesis that intermolecular B cell epitope spreading within multimeric protein complex is influenced by T cell responses to the targeted protein.

DETECTION OF T CELLS REACTIVE TO A-RNP AND T CELLS WITH DUAL REACTIVITIES TO BOTH SmD AND A-RNP IN SmD-IMMUNIZED MICE

To seek evidence to support our hypothesis as stated above, we used the ELISPOT assay to detect antigen-specific T cells in SmD-immunized mice. The readout is for IL-5 production. As shown in TABLE 2, on day 30 after the initial immunization, there were 52 SmD-specific IL-5-producing T cells/1,000,000 splenic cells. By day 90, this number increased to 176. In contrast, 7 A-RNP-specific IL-5-producing T cells/1,000,000 splenic cells were detected, and this number increased to 23. It is of interest to note that there were few splenic T cells that were also seen to be responsive to A-RNP in alum-treated controls. Two other experiments gave similar results.

T–T hybridomas with splenic cells from SmD immunized mice and a HAT-sensitive T cell lymphoma line without expressing TCR as fusion partner. T cell hybridomas capable of responding to both SmD and A-RNP were identified. Some of them have been cloned, and their reactivities to SmD and A-RNP are depicted in FIGURE 3. All four cloned T cell hybridomas responded to both SmD and A-RNP, although all of them responded better to SmD with better secretion of IL-2. In view of the findings that a considerable portion of the $TCR\alpha\beta^+$ cells have two T cell receptor α chains,[9,10] we are in the process of determining whether one or two functional T cell receptor α chains are present in these T–T hybridomas. It is of relevance to note that no T cells reactive to U1RNP were detected in these SmD-immunized

TABLE 2. T cell responses to A-RNP detected in mice immunized with SmD

		SmD		A-RNP	
		Day 30	Day 90	Day 30	Day 90
SmD-immunized	52	176	7	23	
Alum-immunized	0	5	2	3	

NOTE: Each value represents a mean number of IL-5-producing T cells per million spleen cells.

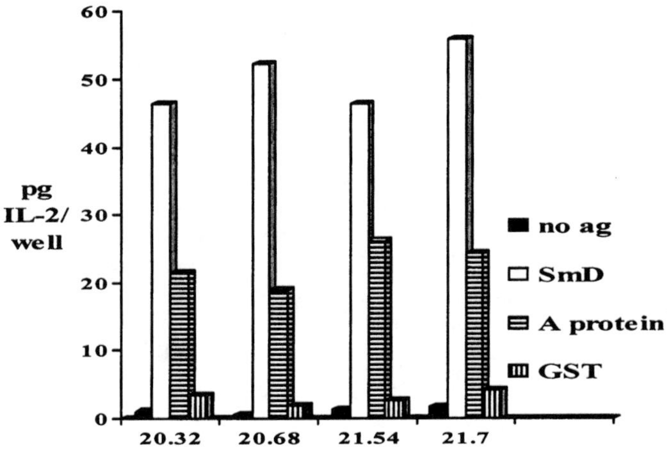

FIGURE 3. T cell cross-reactivity between SmD and A-RNP is revealed by T cell hybridomas. T cell hybridomas reactive with SmD were stimulated with splenic antigen presenting cells in the absence or presence of SmD, A-RNP, and GST for 14 h. Culture supernatants were used in ELISA to estimate IL-2 production.

mice. Recently, we have also obtained similar data in C3H mice, which show a similar immunogen-dependent B epitope spreading pattern in responses to individual proteins in the snRNP complex.

The above observations support our stated hypothesis, and our hypothesis can be further modified to state that dual reactive T cells may play an important role in B cell epitope spreading initially. The interaction of these T cells with B cells specific for the targeted antigen leads to the activation and proliferation of these B cells. The activated antigen-specific B cells would serve as potent antigen-presenting cells to other T cells specific for the targeted antigen, leading to the amplification of a specific B cell response to the target antigen. This mechanism is depicted in FIGURE 4.

CONCLUDING REMARKS

From our studies, the following scenario for autoAb diversification to SLE-related autoantigens emerges. The initiating antigens can be molecular mimics from

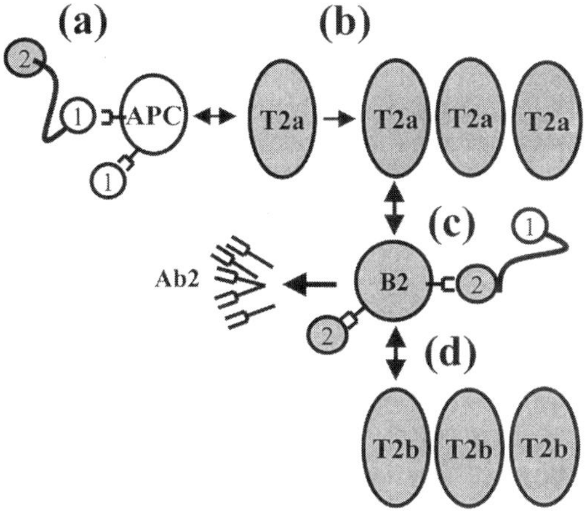

FIGURE 4. T cell responses to individual proteins within a multiprotein complex play a critical role in autoAb diversification. (**a**) Antigen-presenting cells either capture the intact antigenic particle or dissociated antigen 1 and present cross-reactive peptides from antigen 1 to T cells reactive with antigen 2 (T2a). (**b** and **c**) B cells reactive with antigen 2 present peptides from antigen 2 to the activated and amplified population of T2a cells. The resulting T–B interaction results in the generation of antibody reactive with antigen 2. (**d**) B cells activated by T2a cells present novel peptides from antigen 2 to T2b cells, resulting in further amplification of T cell responses. This reciprocal T–B amplification results in the generation of complex autoAb reactivities.

microbes, although the autoimmune response may be initiated by an autoantigen in certain instances. AutoAb diversification can be accomplished by the development of cross-reactive antibodies. The persistence of the autoAb response long after the initial immunizations suggests that endogenous autoantigen is important in sustaining the ongoing autoimmune response. The cross-reactive autoAbs also persist long after the initial immunizations. These cross-reactive autoAbs are of IgM and IgG classes. The role of these cross-reactive Abs in autoAb diversification should be explored further. For intramolecular epitope spreading, interstructural T cell help is necessary and sufficient. For intermolecular epitope spreading to generate specific autoAb to targeted polypeptide within a polymeric protein complex, the presence of cross-reactive T cells is crucial for the initiation and amplification of specific T cell help for the production of specific Ab. The generation of specific T cells for the targeted antigen would provide a mechanism to account for immunogen dependence in patterns of intermolecular epitope spreading. This mechanism together with the interstructural T cell help hypothesis would be sufficient to account for the observed autoAb diversification. The ease of induction of autoAb production with ribonucleoproteins and their peptides suggests that tolerance to these autoantigens is very tenuous. It is also evident that the genetic backgrounds are important determinants for the magnitude of the autoAb production and the degree of autoAb diversification. The emerging scenario provides rationales to design experiments to probe the roles of intrinsic hyperactivity of T cells and antigen processing and presentation in general autoimmunity.

ACKNOWLEDGMENTS

The investigation discussed in this manuscript has been supported partly by grants RO1 AI-45199, RO1 AR-42456, RO1 DE-12544, and P50 AR-45222 from the National Institutes of Health. U.S.D. is partly supported by a scientist development grant from the American Heart Association.

REFERENCES

1. DESHMUKH U.S, J.E. LEWIS, F. GASKIN, *et al.* 1999. Immune responses to Ro60 and its peptides in mice. I. The nature of immunogen and endogenous autoantigen determine the specificities of the induced autoantibodies. J. Exp. Med. **189:** 531–540.
2. DESHMUKH U.S., J.E. LEWIS, F. GASKIN, *et al.* 2000. Ro60 peptides induce autoantibodies to similar epitopes shared among lupus related autoantigens. J. Immunol. **164:** 6655–6661.
3. DESHMUKH U.S., C.C. KANNAPELL & S.M. FU. 2002. Immune responses to small nuclear ribonucleoproteins: antigen-dependent distinct B cell epitope spreading patterns in mice immunized with recombinant polypeptides of small nuclear ribonucleoproteins. J. Immunol. **168:** 5326–5332.
4. LEWIS, J.E. *et al.* 2003. Manuscript in preparation.
5. HARDIN, J.A. 1986. The lupus autoantigens and the pathogenesis of systemic lupus erythematosus. Arthritis Rheum. **29:** 457–460.
6. FATENEJAD, S., M.J. MAMULA & J. CRAFT. 1993. Role of intermolecular/intrastructural B and T cell determinants in the diversification of autoantibodies to ribonucleoprotein particles. Proc. Natl. Acad. Sci. USA **90:** 12010–12014.
7. FATENEJAD, S. & J. CRAFT. 1996. Intrastructural help in diversification of humoral autoimmune responses. Clin. Exp. Immunol. **106:** 1–4.
8. CRAFT, J. & S. FATENEJAD. 1997. Self antigens and epitope spreading in systemic autoimmunity. Arthritis Rheum. **40:** 1374–1382.
9. PADOVAN. E., G. CASORATI, P. DELLABONA, *et al.* 1993. Expression of two T cell receptor a chains: dual receptor T cells. Science **262:** 422–424.
10. CORTHAY, A., K.S. NANDAKUMAR & R. HOLMDAHL. 2001. Evaluation of the percentage of peripheral T cells with two different T cell receptor a-chains and of their potential role in autoimmunity. J. Autoimmun. **16:** 423–429.

Mechanisms Inducing or Controlling CD8+ T Cell Responses against Self- or Non-Self-Antigens

DANIELE ACCAPEZZATO, VITTORIO FRANCAVILLA, ANTONELLA PROPATO, MARINO PAROLI, AND VINCENZO BARNABA

Fondazione "Andrea Cesalpino," Dipartimento di Medicina Interna, Università degli Studi di Roma "La Sapienza," and Istituto Pasteur-Cenci Bolognetti, 00161 Rome, Italy

ABSTRACT: Cytotoxic T lymphocytes (CTLs) generally recognize antigens endogenously synthesized within the cells and presented in the form of peptides on class I molecules. However, a large body of evidence suggests that dendritic cells (DCs) have the capacity to capture and deliver exogenous antigens into the major histocompatibility complex (MHC) class I processing pathway. In this paper, we discuss this function, defined as cross-presentation, and how it is directed, particularly in inducing T cell tolerance, and how it requires special activating signals (such as CD40 ligand) to transform into a mechanism that provides either protective immunity or autoimmunity.

KEYWORDS: CD8+ T cells; cross-presentation; dendritic cells; apoptosis; tolerance; autoimmunity; viral persistence

INTRODUCTION

There have been several studies on the control of CD8+ T cell responses against self- or non-self-antigens, but its underlying molecular mechanisms remain unclear and the subject of controversy. Effector CD8+ cytotoxic T lymphocytes (CTLs), known to be essential for protection against intracellular pathogens,[1] have been shown to cause tissue damage through cytotoxic functions involving perforin, Fas, or cytokines. In particular, perforin-mediated apoptosis or lysis is induced via antigen-specific interaction between T cell receptors (TCRs) on CTLs and major histocompatibility complex (MHC) class I/peptide complexes on target cells. Fas-mediated apoptosis occurs through the interaction of Fas on target cells with Fas ligand (FasL) expressed by both CTLs and interferon-γ-producing CD4+ T helper cells (Th1 cells), and can induce bystander cell damage. Finally, these effector cells can kill target cells by producing pro-apoptotic cytokines, or cytokines promoting nonspecific recruitment of bystander activated T, B, or other inflammatory cells. In

Address for correspondence: Vincenzo Barnaba, M.D., Dipartimento di Medicina Interna, Policlinico "Umberto I," V. le del Policlinico, 155, 00161 Rome, Italy. Voice: +39-06-4453994; fax: +39-06-49383333.
vincenzo.barnaba@uniroma1.it

Ann. N.Y. Acad. Sci. 987: 99–106 (2003). © 2003 New York Academy of Sciences.

contrast to Th cells, which recognize peptides on MHC class II molecules, CTLs recognize those presented by class I molecules.[1] MHC class II molecules present peptides that are derived from both exogenous and endogenous antigens, and that enter the endocytic route of professional antigen-presenting cells (APCs). Class I molecules present peptides generally deriving only from endogenously synthesized antigens (such as viral or self-proteins synthesized within the cell) that have been processed by proteasomes in the cytosol and then transported into the endoplasmic reticulum.[2–4] The selective recognition of endogenous antigens by CTLs is a phenomenon of paramount importance, as it guarantees that CTLs kill infected cells endogenously expressing viral antigens, sparing non-infected bystander cells, which have taken up soluble pathogen-derived proteins via the endocytic route. This phenomenon may, on the other hand, produce a severe side-effect. Since naïve T cells circulate only throughout the secondary lymphoid tissues and require professional APCs to be activated, virus-specific $CD8^+$ T cells should differentiate into effector CTLs only in response to infected APCs. Then, effector CTLs could acquire migratory properties and enter peripheral infected tissues to fight those pathogens that had also infected professional APCs. This phenomenon may constitute a serious hindrance to the protection against viruses selectively infecting peripheral non-lymphoid tissues, and thus not infecting APCs. It may, however, be the price to pay in order to maintain $CD8^+$ T cells specific for self-antigens sequestered in privileged organs in a status of "ignorance," which is one of the main mechanisms controlling the possible generation of autoimmune responses.

In this review, we will consider further issues arising from the above considerations that are still the subject of intense debate and investigation, particularly the mechanisms that induce immunity or tolerance against those intracellular pathogen- or self-antigens sequestered in privileged epithelial tissues that are not endogenously expressed by professional APCs.

THE CROSS-PRESENTATION PHENOMENON: DETERMINANTS THAT ALLOW EFFICIENT CTL CROSS-PRIMING

As emphasized above, antigens must be presented by professional APCs, such as dendritic cells (DCs), in order for antigenic peptides to be recognized on MHC molecules and to activate naïve T cells.

A large body of evidence has demonstrated that DCs display multiple and contrasting properties, according to their stage of maturation, or according to the differential precursors from which they derive.[5–7] The maturational stage model suggests that, under steady-state physiological conditions, immature DCs (imDCs) patrol the various tissues and, because of their high levels of constitutive macropinocytosis and endocytic receptor expression,[7] can capture apoptotic cells, which originate normally during the development of a given tissue. Moreover, imDCs are capable of transporting exogenous antigens from endocytic compartments to cytosol, leading to the presentation on MHC class I molecules to CTLs, which usually recognize endogenous antigens.[8–14] This phenomenon of presenting exogenous antigens derived from other cells to T cells by DCs is defined "cross-presentation," and is normally directed at inducing peripheral self-tolerance.[12,14] As a result of their low co-stimulatory and migratory capacities, a distinct subset of immature or partially mature DCs migrate in small numbers from the normal tissues into the lymph nodes and induce tolerance of naïve T cells specific for self-antigens, including those derived from apoptotic cells (cross-tolerance).[6,7,14] In these conditions, CTLs undergo only a few divisions and then anergy or deletion, without acquiring effector functions.[14–16] Alternatively, imDCs can induce peripheral self-tolerance, because they promote differentiation of both CTLs and Th cells into regulatory/suppressor T cells.[17,18]

Immature DCs promptly become mature DCs (mDCs), upon exposure to severe inflammatory processes, or to those pathogens capable of activating them, through the recognition of selective receptors, such as the Toll-like receptors (TLRs).[19,20] Moreover, cell apoptosis or necrosis are often a consequence of viral infection, and dead cells are in all probability the main source of exogenous viral or self-antigens.[21] In these conditions, the efficient antigen-capturing imDCs are transformed into strongly migratory and stimulatory mDCs by virtue of CCR7, MHC, adhesion, co-stimulatory molecule upregulation, and pro-inflammatory cytokine secretion.[6,22] These changes lead to the migration of continuous waves of stimulatory DCs into lymph nodes. In these compartments, activated Th cells complete the DC maturation program via the CD40 ligand (CD40L) or TRANCE stimuli (interacting with the appropriate receptors on DCs), improving the DC stimulatory capacity that ultimately leads to IL-12 production, essential for CTL differentiation and Th1/Tc1 polarization .[6,22,23] This view is supported by data demonstrating that DCs have to be "conditioned" by CD40L$^+$ cells in order to induce efficient T cell priming.[24–26] In any case, only these terminally differentiated mDCs can efficiently prime and sustain the responses of naïve T cells specific for exogenous antigens transported from the inflamed tissues (cross-priming).[9,27–31]

Recently, we set up an experimental model defining an alternative condition promoting efficient cross-priming. We found that in some pathological conditions, apoptotic cells can express CD40L *per se* (FIG. 1).[32] For instance, CD40L$^+$ apoptotic cells can derive from recently activated CD40L$^+$ T cells that undergo apoptosis once they have performed their effector function in a given inflamed tissue, or upon infection with pro-apoptotic viruses, such as human immunodeficiency virus.[32] Furthermore, we demonstrated that CD40L$^+$ apoptotic cells directly induce DC maturation and CTL cross-priming against apoptotic cell-associated self-antigens (such as vinculin), irrespective of additional exogenous signals (FIG. 1).[32] In contrast, under conditions in which the CD40L$^+$ apoptotic cells were overwhelmed by CD40L$^-$ (that is, in normal conditions or when an inflammatory process is terminated), the latter were unable to provide the appropriate signals to DCs, which deliver tolerance signals to autoreactive CTLs, even though they carry vinculin peptides. Thus, the balance between CD40L$^+$ and CD40L$^-$ apoptotic cells during cross-presentation appears to dictate tolerance or induction of CTL responses. The finding that CD40L$^+$ apoptotic cells induce DC maturation and cross-priming without the addition of exogenous stimuli, including help from CD4$^+$ cells, indicates that the surface phenotype and possibly the lineage of origin of apoptotic cells ultimately dictate the outcome of priming. These data further highlight the critical role played by the CD40L stimulus in DC maturation and efficient cross-priming.[24–26] This has been recently emphasized by evidence that DC stimulation via CD40/CD40L interaction was mandatory for priming CTL responses, while DCs, despite having been matured by a very powerful inflammatory mediator such as the tumor necrosis factor-α, induced T cell tolerance.[33]

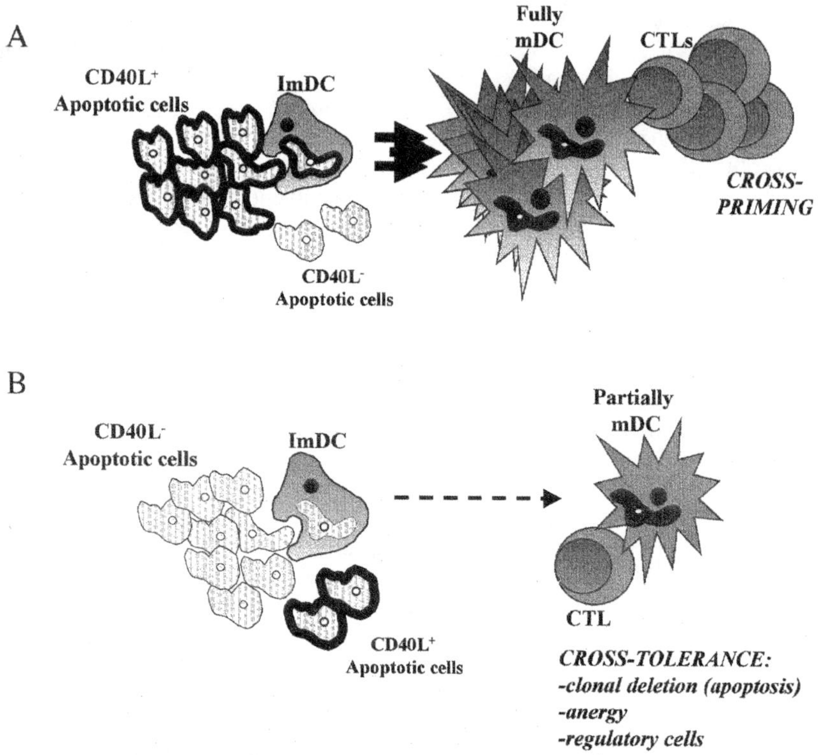

FIGURE 1. The balance between CD40L$^+$ and CD40L$^-$ apoptotic cells during cross-presentation dictates induction or tolerance of CTL responses. (**A**) Apoptotic cells expressing CD40L can bypass the usual requirement for T cell help and directly provide to DCs both antigens and the necessary maturation stimuli for CTL cross-priming. (**B**) Under conditions in which the CD40L$^+$ apoptotic cells are overwhelmed by CD40L$^-$ (that is, in normal conditions or when an inflammatory process is terminated), the latter will be unable to provide the appropriate signals to DCs, which will deliver tolerance signals to autoreactive CTLs, even though they carry apoptotic cell-derived peptides.

Altogether, these data suggest that efficient CTL cross-priming occurs only in the presence of vigorous Th cell responses, high numbers of CD40L$^+$ apoptotic cells, or pathogens activating DCs. Alternatively, CTLs undergo only a few divisions without acquiring effector functions, or even anergy or deletion.[14–16]

ROLE OF CROSS-PRESENTATION IN INDUCING IMMUNITY, TOLERANCE, OR DETRIMENTAL EFFECTS

The cross-priming mechanism might play a pivotal role in inducing CTL immunity, particularly against two types of viruses. First, against those viruses unable to infect and activate DCs, which cannot thus be presented via the conventional endogenous processing pathway leading to efficient formation and presentation of class I/peptide complexes on DC surface.[34,35] Second, cross-priming could be fundamental in generating CTL immunity to viruses that, although capable of infecting DCs, impair their APC functions, thus preventing CTL recognition and/or stimulation.[35–40] Both these cases could, however, be circumvented, as uninfected DCs are capable of capturing infected cells and cross-presenting exogenous viral antigens derived from them on class I molecules, resulting in effector CTL responses (cross-priming).[41] On the other hand, CTL cross-priming might be detrimental, since DCs may cross-present peptides derived from self-antigens upon phagocytosis of apoptotic (that is, viral infected) cells, leading to the generation of autoimmune responses.[21,31,32] Any one of the above described circumstances can occur, and for this reason CTL cross-priming by DCs requires extraordinary stimuli, in particular those derived from Th cells via CD40/CD40L interaction and/or activatory pathogens; otherwise, tolerance is induced.

Taken together, these results reinforce the hypothesis that cross-presentation per se is a system specifically directed at inducing tolerance, given the exceptional requirements of DCs for activation signals (such as CD40L) to induce T cell responses, the poor translocation of exogenous antigens from the endocytic compartment to the cytosol, and the ensuing formation of a lower number of peptide/class I complexes on the DC surface than conventional endogenous antigen presentation. Under these conditions, it is unlikely that DCs could deliver a sufficiently strong and sustained stimulation to induce rapid proliferation of CTLs and their differentiation into effector cells.

If we apply this model to autoimmunity, this poor capacity to mount efficient autoreactive $CD8^+$ T cell responses by the cross-presentation phenomenon may constitute a kind of "protective" mechanism. This could account for the limited occurrence of autoimmune diseases, in spite of self-reactive T cells being physiologically present in our bodies, and other factors such as TCR degeneracy, genetic background, and environment.

This same model could, on the other hand, be detrimental in other pathological conditions. For instance, persisting viruses, such as hepatitis C virus (HCV), that selectively replicate in privileged epithelial tissues (hepatocytes) and elicit poor Th cell responses at the early phase of infection,[42] could take on a status of persistence, as cross-priming (without exogenous help) would be the only mechanism for initiation of CTL responses against them. In agreement with this hypothesis, HCV-infected patients develop both very low frequencies of virus-specific CTLs with effector function and weak antiviral $CD4^+$ T cell responses.[42–46] This suggests that the CTL defect in clearing persisting viruses may be in part caused by the cross-presentation associated with $CD4^+$ T cell failure. This finding further supports the critical need of CD4 help in inducing and maintaining effective cross-priming.[32] On the contrary, pathogens that infect and/or activate DCs require neither cross-presentation nor $CD4^+$ help, owing to their efficient endogenous processing in forming peptide/class I complexes directly on APCs and/or their capacity of inducing full DC maturation.[35] These viruses generally elicit prompt and vigorous CD8 T cell responses, thus hindering a status of viral persistence.[47,48]

This model could be also applied to solid tumors, since it has been demonstrated that the induction of protective CTLs depends on the capacity of sufficient tumor cells directly reaching lymph nodes early and for a long duration, while cross-priming was inefficient and not protective.[49]

CONCLUDING REMARKS

The prevalence of induction or suppression of CTL responses by the cross-presentation phenomenon is correlated to different main factors. One critical factor for inducing cross-priming is the contribution of Th cells: generally, DCs need Th in order to prime CTL. Th cells are further required to induce CTL cross-priming, due to the threshold of MHC/peptide complexes generated during cross-presentation being lower than that of conventional endogenous processing. In this context, CD40L expression on apoptotic cells phagocitized by DCs is instrumental to efficient cross-priming, since the presence or absence of this molecule promotes $CD8^+$ T cell activation or tolerance, respectively. Consequently, viruses or tumor antigens that are poor inducers of $CD40L^+$ T cell responses and/or unable to directly stimulate DCs are more likely to promote inefficient priming, or even cross-tolerance. This may explain the frequent failure of antiviral $CD8^+$ T cell responses to combat persisting viruses such as the HCV, or of anti-tumor immune surveillance. In relation to this, cross-presentation of exogenous antigens to $CD8^+$ T cells by DCs "licensed" by $CD40L^+$ apoptotic cells may provide important building blocks for setting up highly protective, T-cell based vaccination protocols and for devising innovative strategies for the manipulation of anti-microbial, autoimmune, or anti-tumor responses.

ACKNOWLEDGMENTS

This work was supported by the following institutions: Ministero della Sanità-Istituto Superiore di Sanità (Progetti AIDS and Epatite Virale, Ricerche finalizzate su Vaccini); Ministero dell'Università e della Ricerca Scientifica e Tecnologica; Progetto Associazione Italiana Sclerosi Multipla (AISM); Progetto Finalizzato CNR "Biotecnologie"; European Community (contracts BMH4-CT98-3703 and QLK2-CT-2001-01167); Associazione Italiana per la Ricerca sul Cancro (AIRC).

REFERENCES

1. Kagi, D., B. Ledermann, K. Burki, *et al.* 1996. Molecular mechanisms of lymphocyte-mediated cytotoxicity and their role in immunological protection and pathogenesis in vivo. Annu. Rev. Immunol. **14:** 207–232
2. Rock, K.L. 1996. A new foreign policy: MHC class I molecules monitor the outside world. Immunol. Today **17:** 131–137.
3. Watts, C. 1997. Capture and processing of exogenous antigens for presentation on MHC molecules. Annu. Rev. Immunol. **15:** 821–850
4. Yewdell, J.W., C.C. Norbury & J.R. Bennink. 1999. Mechanisms of exogenous antigen presentation by MHC class I molecules in vitro and in vivo: implications for generating CD8+ T cell responses to infectious agents, tumors, transplants, and vaccines. Adv. Immunol. **73:** 1–77
5. Banchereau, J. & R.M. Steinman. 1998. Dendritic cells and the control of immunity. Nature **392:** 245–252.
6. Lanzavecchia, A. & F. Sallusto. 2001. Regulation of T cell immunity by dendritic cells. Cell **106:** 263–266.
7. Lanzavecchia, A. 1996. Mechanisms of antigen uptake for presentation. Curr. Opin. Immunol. **8:** 348–354.

8. BEVAN, M.J. 1976. Cross-priming for a secondary cytotoxic response to minor H antigens with H-2 congenic cells which do not cross-react in the cytotoxic assay. J. Exp. Med. **143:** 1283–1288.
9. ALBERT, M.L., B. SAUTER & N. BHARDWAJ. 1998. Dendritic cells acquire antigen from apoptotic cells and induce class I-restricted CTLs. Nature **392:** 86–89.
10. HEATH, W.R. & F.R. CARBONE. 1999. Cytotoxic T lymphocyte activation by cross-priming. Curr. Opin. Immunol. **11:** 314–318.
11. INABA, K., S. TURLEY, F. YAMAIDE, *et al.* 1998. Efficient presentation of phagocytosed cellular fragments on the major histocompatibility complex class II products of dendritic cells. J. Exp. Med. **188:** 2163–2173.
12. CARBONE, F.R., C. KURTS, S.R. BENNETT, *et al.* 1998. Cross-presentation: a general mechanism for CTL immunity and tolerance. Immunol. Today **19:** 368–373.
13. BARNABA, V., A. FRANCO, A. ALBERTI, *et al.* 1990. Selective killing of hepatitis B envelope antigen-specific B cells by class I-restricted, exogenous antigen-specific T lymphocytes. Nature **345:** 258–260.
14. HEATH, W.R. & F.R. CARBONE. 2001. Cross-presentation, dendritic cells, tolerance and immunity. Annu. Rev. Immunol. **19:** 47–64
15. HAWIGER, D., K. INABA, Y. DORSETT, *et al.* 2001. Dendritic cells induce peripheral T cell unresponsiveness under steady state conditions in vivo. J. Exp. Med. **194:** 769–779.
16. HERNANDEZ, J., S. AUNG, W.L. REDMOND, *et al.* 2001. Phenotypic and functional analysis of CD8(+) T cells undergoing peripheral deletion in response to cross-presentation of self-antigen. J. Exp. Med. **194:** 707–717.
17. JONULEIT, H., E. SCHMITT, G. SCHULER, *et al.* 2000. Induction of interleukin 10-producing, nonproliferating CD4(+) T cells with regulatory properties by repetitive stimulation with allogeneic immature human dendritic cells. J. Exp. Med. **192:** 1213–1222.
18. DHODAPKAR, M.V., R.M. STEINMAN, J. KRASOVSKY, *et al.* 2001. Antigen-specific inhibition of effector T cell function in humans after injection of immature dendritic cells. J. Exp. Med. **193:** 233–238.
19. SCHNARE, M., G.M. BARTON, A.C. HOLT, *et al.* 2001. Toll-like receptors control activation of adaptive immune responses. Nat. Immunol. **2:** 947–950.
20. KADOWAKI, N., S. HO, S. ANTONENKO, *et al.* 2001. Subsets of human dendritic cell precursors express different toll-like receptors and respond to different microbial antigens. J. Exp. Med. **194:** 863–869.
21. BARNABA, V. 1996. Viruses, hidden self-epitopes and autoimmunity. Immunol. Rev. **152:** 47–66.
22. MELLMAN, I. & R.M. STEINMAN. 2001. Dendritic cells: specialized and regulated antigen processing machines. Cell **106:** 255–258.
23. CELLA, M., D. SCHEIDEGGER, K. PALMER-LEHMANN, *et al.* 1996. Ligation of CD40 on dendritic cells triggers production of high levels of interleukin-12 and enhances T cell stimulatory capacity: T-T help via APC activation. J. Exp. Med. **184:** 747–752.
24. SCHOENBERGER, S.P., R.E. TOES, E.I. VAN DER VOORT, *et al.* 1998. T-cell help for cytotoxic T lymphocytes is mediated by CD40-CD40L interactions. Nature **393:** 480–483.
25. RIDGE, J.P., F. DI ROSA & P. MATZINGER. 1998. A conditioned dendritic cell can be a temporal bridge between a CD4+ T-helper and a T-killer cell. Nature **393:** 474–478.
26. BENNETT, S.R., F.R. CARBONE, F. KARAMALIS, *et al.* 1998. Help for cytotoxic-T-cell responses is mediated by CD40 signalling. Nature **393:** 478–480.
27. SIGAL, L.J., S. CROTTY, R. ANDINO, *et al.* 1999. Cytotoxic T-cell immunity to virus-infected non-haematopoietic cells requires presentation of exogenous antigen. Nature **398:** 77–80.
28. ROVERE, P., C. VALLINOTO, A. BONDANZA, *et al.* 1998. Bystander apoptosis triggers dendritic cell maturation and antigen-presenting function. J. Immunol. **161:** 4467–4471.
29. IGNATIUS, R., M. MAROVICH, E. MEHLHOP, *et al.* 2000. Canarypox virus-induced maturation of dendritic cells is mediated by apoptotic cell death and tumor necrosis factor alpha secretion. J. Virol. **74:** 11329–11338.
30. HUANG, F.P., N. PLATT, M. WYKES, *et al.* 2000. A discrete subpopulation of dendritic cells transports apoptotic intestinal epithelial cells to T cell areas of mesenteric lymph nodes. J. Exp. Med. **191:** 435–444.

31. DI ROSA, F. & V. BARNABA. 1998. Persisting viruses and chronic inflammation: understanding their relation to autoimmunity. Immunol. Rev. **164:** 17–27.
32. PROPATO, A., G. CUTRONA, V. FRANCAVILLA, *et al.* 2001. Apoptotic cells overexpress vinculin and induce vinculin-specific cytotoxic T-cell cross-priming. Nat. Med. **7:** 807–813.
33. ALBERT, M.L., M. JEGATHESAN & R.B. DARNELL. 2001. Dendritic cell maturation is required for the cross-tolerization of CD8+ T cells. Nat. Immunol. **2:** 1010–1017.
34. CELLA, M., M. SALIO, Y. SAKAKIBARA, *et al.* 1999. Maturation, activation, and protection of dendritic cells induced by double-stranded RNA. J. Exp. Med. **189:** 821–829.
35. RESCIGNO, M. & P. BORROW. 2001. The host-pathogen interaction: new themes from dendritic cell biology. Cell **106:** 267–270.
36. TORTORELLA, D., B.E. GEWURZ, M.H. FURMAN, *et al.* 2000. Viral subversion of the immune system. Annu. Rev. Immunol. **18:** 861–926
37. SALIO, M., M. CELLA, M. SUTER, *et al.* 1999. Inhibition of dendritic cell maturation by herpes simplex virus. Eur. J. Immunol. **29:** 3245–3253.
38. ENGELMAYER, J., M. LARSSON, M. SUBKLEWE, *et al.* 1999. Vaccinia virus inhibits the maturation of human dendritic cells: a novel mechanism of immune evasion. J. Immunol. **163:** 6762–6768.
39. FUGIER-VIVIER, I., C. SERVET-DELPRAT, P. RIVAILLER, *et al.* 1997. Measles virus suppresses cell-mediated immunity by interfering with the survival and functions of dendritic and T cells. J. Exp. Med. **186:** 813–823.
40. SEVILLA, N., S. KUNZ, A. HOLZ, *et al.* 2000. Immunosuppression and resultant viral persistence by specific viral targeting of dendritic cells. J. Exp. Med. 192: 1249–1260.
41. HEATH, W.R. & F.R. CARBONE. 2001. Cross-presentation in viral immunity and self-tolerance. Nat. Rev. Immunol. **1:** 126–135
42. MISSALE, G., R. BERTONI, V. LAMONACA, *et al.* 1996. Different clinical behaviors of acute hepatitis C virus infection are associated with different vigor of the anti-viral cell-mediated immune response. J. Clin. Invest. **98:** 706–714.
43. PREZZI, C., M.A. CASCIARO, V. FRANCAVILLA, *et al.* 2001. Virus-specific CD8(+) T cells with type 1 or type 2 cytokine profile are related to different disease activity in chronic hepatitis C virus infection. Eur. J. Immunol. **31:** 894–906.
44. LECHNER, F., D.K. WONG, P.R. DUNBAR, *et al.* 2000. Analysis of successful immune responses in persons infected with hepatitis C virus. J. Exp. Med. **191:** 1499–1512.
45. THIMME, R., D. OLDACH, K.-M. CHANG, *et al.* 2001. Determinants of viral clearance and persistence during acute hepatitis C infection. J. Exp. Med. **194:** 1395–1406
46. GRUENER, N.H., F. LECHNER, M.C. JUNG, *et al.* 2001. Sustained dysfunction of antiviral CD8+ T lymphocytes after infection with hepatitis C virus. J. Virol. **75:** 5550–5558.
47. TAN, L. C., N. GUDGEON, N.E. ANNELS, *et al.* 1999. A re-evaluation of the frequency of CD8+ T cells specific for EBV in healthy virus carriers. J. Immunol. **162:** 1827–1835.
48. GILLESPIE, G.M., M.R. WILLS, V. APPAY, *et al.* 2000. Functional heterogeneity and high frequencies of cytomegalovirus-specific CD8(+) T lymphocytes in healthy seropositive donors. J. Virol. **74:** 8140–8150.
49. OCHSENBEIN, A.F., S. SIERRO, B. ODERMATT, *et al.* 2001. Roles of tumour localization, second signals and cross priming in cytotoxic T-cell induction. Nature **411:** 1058–1064.

Role of Homeostatic Chemokine and Sphingosine-1-Phosphate Receptors in the Organization of Lymphoid Tissue

GERD MÜLLER, PHILLIP REITERER, UTA E. HÖPKEN, SVEN GOLFIER, AND MARTIN LIPP

Department of Molecular Tumor Genetics and Immunogenetics, Max Delbrück Center for Molecular Medicine (MDC), Berlin, Germany

ABSTRACT: Chemokines regulate both homeostatic leukocyte recirculation and trafficking to sites of infection and inflammation. Apart from the well-established physiological functions, chemokines receive growing interest for their role in pathophysiological processes such as autoimmune diseases, cancer, and allograft rejection. The chemokine receptor CCR7, which is responsible for directing T cells, B cells, and dendritic cells (DCs) into secondary lymphoid organs and their precise positioning therein, has already been implicated in lymphoid organ infiltration by neoplastic cells and the localization of metastasis formation. We have shown that the differential expression of CCR7 by neoplastic cells in two entities of Hodgkin's disease (HD), classic HD (cHD) and the nodular lymphocyte predominant HD (NLPHD), may account for the differences observed in tumor cell dissemination within the affected lymph nodes. Because of the prominent role of the chemokine receptors CCR7 and CXCR5 in lymphocyte homing to secondary lymphoid organs, we hypothesized that they may also be involved in the action of FTY720, a synthetic immunosuppressant inducing lymphopenia. By using CXCR5 and CCR7 knockout mice, we have tested for a possible function of these receptors in the FTY720-induced migration of lymphocytes into Peyer's patches (PPs) and peripheral lymph nodes (PLNs). Lymphopenia is noticeably delayed in mice lacking CCR7, whereas CXCR5 knockout mice show a significant reduction of lymphocyte accumulation in secondary lymphoid organs that are infrequently present in these mice. However, FTY720-induced lymphocyte sequestration appears to be essentially independent of CCR7 and CXCR5.

KEYWORDS: CCR7; CXCR5; Hodgkin's disease; FTY720; EDG6; sphingosine-1-phosphate

INTRODUCTION

Chemokines are indispensable regulators of leukocyte migration throughout the body. They participate in lymphoid organ development, lymphoid system homeostasis, and the recruitment of leukocytes at the sites of infection and inflammation.

Address for correspondence: Dr. Martin Lipp, Department of Molecular Tumor Genetics and Immunogenetics, Max Delbrück Center for Molecular Medicine (MDC), Robert-Rössle-Straße 10, 13092 Berlin, Germany. Voice: +49 30 9406-2886; fax: +49 30 9406-2887.
mlipp@mdc-berlin.de

Ann. N.Y. Acad. Sci. 987: 107–116 (2003). © 2003 New York Academy of Sciences.

Chemokines are cytokine-like molecules, about 8–14 kDa in size, that bind to G protein-coupled receptors characterized by seven transmembrane-spanning domains. A conserved motif of cysteine residues is used to classify chemokines. Depending on the number and spacing of the first two cysteines, they are designated as CC, CXC, CX_3C, or C chemokines. Most chemokines bind several chemokine receptors and vice versa, with the reservation that the ligand-binding profile of a given chemokine receptor is generally restricted to a single class of chemokines. Accordingly, chemokine receptors are classified as CCR, CXCR, CX_3CR, or XCR receptors.

The first chemokines to be identified were pro-inflammatory cytokines such as CXCL8 (interleukin-8), CCL2 (MCP-1), CCL3 (MIP-1α), and CCL4 (MIP-1β).[1] Expression of these chemokines can be induced in inflammation resulting in the attraction of leukocyte subsets mediating inflammatory reactions. Chemokines involved in lymphoid system homeostasis such as CCL19 (ELC), CCL21 (SLC), CXCL12 (SDF-1), and CXCL13 (BCA-1/BLC) were a later discovery.[2] They are constitutively expressed in lymphoid tissues, thereby enabling basal leukocyte trafficking and homing, which are a prerequisite for adaptive immune responses.

The homeostatic chemokines CCL19 and CCL21, ligands for CCR7, are involved in the homing of lymphocytes and DCs to secondary lymphoid organs.[3] Knockout mice have proved the importance of CCR7 and its ligands in lymphoid organ entry.[4] Migration of T cells, B cells, and dendritic cells (DCs) into Peyer's patches (PPs) and peripheral lymph nodes (PLNs) is impaired in these mice. They show increased numbers of T cells in the peripheral blood and at the same time reduced numbers of T cells within lymph nodes. Mice harboring the paucity of lymph node T cell (*plt*) mutation are defective in the expression of CCL19 and CCL21 in lymphatic organs and display a phenotype similar to $CCR7^{-/-}$ mice.[5,6] The precise positioning of lymphocytes within secondary lymphoid organs is controlled by the responsiveness of these cells towards overlapping gradients of homeostatic chemokines, including CCL19, CCL21, and CXCL13, that are expressed in separate but adjacent areas of the organ.[7] These data correspond with our earlier observation that $CXCR5^{-/-}$ mice show a severe defect in the segregation of B cells and T cells into functionally distinct areas within the spleen and PPs.[8] In addition, PLNs and PPs are either absent or present at low frequency in $CXCR5^{-/-}$ mice.

During the last several years, it became clear that homeostatic chemokines are not only involved in physiological but also in a variety of pathophysiological processes, including autoimmune diseases, bacterial and viral infections, neoplasia, and graft rejection.[9–11] The chemokines CCL19, CCL21, and CXCL13 are particularly interesting in the development of lymphoproliferative diseases affecting secondary lymphoid tissue or diseases involving the formation of ectopic follicles. CCR7 has been associated with adult T cell leukemia, where expression levels for CCR7 correlate with the probability of lymphoid organ infiltration.[12] CCR7 has also been connected with breast cancer metastasis in a mouse model.[13] The expression of the chemokine receptor CXCR4, and less frequently CCR7, in breast cancer cells is consistent with the fact that typical sites of metastasis formation, lymph node and lung, are abundant sources of the corresponding ligands CXCL12, CCL19, and CCL21. From this point of view, the histopathological features and the pattern of cytokines expressed in HD prompted us to test for a possible involvement of homeostatic chemokines and their receptors in HD.

TABLE 1. Chemokine receptor expression in Hodgkin's disease

	Proportion of positive tumor cells					
Entity	CCR7		CXCR4		CXCR5	
HD-MC	3/10	+++	5/10	+++	4/10	±
	2/10	++	3/10	++	3/10	negative
	2/10	+	2/10	±		
	3/10	±				
HD-NS	7/18	+++	9/18	+++	2/18	+++
	3/18	++	4/18	++	8/18	++
	6/18	+	2/18	+	4/18	±
	2/18	±	2/18	±	4/18	n.a.
			1/18	n.a.		
HD-NLP	1/7	±	3/7	++	1/7	++
	6/7	negative	2/7	+	2/7	+
			2/7	negative	4/7	n.a.

ABBREVIATIONS: HD, Hodgkin's disease; MC, mixed cellularity; NS, nodular sclerosis; NLP, nodular lymphocyte predominant; +++, 50 to 100% positive tumor cells; ++, 20 to 50% positive tumor cells; +, 10 to 20% positive tumor cells; ±, less than 10% positive tumor cells or single positive tumor cells; n.a., not analyzed.

EXPRESSION OF THE CHEMOKINE RECEPTOR CCR7 CORRELATES WITH THE DISSEMINATION OF NEOPLASTIC CELLS IN CLASSIC HODGKIN'S DISEASE

Hodgkin's disease (HD) comprises a group of lymphomas in which a small proportion of tumor cells is present in a large background of reactive cells, including T cells, eosinophils, histiocytes, and plasma cells. Although tumor cells in HD express surface markers of distinct hematopoietic lineages, they are most likely derived from germinal center (GC) or post-GC B cells, since they frequently show clonal immunoglobulin gene rearrangements.[14] The tumor cells express various cytokines and chemokines, including interleukin (IL)-1, IL-7, CCL17 (TARC), and CCL22 (MDC), that may explain the characteristic inflammatory infiltrate in the affected tissue.[15–18] Lymphotoxin-α and tumor necrosis factor-α, which are critical factors for the expression of CCL19, CCL21, and CXCL13,[19] are also expressed in HD tissue.[20] By testing for the expression of homeostatic chemokines and their receptors in classic HD (cHD) and nodular lymphocyte predominant HD (NLPHD), two entities of HD with different histologic features, we found that only tumor cells in cHD, Hodgkin/Reed-Sternberg (HRS) cells, strongly express CCR7 (TABLE 1).[21] HRS cells are predominantly located in the interfollicular zone or the follicular mantle zone of partially infiltrated lymph nodes, whereas tumor cells in the NLPHD, L&H cells that lack expression of CCR7, frequently reside within follicular structures. Expression of the corresponding chemokines CCL19 and CCL21 was restricted to reactive cells and detected in practically all tumor infiltrates in cHD, whereas tumor nodules in NLPHD are generally devoid of these chemokines. In comparison, expression of CXCR5 was less prominent in cHD and essentially restricted to the nodular sclerosis subtype. Expression of CXCR5 on L&H cells in NLPHD could not be reliably determined since most reactive cells in the tumor nodules express CXCR5. These most likely comprise B cells but also $CD4^+CD57^+$ T cells expressing CXCR5[22] and rosetting the lymphocytic and histiocytic (L&H) cells.[23]

Taken together, our results point towards a crucial role for CCR7 in the dissemination of tumor cells in cHD, because $CCR7^+$ HRS cells but not $CCR7^-$ L&H cells co-localize with CCL19 and CCL21-secreting reactive cells. In support of this hypothesis, it has now been shown that the activation-dependent positioning of B cells within microcompartments of the spleen is controlled by their responsiveness towards chemokines such as CCL19, CCL21, and CXCL13.[7] Therefore, the question arises to what extent CXCL13 might be responsible for the localization of L&H cells and the organization of tumor nodules in NLPHD. In this connection, it is appealing to investigate the interaction of L&H cells with the surrounding $CD4^+CD57^+$ T cells, because these T cells normally exert a classic T helper function for B cells in GCs by stimulating (for example) the production of immunoglobulins.[22,24]

CHEMOKINE AND SPHINGOSINE-1-PHOSPHATE (S1P) RECEPTORS IN LYMPHOCYTE ORGAN ENTRY: LONE PLAYERS OR TEAM PLAYERS?

The synthetic immunosuppressant FTY720 (2-amino-2-[2-(4-octylphenyl)-ethyl]-1,3-propanediol), a derivative of myriocin that can be isolated from *Myriococcum albomyces* and *Isaria sincalide*, induces lymphopenia by modulating lymphocyte recirculation (FIG. 1). Administration of FTY720 leads to a rapid sequestration of lymphocytes from peripheral blood to the lymph nodes and PPs.[25] It neither affects the cellular or humoral immunity to systemic viral infection,[26] nor the induction and expansion of immune responses in secondary lymphoid tissue,[27–29] but it efficiently reduces the number of effector T cells in the periphery. The sequestration of lymphocytes, in opposition to earlier reports, is unrelated to apoptosis, and it is completely reversible.[25,26] FTY720-induced lymphocyte trafficking is sensitive to pertussis toxin and can be inhibited by a mixture of antibodies against the adhesion molecules L-selectin, leukocyte function-associated antigen 1 (LFA-1), and very late antigen 4 (VLA-4).[25,30]

Since $CCR7^{-/-}$ and *plt* mice show severe defects in the homing of T cells to PPs and PLNs, compartments where lymphocytes are sequestered following treatment with FTY720, we tested for a possible interplay of CCR7 and FTY720 in lymphocyte homing. Surprisingly, administration of FTY720 was still effective on lymphocytes in $CCR7^{-/-}$ and *plt* mice. Lymphocyte counts are reduced in the periphery, whereas elevated numbers of lymphocytes could be detected in PPs, mesenteric lymph nodes, and PLNs. Nevertheless, the kinetics of lymphocyte sequestration is delayed, and lymphocyte clearance in the periphery is not as complete as in wild-type animals.[31] In comparison, *plt* mice resemble $CCR7^{-/-}$ mice with respect to FTY720 kinetics and sequestration efficiency. In a similar study using *plt* mice, Chen *et al.*[32] point out that FTY720 is significantly less effective in the sequestration of T lymphocytes in *plt* mice compared to wild-type mice, suggesting an essential

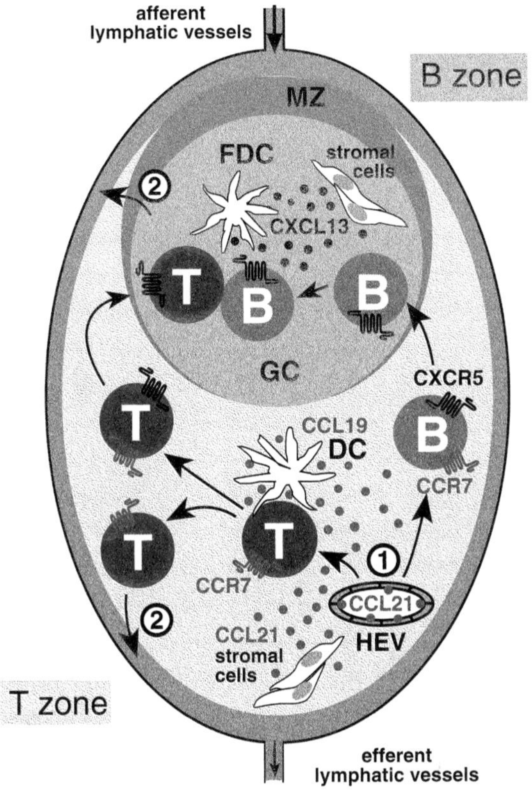

FIGURE 1. Chemokine-directed trafficking of lymphocytes through secondary lymphoid organs: modulation by FTY720. Chemokines, such as CCL19 and CCL21, which are ligands for the chemokine receptor CCR7, enable lymphocytes to cross the endothelial cell layer of high-endothelial venules (HEVs) to enter the T cell zone of lymph nodes. B cells expressing CXCR5 migrate through the T cell zone into B cell follicles where BLC (CXCL13), the ligand for CXCR5, is produced by follicular stromal cells. Activated B cells may give rise to the formation of germinal centers (GCs). A subset of activated T cells upregulates CXCR5, consequently entering a follicle and participating in the germinal center reaction. The precise mechanism by which FTY720 induces lymphopenia still has to be identified. FTY720 may affect lymphocyte entry into lymph nodes in combination with or independent of the chemokine system (1). FTY720 may also interfere with the exit of lymphocytes from lymph nodes via the lymphatic system (2). ABBREVIATIONS: FDC, follicular dendritic cell; MZ, mantle zone.

function for the chemokine system in the action of FTY720. Considering the relatively short observation periods by Chen and co-workers, CCR7-independent lymphopenia may have escaped the notice of the authors.

In addition, we tested for a possible interaction of FTY720 and CXCR5 by using CXCR5$^{-/-}$ mice (FIG. 2). Administration of FTY720 induces lymphopenia in wild-type as well as in CXCR5$^{-/-}$ mice. T cell and B cell counts in peripheral blood are reduced by at least 50% following 3 h of incubation. At the same time, increasing

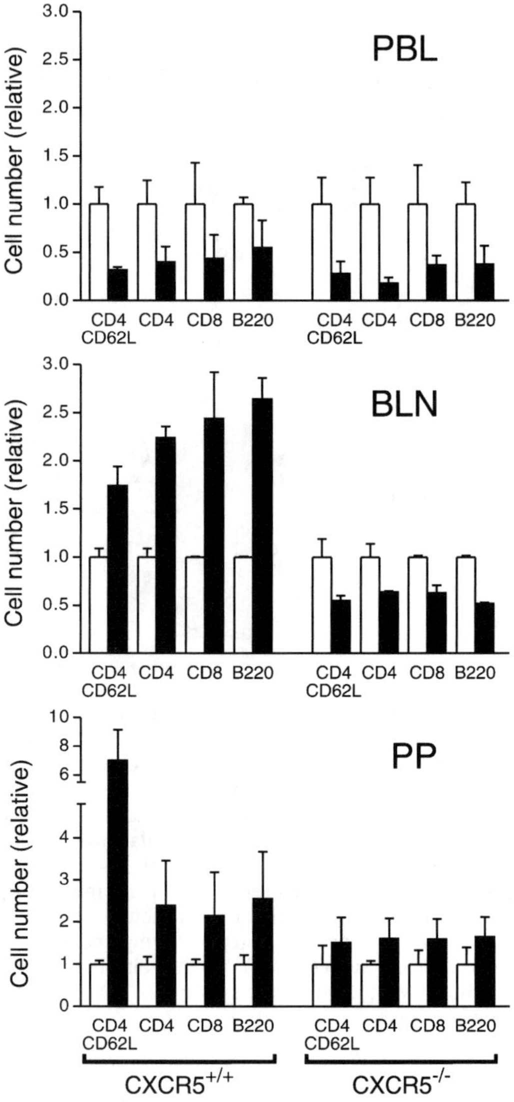

FIGURE 2. FTY720-induced lymphopenia in wild-type and CXCR5$^{-/-}$ mice. FTY720 (0.3 mg/kg) was dissolved in distilled water and administered orally. Control mice received water only. After 3 h, lymphocytes were isolated from peripheral blood leukocytes (PBL), brachial lymph nodes (BLNs), and Peyer's patches (PPs). Cells were counted and stained for the surface markers indicated. Each value represents a mean ± SD from at least four animals in two independent experiments.

numbers of lymphocytes appear in PLNs and PPs of wild-type mice. On average, B and T cell numbers increased with comparable efficiency by a factor of 2.5. Interestingly, there seems to be a preferential sequestration of naïve $CD4^+CD62L^{high}$ T cells into PPs as compared to brachial lymph nodes (BLNs). The increase in the number of naïve $CD4^+$ T cells in PPs accounts for almost all of the increase in the total number of $CD4^+$ T cells in this organ. In comparison, $CXCR5^{-/-}$ mice showed a severe defect in the FTY720-mediated sequestration of lymphocytes in BLNs and PPs, lymphoid organs that are either absent or present in low numbers in these mice. FTY720 induced only a moderate increase in the number of B and T cells in PPs and, surprisingly, a slight decrease in BLNs. In addition, we could no longer observe a preferential accumulation of naïve $CD4^+$ T cells in PPs. As there is accumulating evidence for an involvement of CXCR5 and CXCL13 in lymphoid organogenesis,[33] BLNs and PPs may have intrinsic defects impeding the sequestration of lymphocytes.

Only recently the phosphate ester metabolite of FTY720, a close structural homologue to S1P, has been identified as a high-affinity agonist for four out of five known S1P receptors.[34,35] A nonhydrolyzable analogue of this phosphate ester of FTY720 induces peripheral blood lymphopenia comparable to FTY720. S1P exerts the same effect on lymphocytes but with lower efficiency. It is especially interesting that the affinity of the S1P receptor EDG6 for the phosphate ester metabolite of FTY720 tested to be 20-fold higher compared to S1P. Expression of EDG6 is restricted to lymphoid tissues including cell populations such as peripheral lymphocytes and DCs.[36,37] It is therefore tempting to speculate that EDG6 may act in cooperation with chemokine receptors in the course of lymphocyte emigration from the peripheral blood into PPs and PLNs upon binding to the phosphate ester metabolite of FTY720. Cooperation may include chemokine receptors other than CCR7 since FTY720-induced lymphopenia can still be observed in $CCR7^{-/-}$ and *plt* mice. In favor of this hypothesis is the finding that FTY720 enhances the chemotaxis of lymphocytes towards several chemokines, including CCL19, CCL21, and CXCL13 *in vitro*.[32] The histological finding that lymphocytes disappear from subcapsular and medullary sinuses of mesenteric lymph nodes following application of FTY720 points towards an alternative mode of action for this substance. FTY720 may selectively inhibit the emigration of lymphocytes from lymph nodes and PPs, thereby accelerating and reinforcing the lymphopenia (FIG. 1). It still needs to be shown if the sequestration of lymphocytes within secondary lymphoid organs occurs independent of chemokines, thereby proposing a comparable function for EDG and chemokine receptors in the emigration of lymphocytes from peripheral blood. Other possible scenarios include a cooperative effect between S1P and chemokine receptors in lymphocyte migration into lymph nodes and PPs complemented by an inhibition of lymphocyte emigration from these organs. Regarding the efficiency and specificity of lymphocyte sequestration into lymphoid organs, it will be interesting to test for the selective involvement of S1P receptors and adhesion molecules in the action of FTY720.

CONCLUSIONS

In consideration of the importance of chemokines in leukocyte trafficking, the chemokine system is an attractive target for therapeutic interventions in diseases associated with an errant cell trafficking or localization. However, due to the diverse nature of these diseases, tailor-made strategies are necessary in combating them. Results from a mouse model for breast cancer suggest that chemokine receptor blocking antibodies may inhibit cancer metastasis.[13] In diseases such as HD with a high background of reactive cells, a change in the local cytokine and chemokine milieu may help to change a favorable environment sustaining tumor growth into a destructive environment for the neoplastic cells. On the other hand, a naturally occurring immune response towards the grafted tissue needs to be suppressed to prevent allograft rejection in transplantation. The latter may be achieved by FTY720 in a deceptively simple procedure—by excluding effector T cells from recirculation, thereby impeding their recruitment to peripheral sites. Contrary to expectation, it seems FTY720 does not directly interact with the chemokine receptors CCR7 or CXCR5, but instead interacts with S1P receptors. However, the precise mode of action of FTY720 and its possible dependency on the chemokine system still needs to be evaluated to improve disease- and case-specific therapies.

REFERENCES

1. BAGGIOLINI, M., B. DEWALD & B. MOSER. 1994. Interleukin-8 and related chemotactic cytokines—CXC and CC chemokines. Adv. Immunol. **55:** 97–179.
2. KIM, C.H. & H.E. BROXMEYER. 1999. Chemokines: signal lamps for trafficking of T and B cells for development and effector function. J. Leukocyte Biol. **65:** 6–15.
3. CYSTER, J.G. 1999. Chemokines and cell migration in secondary lymphoid organs. Science **286:** 2098–2102.
4. FÖRSTER, R., A. SCHUBEL, D. BREITFELD, *et al.* 1999. CCR7 coordinates the primary immune response by establishing functional microenvironments in secondary lymphoid organs. Cell **99:** 23–33.
5. MORI, S., H. NAKANO, K. ARITOMI, *et al.* 2001. Mice lacking expression of the chemokines CCL21-ser and CCL19 (plt mice) demonstrate delayed but enhanced T cell immune responses. J. Exp. Med. **193:** 207–218.
6. GUNN, M.D., S. KYUWA, C. TAM, *et al.* 1999. Mice lacking expression of secondary lymphoid organ chemokine have defects in lymphocyte homing and dendritic cell localization. J. Exp. Med. **189:** 451–460.
7. REIF, K., E.H. EKLAND, L. OHL, *et al.* 2002. Responsiveness to chemoattractants from adjacent microenvironments determines B-cell position. Nature **416:** 94–99.
8. FÖRSTER, R., A.E. MATTIS, E. KREMMER, *et al.* 1996. A putative chemokine receptor, BLR1, directs B cell migration to defined lymphoid organs and specific anatomic compartments of the spleen. Cell **87:** 1037–1047.
9. VICARI, A.P. & C. CAUX. 2002. Chemokines in cancer. Cytokine Growth Factor Rev. **13:** 143–154.
10. GODESSART, N. & S.L. KUNKEL. 2001. Chemokines in autoimmune disease. Curr. Opin. Immunol. **13:** 670–675.
11. HJELMSTROM, P. 2001. Lymphoid neogenesis: de novo formation of lymphoid tissue in chronic inflammation through expression of homing chemokines. J. Leukocyte Biol. **69:** 331–339.
12. HASEGAWA, H., T. NOMURA, M. KOHNO, *et al.* 2000. Increased chemokine receptor CCR7/EBI1 expression enhances the infiltration of lymphoid organs by adult T-cell leukemia cells. Blood **95:** 30–38.
13. MÜLLER, A., B. HOMEY, H. SOTO, *et al.* 2001. Involvement of chemokine receptors in breast cancer metastasis. Nature **410:** 50–56.
14. STAUDT, L.M. 2000. The molecular and cellular origins of Hodgkin's disease. J. Exp. Med. **191:** 207–212.
15. RUCO, L.P., D. POMPONI, R. PIGOTT, *et al.* 1990. Cytokine production (IL-1 alpha, IL-1 beta, and TNF alpha) and endothelial cell activation (ELAM-1 and HLA-DR) in

reactive lymphadenitis, Hodgkin's disease, and in non-Hodgkin's lymphomas. An immunocytochemical study. Am. J. Pathol. **137:** 1163–1171.

16. FOSS, H.D., M. HUMMEL, S. GOTTSTEIN, *et al.* 1995. Frequent expression of IL-7 gene transcripts in tumor cells of classical Hodgkin's disease. Am. J. Pathol. **146:** 33–39.
17. VAN DEN BERG, A., L. VISSER & S. POPPEMA. 1999. High expression of the CC chemokine TARC in Reed-Sternberg cells. A possible explanation for the characteristic T-cell infiltratein Hodgkin's lymphoma. Am. J. Pathol. **154:** 1685–1691.
18. HEDVAT, C.V., E.S. JAFFE, J. QIN, *et al.* 2001. Macrophage-derived chemokine expression in classical Hodgkin's lymphoma: application of tissue microarrays. Mod. Pathol. **14:** 1270–1276.
19. NGO, V.N., H. KORNER, M.D. GUNN, *et al.* 1999. Lymphotoxin α/β and tumor necrosis factor are required for stromal cell expression of homing chemokines in B and T cell areas of the spleen. J. Exp. Med. **189:** 403–412.
20. CLODI, K. & A. YOUNES. 1997. Reed-Sternberg cells and the TNF family of receptors/ligands. Leuk. Lymphoma **27:** 195–205.
21. HÖPKEN, U.E., H.D. FOSS, D. MEYER, *et al.* 2002. Up-regulation of the chemokine receptor CCR7 in classical but not in lymphocyte-predominant Hodgkin disease correlates with distinct dissemination of neoplastic cells in lymphoid organs. Blood **99:** 1109–1116.
22. KIM, C.H., L.S. ROTT, I. CLARK-LEWIS, *et al.* 2001. Subspecialization of $CXCR5^+$ T cells: B helper activity is focused in a germinal center-localized subset of $CXCR5^+$ T cells. J. Exp. Med. **193:** 1373–1381.
23. POPPEMA, S. 1989. The nature of the lymphocytes surrounding Reed-Sternberg cells in nodular lymphocyte predominance and in other types of Hodgkin's disease. Am. J. Pathol. **135:** 351–357.
24. BREITFELD, D., L. OHL, E. KREMMER, *et al.* 2000. Follicular B helper T cells express CXC chemokine receptor 5, localize to B cell follicles, and support immunoglobulin production. J. Exp. Med. **192:** 1545–1552.
25. CHIBA, K., Y. YANAGAWA, Y. MASUBUCHI, *et al.* 1998. FTY720, a novel immunosuppressant, induces sequestration of circulating mature lymphocytes by acceleration of lymphocyte homing in rats. I. FTY720 selectively decreases the number of circulating mature lymphocytes by acceleration of lymphocyte homing. J. Immunol. **160:** 5037–5044.
26. PINSCHEWER, D.D., A.F. OCHSENBEIN, B. ODERMATT, *et al.* 2000. FTY720 immunosuppression impairs effector T cell peripheral homing without affecting induction, expansion, and memory. J. Immunol. **164:** 5761–5770.
27. CHIBA, K., Y. HOSHINO, C. SUZUKI, *et al.* 1996. FTY720, a novel immunosuppressant possessing unique mechanisms. I. Prolongation of skin allograft survival and synergistic effect in combination with cyclosporine in rats. Transplant. Proc. **28:** 1056–1059.
28. YANAGAWA, Y., K. SUGAHARA, H. KATAOKA, *et al.* 1998. FTY720, a novel immunosuppressant, induces sequestration of circulating mature lymphocytes by acceleration of lymphocyte homing in rats. II. FTY720 prolongs skin allograft survival by decreasing T cell infiltration into grafts but not cytokine production in vivo. J. Immunol. **160:** 5493–5499.
29. WANG, M.E., N. TEJPAL, X. QU, *et al.* 1998. Immunosuppressive effects of FTY720 alone or in combination with cyclosporine and/or sirolimus. Transplantation **65:** 899–905.
30. LI, X.K., S. ENOSAWA, T. KAKEFUDA, *et al.* 1997. FTY720, a novel immunosuppressive agent, enhances upregulation of the cell adhesion molecule ICAM-1 in TNF-α treated human umbilical vein endothelial cells. Transplant. Proc. **29:** 1265–1266.
31. HENNING, G., L. OHL, T. JUNT, *et al.* 2001. CC chemokine receptor 7-dependent and -independent pathways for lymphocyte homing: modulation by FTY720. J. Exp. Med. **194:** 1875–1881.
32. CHEN, S., K.B. BACON, G. GARCIA, *et al.* 2001. FTY720, a novel transplantation drug, modulates lymphocyte migratory responses to chemokines. Transplant. Proc. **33:** 3057–3063.
33. HONDA, K., H. NAKANO, H. YOSHIDA, *et al.* 2001. Molecular basis for hematopoietic/mesenchymal interaction during initiation of Peyer's patch organogenesis. J. Exp. Med. **193:** 621–630.

34. BRINKMANN, V., M.D. DAVIS, C.E. HEISE, *et al.* 2002. The immune modulator FTY720 targets sphingosine 1-phosphate receptors. J. Biol. Chem. **277:** 21453–21457.
35. MANDALA, S., R. HAJDU, J. BERGSTROM, *et al.* 2002. Alteration of lymphocyte trafficking by sphingosine-1-phosphate receptor agonists. Science **296:** 346–349.
36. IDZKO, M., E. PANTHER, S. CORINTI, *et al.* 2002. Sphingosine 1-phosphate induces chemotaxis of immature and modulates cytokine-release in mature human dendritic cells for emergence of T_H2 immune responses. FASEB J. **16:** 625–627.
37. GRÄLER, M.H., G. BERNHARDT & M. LIPP. 1998. EDG6, a novel G-protein-coupled receptor related to receptors for bioactive lysophospholipids, is specifically expressed in lymphoid tissue. Genomics **53:** 164–169.

The Human Marginal Zone B Cell

MARIELLA DONO,[a] SIMONA ZUPO,[a] MONICA COLOMBO,[a] ROSANNA MASSARA,[a] GIANLUCA GAIDANO,[b] GIUSEPPE TABORELLI,[c] PAOLA CEPPA,[d] VITO L. BURGIO,[e] NICHOLAS CHIORAZZI,[f] AND MANLIO FERRARINI[a,g]

[a]*Oncologia Medica C, Istituto Nazionale per la Ricerca sul Cancro, IST, Genova, Italy*

[b]*Dipartimento di Medicina, Università del Piemonte Orientale, Novara, Italy*

[c]*Divisione ORL, Istituto Giannina Gaslini, Genova, Italy*

[d]*Sezione di Anatomia Pat., DICMI, Università di Genova, Genova, Italy*

[e]*Istituto 1[a] Clinica Medica Generale e Terapia Medica, Università di Roma, La Sapienza, Roma, Italy*

[f]*Department of Medicine, North Shore University Hospital and NYU School of Medicine, Manhasset, New York, USA*

[g]*Dipartimento di Oncologia, Biologia e Genetica, Università degli Studi di Genova, Genova, Italy*

ABSTRACT: This study describes the features of the marginal zone (MZ) B cells of human tonsils and spleens and compares them with those of the follicular mantle (FM) B cells from the same tissues. The two B cell subpopulations displayed marked differences in phenotype, in response capacity to T cell-independent antigens and polyclonal B cell activators, and in presentation of antigens to T cells. FM B cells expressed surface CD5, and hence should be considered as B1 cells by current nomenclature. Fractionation of MZ B cells according to the presence or absence of surface IgD revealed the presence of two subsets. These subsets were characterized by different properties, including the presence of Ig V_H gene mutations and the response capacity to TI-2 antigens, this latter property being associated with IgD-positive cells. Comparison of the data with those reported for mice revealed that human MZ B cells had strong analogies with both the murine MZ and B1 cells. In contrast, human B1 cells (that is, CD5-positive FM cells) were considerably different, an observation that should prompt further studies. Indeed, B cells with characteristics analogous to those of murine B1 cells were detected in small but definite proportions in the peripheral blood and tonsils. If the current distinction into B1 and B2 cells has to be maintained also for humans, it is likely that only these CD5-positive cells rather than the FM B cells should be called B1 cells.

KEYWORDS: marginal zone; B1 cells; heterogeneity

Address for correspondence: Mariella Dono, Ph.D., Oncologia Medica C, Istituto Nazionale per la Ricerca sul Cancro, IST, L.go Rosanna Benzi, n. 10, 16132 Genova GE, Italy. Voice: +39.010.5600.263 or +39.010.5600.271; fax: +39.010.5600.264.
mariella.dono@istge.it

Ann. N.Y. Acad. Sci. 987: 117–124 (2003). © 2003 New York Academy of Sciences.

INTRODUCTION

After B cells mature, they emigrate from the bone marrow to the peripheral lymphoid organs, where they seed into the so-called B cell-dependent areas.[1] Based upon morphological criteria, these B cell areas have been broadly subdivided into follicles and extrafollicular zones. The follicles are in turn subdivided into germinal centers and follicular mantles. The extrafollicular area is often referred to as the marginal zone or the marginal zone equivalent area. Consequently, the B cells found in the different areas are referred to as follicular mantle (FM), germinal center (GC), or marginal zone (MZ) B cells. Since the advent of *in situ* staining methods with monoclonal antibodies, it has been determined that these cells in the different lymphoid areas express different markers. With this information, it has also been possible to isolate these cells in suspension and test them for their functional features *in vitro*. It is widely accepted that germinal centers are the sites of origin of memory B cells, where selection of B cells that have generated new antibody specificities following antigenic stimulation takes place.[1] In contrast, much less is known about the two other subsets and their mutual relationships, although the most widely held view is that FM B cells are mainly virgin B cells, while the MZ B cells are mostly memory B cells. In this paper, we shall review our work on the features of MZ B cells and compare their characteristics with those of other human and murine B cell subsets.

Another approach to B cell subsetting, utilized primarily in mice, has been that of defining certain subpopulations based upon surface markers and functional features, irrespective of the cell's homing properties. These studies have led to the definition of two major subsets, called B1 and B2 cells, the main distinction being the presence or absence, respectively, of CD5 on the cell surface.[2] Such a distinction is not easily applicable to human cell subsets, and this problem will also be briefly dealt with in this article.

ISOLATION OF MZ B CELLS

Extrafollicular B cells have distinctive homing properties. They are detected in the splenic marginal zone, in the subepithelial area of tonsils, in the subcapsular area of lymph nodes, in the dome region of Peyer's patches, in the thymic medulla, and in the newly formed mucosa-associated lymphoid tissue (MALT). This has led to the use of various terminologies to indicate these cells; however, for the sake of simplicity, in this article they shall be referred to as MZ B cells.

In our studies, we have used mainly tonsillar MZ B cells.[3,4] The strategy employed included an assessment of the *in situ* morphological and phenotypic features of the cells, a fractionation of the tonsillar cell suspensions by immunomagnetic beads, and a subsequent check of the phenotypic and morphological features of the cells isolated in suspensions. MZ B cells were detected in the Percoll fractions of highest density and separated from FM B cells found in the same fractions by cell sorting or magnetic beads. Separation was possible since, unlike MZ B cells, FM B cells invariably express CD5 and CD39. Reanalysis of the phenotypic and morphological features of MZ and FM cells obtained in suspension confirmed identity with those determined *in situ*. Although a number of MZ B cells were also found in lighter Percoll fractions (a finding that likely indicates that they were activated *in vivo*), the

TABLE 1. Summary of the main phenotypic features of human and murine MZ and B1 B cells

	CD5	CD23	CD21	IgM	IgD	CD9
MZ B cells						
Human tonsil	–	–	±	++	±	–
Human spleen	–	–	++	++	±	±
Murine spleen	–	–	++	++	±	±
B1 B cells						
Human tonsil	+	++	+	++	++	n.d.
Human spleen	+	++	+	++	++	n.d.
Murine peritoneal cavity	+	–	+	++	±	n.d.

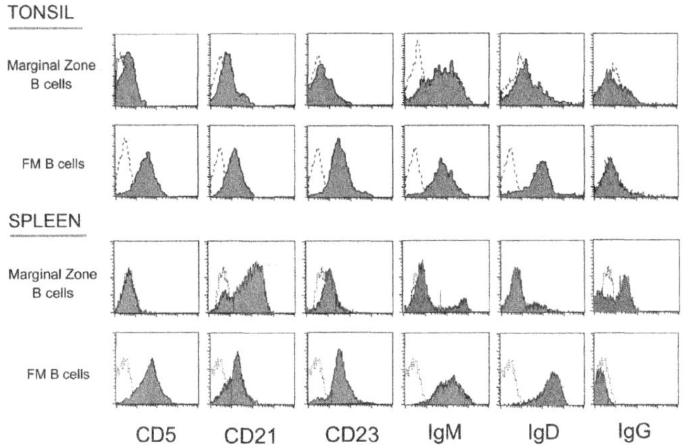

FIGURE 1. Immunofluorescence analysis of human MZ B cells. MZ and FM B cells isolated from tonsil (*top*) and spleen (*bottom*) were stained with the indicated mAb (*gray areas*) or with an unrelated (control) mAb (*dotted lines*) and analyzed by flow cytometry.

functional studies were carried out on resting cells to avoid interference by the previous signals received by the cells *in vivo*.

Unlike the observations in tonsils, MZ B cells were not found in the high-density cellular fractions of the spleen, although they were abundant in the less dense (50%) Percoll fractions, possibly indicating that MZ B cells are constantly activated in the spleen. Alternatively, the pathological conditions that led to splenectomy (thrombocytopenia and gastric cancer) in the two patients studied may have been responsible for the observed phenomenon. As is apparent from FIGURE 1, the phenotypic features of tonsillar and splenic MZ B cells are similar, with the remarkable exception that surface CD21 is high in the spleen and low in tonsils. This may reflect an intrinsic difference between the two cells, or it may be related to the different state of *in vivo* activation discussed above. Notably, studies in mice have also demonstrated high levels of surface CD21on splenic MZ B cells.[5,6] Another difference between tonsil and spleen relates to the high expression of IgG by the splenic MZ B cells. This feature is likely to be related to cell activation *in vivo* since it is observed also in the tonsillar MZ B cells from the same Percoll fraction. It is important to note, however, that multicolor analyses of these suspensions (not reported in detail) showed that the IgM-positive cells in the suspensions had a surface phenotype characterized by low or absent IgD and CD23.

CONSIDERATIONS ON THE PHENOTYPIC FEATURES OF MZ B CELLS AND B1 CELLS

TABLE 1 shows the phenotypic features of human MZ B cells in comparison with those reported for murine MZ B cells. Apart from the already-mentioned differences in CD21 expression, their phenotypes are remarkably similar. The same table also shows a comparison between the features of the murine and human B1 cells. The definition of the B1 cell is provided by the constitutive expression of surface CD5; consequently, human B1 cells have the phenotype of FM B cells, which, as discussed earlier, represent the most numerous population of B cells expressing the CD5 marker. The phenotype of murine B1 cells is deduced from literature data and refers to peritoneal cavity B cells that express CD5.[2] Clearly, the human and murine cells differ phenotypically, and the murine B1 cells closely resemble the MZ cells of both species.

FUNCTIONAL PROPERTIES OF MZ B CELLS

TABLE 2 reports the major functional features of human tonsillar MZ B cells in comparison with those of the corresponding murine B cells. Distinctive features of MZ B cells from the two species appear to be the capacity of responding to TI-2 antigens and of presenting antigens to T cells very efficiently.[3,4,7] Despite some controversies existing in the mouse field concerning the response to polyclonal B cell activators, the properties of human and murine B cells appear to be remarkably sim-

TABLE 2. Summary of the functional properties of MZ and B1 B cells

	Human		Mouse	
	MZ	B1	MZ	B1
Response to:				
IgM cross-linking	–	++	–	–
Polyclonal B cell activators	–	++	++ or –	+
TI-1 antigens	–	–	+	+
TI-2 antigens	+++	–	+++	++
T cell–dependent antigens	++	+	++	±?
Antigen presentation	++	±	++	+++
Propensity to apoptosis	–	–	–	–

ilar. These features differ considerably from those of human FM B cells that, for the aforementioned reasons, should be considered as B1 cells. In contrast, when the functional features of MZ B and B1 cells have been compared in the mouse, more similarities than differences have been found. These considerations reiterate the concept that a large fraction of CD5-expressing cells in human lymphoid organs do not correspond to the CD5-bearing cells from the peritoneal cavity, and they may not correspond to those from the spleen of mice, either.

HETEROGENEITY OF TONSILLAR MZ B CELLS

Staining of tonsillar MZ B cells for surface IgM demonstrated that virtually all of the cells displayed a uniformly bright fluorescence; in contrast, staining for surface IgD showed a bimodal distribution, with the majority of the cells negative and a small but definite fraction displaying a weak fluorescence.[8] Sorting and re-staining of these cells unequivocally demonstrated that they were IgM-IgD-low, and were thus clearly different from the majority of MZ B cells that were IgM-only. The existence of a subset of MZ B cells characterized by weak expression of surface IgD was confirmed by *in situ* staining experiments, which demonstrated that a fraction of the B cells in the subepithelial region expressed IgD.[9] Analyses of the properties of the two subpopulation of MZ B cells (that is, IgM-only and IgM-IgD-low) demonstrated the following:

1. IgM-only MZ B cells utilized mainly unmutated V_H gene segments. This indicates that these cells are either virgin B cells or B cells that have probably expanded following T cell-independent antigenic stimulation. The latter possibility seems to be supported by the finding of multiple cells expressing the same V_H gene segment and the same CDR3, an observation that strongly suggests *in situ* expansion of a few selected clonal populations of B cells.

2. IgM-IgD-low MZ B cells utilized both unmutated and mutated V_H gene segments, indicating that they consisted of virgin and memory B cells, although the latter appeared to be more abundant. This concept was reinforced by the observation that a large fraction of these cells expressed surface CD27, a memory B cell marker. Moreover, enrichment of CD27-positive cells in the suspensions resulted in a concomitant increase in cells that utilized mutated V_H gene segments. Taking all the above into consideration, it is possible that a fair proportion of IgM-IgD-low B cells are cells that, following stimulation/selection by antigen, have exited from the germinal centers and have seeded into the MZ. Alternatively, these cells may have undergone V_H gene mutations within the MZ, outside the germinal centers, possibly following T cell-independent antigenic stimulation. This apparently unorthodox explanation seems supported by observations in patients with the X-linked hyper-IgM syndrome; in these patients, IgM-IgD-low B cells are seen in the absence of germinal centres.[10]

3. Functional *in vitro* studies have led to the striking observation that the responses to TI-2 antigens were mediated primarily by IgM-IgD-low B cells (FIG. 2). These cells also appeared capable of a weak but definite response to polyclonal B cell activators.

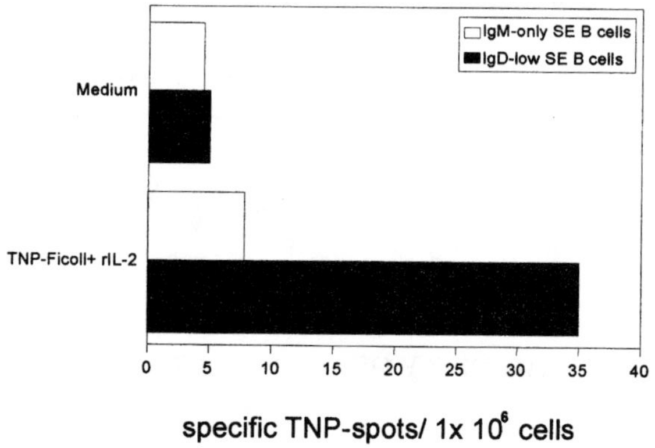

FIGURE 2. Production of anti-TNP antibodies in response to 2,4,6-teiniteophenyl (TNP)-Ficoll by tonsillar MZ B cell subsets. MZ B cells were purified from the high-density (60%) Percoll fraction and further separated into IgD-positive (IgD-low) and IgD-negative (IgM-only) cells by magnetic separation. One million cells from the two purified subepithelial (SE) B cell populations were cultured with TNP-Ficoll (1 μg/mL) in the presence of rIL-2 (50 U/mL). Cells were harvested after 6 days and subjected to an ELI-spot assay so that cells producing anti-TNP antibody could be detected.

IDENTIFICATION OF A HUMAN B CELL SUBSET SHARING SOME FEATURES WITH MURINE B1 CELLS

As alluded to before, the CD5-positive cells encountered so far in our studies on B cell subsets had the phenotype of FM B cells and were considerably different from the murine B1 cells that share many phenotypic and functional features with MZ B cells. In an attempt to find cells that resembled murine B1 cells, we analyzed the CD5-positive tonsillar B cells found in the 50% Percoll gradient. The majority of these cells were activated *in vivo*, as demonstrated by their buoyant properties as well as by the surface expression of CD69. Staining for CD38 revealed a bimodal distribution of the cells, possibly related to a different state of maturation/activation. When the CD38-positive cells were separated and further analyzed for co-expression of CD5 and other surface markers, it was found that the majority of the CD5-positive cells had the phenotype of FM B cells (data not shown). In contrast, when the CD38-negative cell fraction was analyzed using the same approach, it was found that a variable (but substantial) proportion of the CD5-positive cells had a phenotype resembling that of a subset of the human MZ B cells described above, that is, IgM-IgD-low. FIGURE 3 reports typical flow-cytometry profiles that illustrate the phenotypic properties of these cells. Analyses of the sequences of the V_H3 genes utilized by the CD5-positive cells present in the CD38-negative B cell fraction revealed the presence of both mutated and unmutated sequences (data not shown). The proportion of these two types of sequences varied in the different tonsils and correlated broadly with the proportion of CD5-CD27-positive cells that were consistently present in the suspensions (see FIG. 3). Notably, when the CD5-CD27-positive B cells were purified and analyzed for the V_H3 gene sequence, it was found that there was a consid-

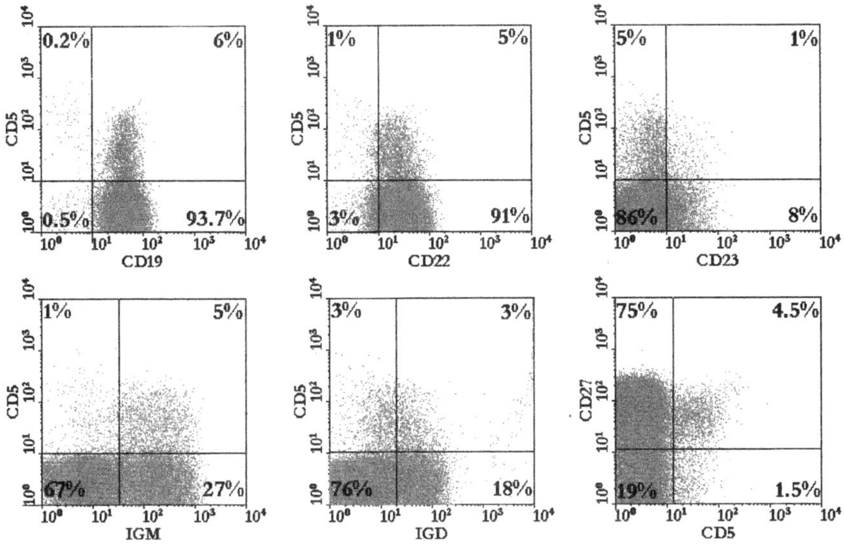

FIGURE 3. Phenotypic profiles of tonsillar CD38-negative B cells purified from tonsils. CD38-negative B cells were isolated from the 50% Percoll fractions by negative selection, double-stained with anti-CD5 mAb in combination with the indicated mAbs, and analyzed by flow cytometry. Indicated are the percentages of cells in each of the four quadrants.

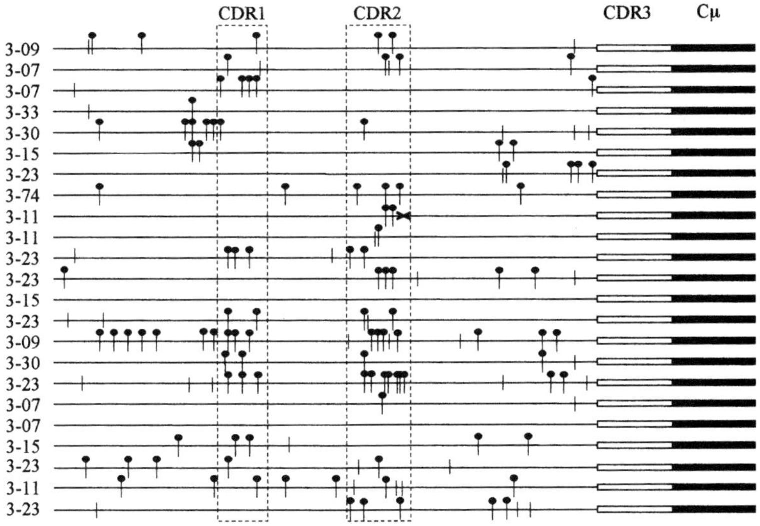

FIGURE 4. V_H 3-μ sequences from purified CD5-positive, CD27-positive, CD38-negative tonsillar B cells. CD38-negative B cells isolated from the 50% Percoll fractions were double-stained with anti-CD5 CyChrome-conjugated and anti-CD27 PE-conjugated mAbs and analyzed by flow cytometry. CD5-CD27-positive cells were then sorted, and V_H 3-μ transcripts were amplified and analyzed for the presence of mutations. Replacement mutations (*closed circle with stem*) and silent mutations (*stem*) are indicated. Each line represents one V_H3 sequence, and the corresponding germ line gene segment is reported on the right.

erable enrichment in mutated V_H gene sequences compared to the unfractionated CD5-positive, CD38-negative B cells that contained both CD27-positive and CD27-negative cells (FIG. 4). Taken together, these data suggest strong similarities between these newly identified cells and IgM-IgD-low MZ B cells. In addition, these B cells share a number of phenotypic features with the murine B1 cells (TABLE 1).

In a subsequent series of experiments, we searched for peripheral blood cells that shared similar features with MZ B cells and expressed CD5. Indeed, these cells were detected within the CD38-negative B cells and, like the cells isolated from the tonsillar B cells suspensions, had a CD23-low, CD22-positive, IgM-IgD-low surface phenotype (data not shown). Like the tonsillar B cells, the peripheral blood cells consisted of cells that expressed mutated and unmutated V_H genes, a property that closely correlated with the cell capacity of expressing surface CD27. These data therefore demonstrate that these CD5-positive cells have the capacity of recirculating and represent a cell subpopulation possibly present in all of the lymphoid organs. IgM-positive, IgD-positive, CD27-positive B cells without surface CD5 have been also described by Küppers *et al.*[11] Notably, although B cells with this phenotype were found to be present in the peripheral blood of a patient with activation-induced deaminase (AID) deficiency, they did not display point mutations in their V_H genes (data not shown).

REFERENCES

1. RAJEWSKY, K. 1996. Clonal selection and learning in the antibody system. Nature **381:** 751–758.
2. HARDY, R.R. & K. HAYAKAWA. 2001. B cell development pathways. Annu. Rev. Immunol. **19:** 595–561.
3. DONO, M. *et al.* 1996. Subepithelial B cells in the human palatine tonsil. I. Morphologic, cytochemical and phenotypic characterization. Eur. J. Immunol. **26:** 2035–2042.
4. DONO, M. *et al.* 1996. Subepithelial B cells in the human palatine tonsil. II. Functional characterization. Eur. J. Immunol. **26:** 2043–2050.
5. GUINAMARD, R. *et al.* 2000. Absence of marginal zone B cells in Pyk-2-deficient mice define their role in the humoral response. Nature Immunol. **1:** 31–39.
6. MARTIN, F., A.M. OLIVER & J.F. KEARNEY. 2001. Marginal zone and B1 B cells unite in the early response against T-independent blood-borne particulate antigens. Immunity **14:** 617–629.
7. MARTIN, F. & J.F. KEARNEY. 2001. B1 cells: similarities and differences with other B cell subsets. Curr. Opinion Immunol. **13:** 195–201.
8. DONO, M. *et al.* 2000. Heterogeneity of tonsilar subepithelial B lymphocytes, the splenic marginal zone equivalents. J. Immunol. **164:** 5596–5604.
9. WELLER, S. *et al.* 2001. CD40-CD40L-independent Ig gene hypermutation suggests a second B cell diversification pathway in humans. Proc. Natl. Acad. Sci. USA **98:** 1166–1170.
10. DONO, M. *et al.* 1997. Phenotypic and functional characterization of human tonsillar subepithelial (SE) B cells. Ann. N.Y. Acad. Sci. **815:** 171–178.
11. KLEIN, U., K. RAJEWSKY & R. KÜPPERS. 1998. Human immunoglobulin (Ig)M+IgD+ peripheral blood B cells expressing the CD27 cell surface antigen carry somatically mutated variable region genes: CD27 as a general marker for somatically mutated (memory) B cells. J. Exp. Med. **188:** 1679–1689.

B Cell Immunity Regulated by the Protein Kinase C Family

KAORU SAIJO,[a] INGRID MECKLENBRÄUKER,[a] CHRISTIAN SCHMEDT, AND ALEXANDER TARAKHOVSKY

Laboratory of Lymphocyte Signaling, The Rockefeller University, New York, New York 10021, USA

ABSTRACT: Protein kinase C (PKC) is a family of serine/threonine kinases which mediate essential cellular signals required for activation, proliferation, differentiation, and survival. Several PKC members are expressed in B lineage cells and activated by stimulation of the B cell receptor (BCR), thus suggesting a contribution of PKCs to the B cell–mediated immune response. To understand the individual roles of PKCs for B cell immunity, mice deficient for PKC-βI/II (PKCβ) or PKCδ were analyzed. PKCβ and PKCδ play essential but distinctive roles in B cell immunity. In addition to its role in B cell activation and humoral immunity, PKCβ was recently shown to control NF-κB activation and survival of mature B cells. PKCδ on the other hand specifically regulates the induction of tolerance in self-reactive B cells. Thus, individual PCKs regulate B cell immunity specifically.

KEYWORDS: protein kinase C; cell signaling; B cell immunity

INTRODUCTION

The family of PKC is composed of 9 isoforms, which share a highly homologous kinase domain and some regulatory domains, including C1-, C2-, and pseudosubstrate domains.[1] The C1 domain is important for the binding to phorbol ester and the C2 domain is responsible for the binding to calcium and phospholipids. All PKC isoforms contain a pseudosubstrate domain, which binds to the substrate binding site inside the kinase domain and prevents the activation of kinase in the absence of cofactors such as Ca^{2+} and diacylglycerol (DAG). The PKC family can be subdivided into three groups based on their structure and requirement of activation (FIG. 1). Conventional PKCs (cPKC) represented by PKCα, PKCβI/II and PKCγ are dependent on Ca^{2+} and diacylglycerol (DAG) for their activation. Indeed, the C2 domain is found only in cPKC but not in other subfamilies. PKCβI and PKCβII are produced by alternative splicing, resulting in different C-termini of the protein.[2] Novel PKCs (nPKC), which include PKCδ, PKCε, PKCη and PKCθ, require only DAG but not Ca^{2+} to be activated, and atypical PKCs (aPKC) PKCζ and PKCλ/ι are entirely independent of

[a]K.S. and I.M. contributed equally to this work.

Address for correspondence: Dr. Alexander Tarakhovsky, Laboratory of Lymphocyte Signaling, Rockefeller University, Box 301, 1230 York Avenue, New York, NY 10021. Voice: 212-327-8256; fax: 212-327-8258.

tarakho@mail.rockefeller.edu

Ann. N.Y. Acad. Sci. 987: 125–134 (2003).

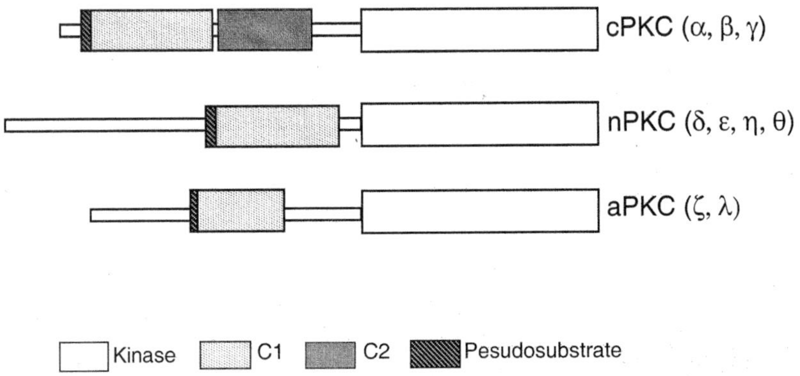

FIGURE 1. Domain structure of PKC. A comparison of domain architecture of PKC subgroup is shown. Each rectangle indicates the different domain.

DAG and Ca^{2+} for their activation. aPKCs also contain a C1 domain, although it is shorter than the C1 domain found in other PKC subgroups. In B cells, crosslinking of BCR induces the translocation and activation of PKC from the cytoplasm to the plasma membrane. PKC activation eventually leads the induction of various transcriptional factors in B cell.[3] In B-lineage cells, PKCα, βI/II, δ, ε, η, ζ and λ isoforms are expressed.[4]

Despite broad comprehension of the enzymatic properties, little is known about physiological and specific roles of individual PKC isoforms in B cell–mediated immune responses. To understand the contribution of individual PKC isoforms in B cell immunity, PKCβI/II- and PKCδ-deficient mice were generated by the target disruption of the corresponding genes in ES cells. Here we discuss the essential and distinctive roles of two isoforms of PKC family, PKCβI/II and PKCδ in B cell immunity by the analysis of mice deficient for these enzymes.

PKCβ CONTROLS B CELL SURVIVAL BY ACTIVATION OF NF-κB

The defective immunity of PKCβ-deficient mice is characterized by impaired B cell activation and ineffective T-independent immune response.[5] These defects are similar to those seen in mice deficient for Bruton's tyrosine kinase (Btk) or X-linked immunodeficient mice (Xid) which are carrying a point mutation in the pleckstrin-homology (PH) domain of Btk. B cells of either Btk-deficient mice or Xid mice are also characterized by a drastic reduction of the life span *in vivo* and *in vitro.*[6] Similar defects are observed in PKCβ-deficient B cells. *Ex vivo* isolated PKCβ-deficient B cells die rapidly if no exogenous stimulus like IL-4 is added to promote survival (FIG. 2A). PKCβ-deficient B cells also show poor proliferative responses to antibody-mediated BCR stimulation. This impairment is also rescued by the provision of additional survival signals (i.e., IL-4).[7] The poor survival of PKCβ-deficient B cells is associated with the inability to upregulate the essential survival factor Bcl-xL and Bcl-2 upon BCR crosslinking (FIG. 2B). Because the expression of Bcl-xL

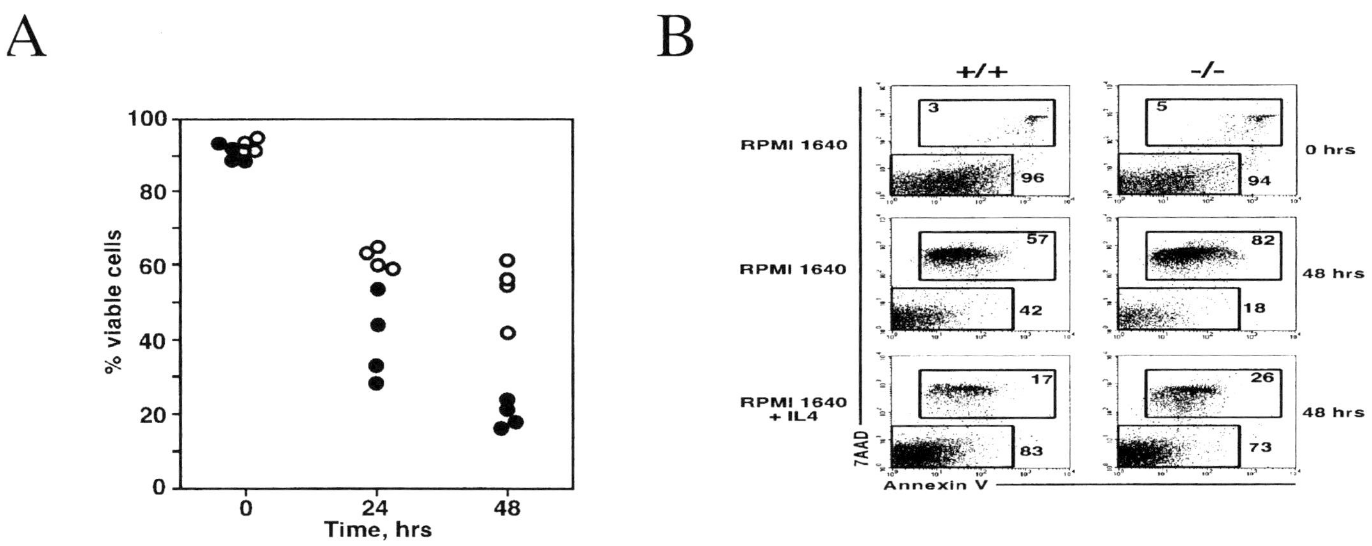

FIGURE 2. Impaired survival of PKCβ-deficient B cells. (**A**) Accelerated cell death of B cell from PKCβ$^{-/-}$ mice cultured with medium alone. Percentage of cell viability of cultured purified splenic B cells is plotted. *Solid* and *open circles* correspond to PKCβ$^{-/-}$ B cell and wild-type controls, respectively. (**B**) The addition of IL-4 promotes the cell survival of both PKCβ$^{-/-}$ B cells (–/–) and wild-type control (+/+). Cell death is determined by FACS analysis of Annexin-V and 7AAD expression level after the culturing of the cells in RPMI medium with or without IL-4. Numbers indicate percentage of gated cells.

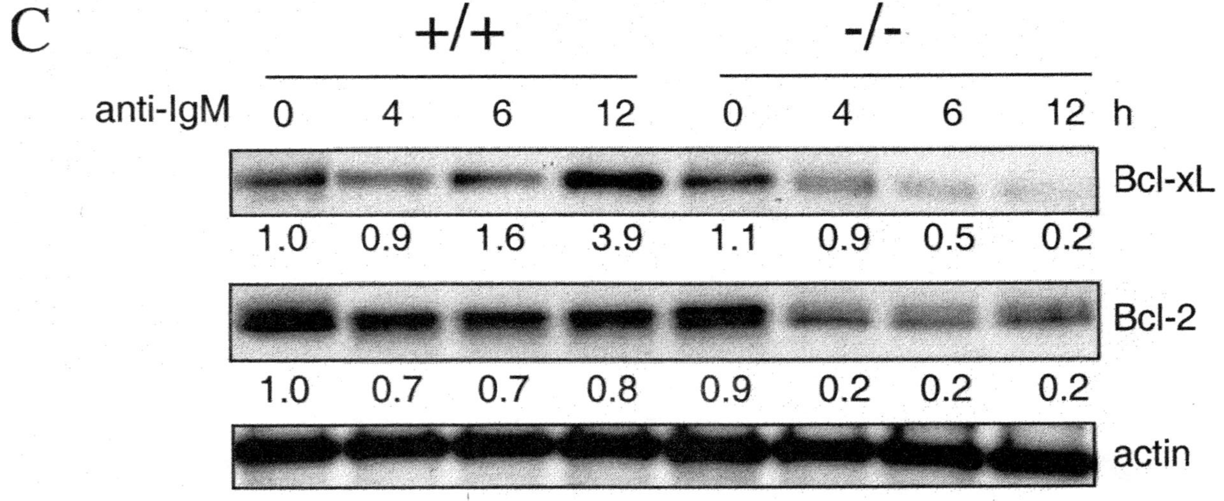

FIGURE 2. **(C)** Reduced expression of Bcl-xL and Bcl-2 protein after IgM-mediated stimulation of PKCβ$^{-/-}$ B cells. Purified splenic B cells from PKCβ$^{-/-}$ mice (–/–) and wild-type control (+/+) are incubated with 10 mg/ml F(ab′)$_2$ fragment of anti-IgM antibody for indicated time (h). Bcl-xL and Bcl-2 protein expression level are examined by Western blotting. For quantification, band intensity is first normalized to respective actin signal and then calculated as fold-change relative to unstimulated PKCβ$^{+/+}$, which is set to 1.0.

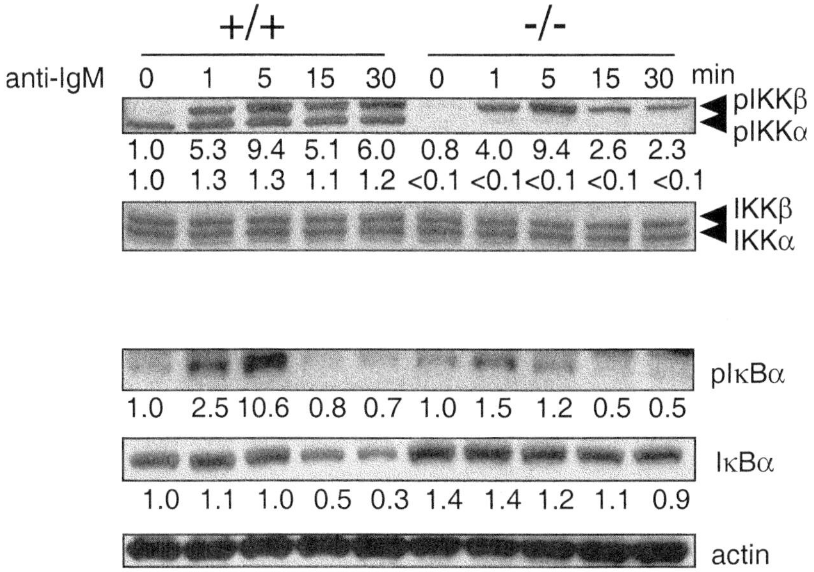

FIGURE 3. Impaired activation of NF-κB in PKC$\beta^{-/-}$ B cells. Absence of IKKα activation and shortening of the duration of IKKβ activation in PKC$\beta^{-/-}$ B cells (–/–) is observed (*top panel*) when purified splenic B cells from PKC$\beta^{-/-}$ mice (–/–) and wild-type control mice (+/+) are stimulated with 20 mg/ml of F(ab′)$_2$ fragment of anti-IgM antibody for indicated time (min). Activation of IKKα and IKKβ is determined by Western blotting with antibody specific for phospho-Ser180 and phospho-Ser181 in IKKα and IKKβ, respectively. The membrane is stripped and equal protein loading is examined by Western blotting with antibodies specific for nonphosphorylated IKKα and IKKβ. IkBα phosphorylation at Ser-32 and expression level after anti-IgM-mediated stimulation are analyzed (*bottom panel*). Equal protein loading is determined by the western blotting with anti-actin antibody. Numbers indicate the fold-change as described in FIGURE 2.

in B cells is dependent on the activation of transcription factor NF-κB, the involvement of PKCβ in NF-κB activation upon BCR stimulation is suggested.

Mechanisms of NF-κB activation have been reviewed in depth elsewhere.[8] In brief: successful NF-κB activation and transcription of target genes requires the phosphorylation and degradation of IκB protein . The phsophorylation of IκB is mediated by the multi-component IκB kinase (IKK) complex containing two catalytic subunits, IKKα and IKKβ, and one regulatory subunit IKKγ. NF-κB is rapidly induced upon BCR stimulation, but no detailed understanding exists about the signaling pathway, particularly upstream of the IKK signaling complex. The involvement of Btk in this signaling has been suggested,[9,10] but the substrate specificity of Btk for tyrosines precludes a direct connection between Btk and the IKK complex. Recently, a functional link between Btk and the classical PKCβI/II was demonstrated.[11] Therefore, we hypothesized that PKCβ may be involved in Btk-mediated NF-κB activation upon BCR crosslinking.

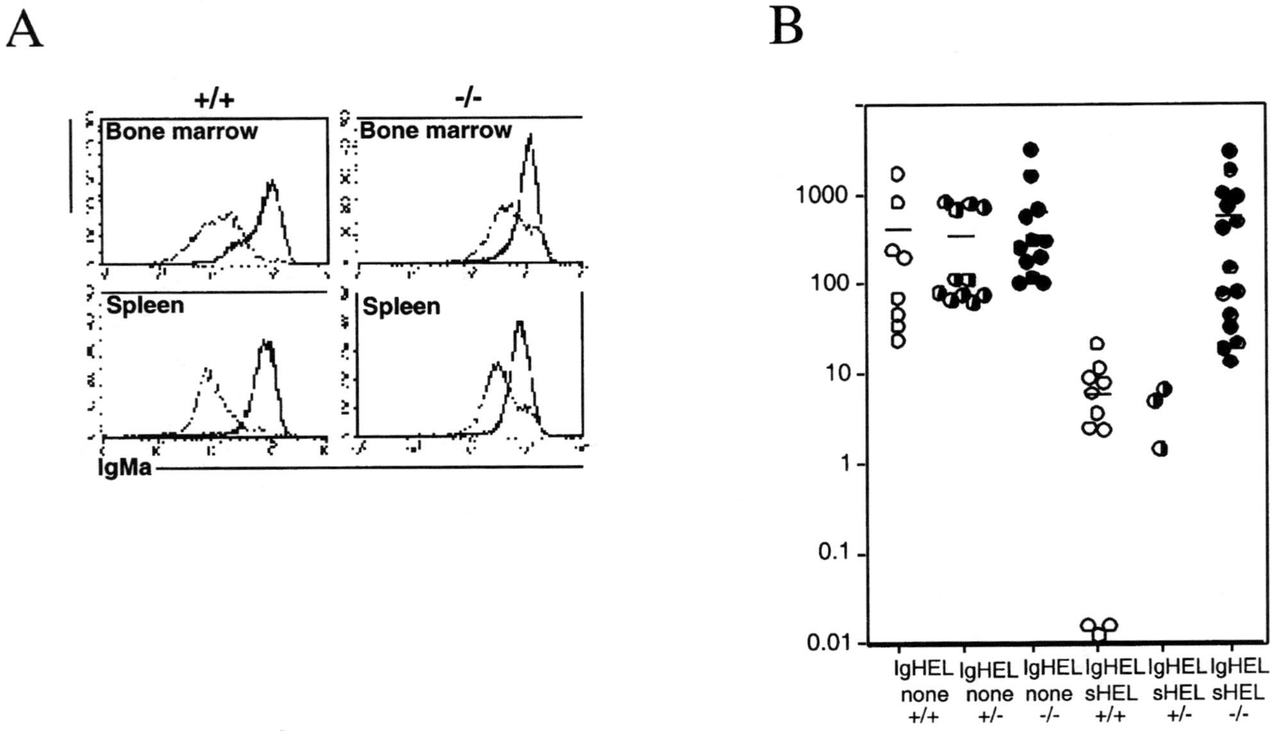

FIGURE 4. Defective tolerance induction in the absence of PKCδ. (**A**) Histogram shows expression level of surface IgM on bone marrow B cells (*upper panel*) or splenic B cells (*lower panel*) of PKCδ$^{+/+}$ mice (*left column*) or PKCδ$^{-/-}$ (*right column*) IgHEL (*thick line*) or IgHEL;sHEL (*thin line*) mice. (**B**) Anti-HEL antibody titer in the serum of PKCδ$^{+/+}$ (*open circle*), PKCδ$^{+/-}$ (*half-filled circle*), and PKCδ$^{-/-}$ (*solid circle*) mice carrying HEL alone or in combination with sHEL are shown. *Horizontal bar* indicates mean value.

Rapid and sustained BCR crosslinking-induced activation of IKKβ was observed while the activation of IKKα seemed stimulation-independent and constitutive in wild-type splenic B cells (FIG. 3). In PKCβ-deficient B cells, IKKβ activation upon BCR crosslinking was observed, but was not sustained, unlike the case in wild-type controls. Surprisingly, in PKCβ-deficient B cell, the activation of IKKα with or without BCR stimulation is effectively diminished (FIG. 3). The phosphorylation of IκBα is decreased and little degradation is observed. As a consequence, nuclear translocation and DNA-binding activity of NF-κB is reduced in PKCβ-deficient B cells (data not shown). When PKCβ-deficient B cells were crosslinked on CD40, these B cells show NF-κB activation similar to wild-type B cells. These data suggest that PKCβ is involved in BCR-mediated NF-κB activation, particularly via controlling IKKα. Although, it is not clear that IKKα is a direct substrate of PKCβ, the amino acid sequence analysis shows a potential serine phosphorylation site by PKCβ in IKKα, but not in IKKβ. Collectively, PKCβ is important for NF-κB activation upon BCR stimulation by controlling IKKα activity.

THE ROLE OF PKCδ IN TOLERANCE INDUCTION IN B CELLS

Upon BCR crosslinking, PKCδ is rapidly phosphorylated on tyrosine and serine/threonine residues (data not shown). These data suggest that PKCδ is involved in BCR-mediated signaling. To understand the physiological role of PKCδ *in vivo*, mice deficient for this enzyme were generated and their immune system analyzed.[12]

Although the expression of PKCδ protein is higher in bone marrow immature B cells, no noteworthy alteration in B cell development in bone marrow is observed in PKCδ-deficient mice (data not shown). Interestingly PKCδ-deficient mice show significant splenomegaly and lymphadenopathy because of increased the numbers of B cells. Mice die prematurely by severe autoimmune disease, which is characterized by the detection of autoreactive antibodies such as antinuclear antibody or anti-DNA antibody. These findings suggest autoimmune disease in PKCδ-deficient mice.

Interaction of a B cells expressing self-reactive BCR to autoantigen normally induce either clonal deletion or functional inactivation (anergy) *in vivo*.[13] Both of these processes lead B cells to tolerance, which is essential for the prevention of autoimmune disease. Whereas clonal deletion results the death of developing autoreactive B cell, anergy leads to a complex phenotypical change in autoreactive B cells. To determine whether PKCδ is involved in negative selection of autoreactive B cells, we employed the well-established immunoglobulin-transgenic system harboring hen egg lysozyme-specific BCR (IgHEL). Mice in which the IgHEL transgene was bred together with a transgene encoding either a membrane-bound form (mHEL) or a soluble form of HEL (sHEL) show clonal deletion or anergy of autoreactive B cells, respectively.[14]

Encounter of B cells from PKCδ$^{+/+}$IgHEL with sHEL in bone marrow leads the 7–11-fold reduction of surface expression of HEL-specific receptor. In contrast, interaction of B cells from PKCδ$^{-/-}$IgHEL with sHEL in bone marrow as well as spleen results in only 2-fold reduction of the surface expression of HEL-specific receptor (FIG. 4A). These data suggests that PKCδ-deficient B cells might be refractory to tolerance-inducing signals generated in B cells exposed to self-antigen.

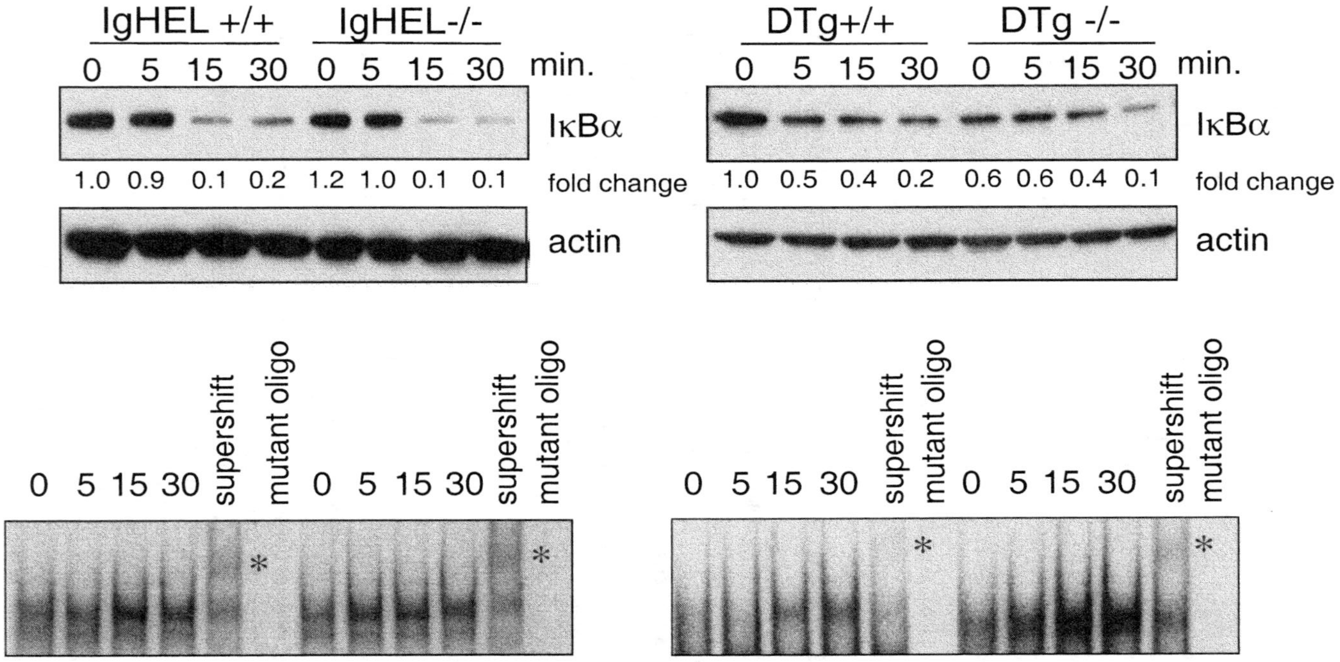

FIGURE 5. BCR-mediated NF-κB activation in B cells from PKCδ$^{-/-}$ mice. Purified splenic B cells from IgHEL PKCδ$^{+/+}$ (IgHEL$^{+/+}$), IgHEL PKCδ$^{-/-}$ (IgHEL$^{-/-}$), IgHEL;sHEL PKCδ$^{+/+}$ (DTγ$^{+/+}$) and IgHEL;sHEL PKCδ$^{-/-}$ (DTγ$^{-/-}$) mice are stimulated as described in FIGURE 3. IkBα degradation is determined by Western blotting with anti-IkBα antibody (*upper panel*). Equal protein loading is tested by the Western blotting with anti-actin antibody. Nuclear translocation and DNA-binding activity is examined by EMSA (*bottom panel*).

Chronic exposure of PKCδ$^{+/+}$ or PKCδ$^{+/-}$ IgHEL B cells to sHEL leads to the reduction of peripheral B cells compared to IgHEL single transgenic mice. In the absence of PKCδ, chronic exposure to self-antigen does not induce the reduction of HEL-specific B cells. IgHEL B cells remain in peripheral lymphoid organs of IgHEL; sHEL mice at numbers equal to those observed in single IgHEL transgenic mice. Defective tolerance induction in the absence of PKCδ is further supported by 80-fold increase of serum titer of HEL-specific antibody (FIG. 4B). These data indicate that in the absence of PKCδ self-reactive B cells do not induce tolerance, but rather grow, accumulate, and terminally differentiate.

The defective tolerance induction is not because of general defect of BCR-mediated signaling in the absence of PKCδ, because B cells from PKCδ$^{-/-}$ in Non-Tg background or IgHEL normally respond to the antibody-mediated BCR stimulation *in vitro*. PKCδ$^{-/-}$ B cells normally respond to the antigen stimulation *in vivo*, as measured by immunization with T-independent or T-dependent antigen (data not shown). Despite unaltered normal response of antigen receptor, PKCδ$^{-/-}$ B cells are unable to induce the tolerance to chronic exposure to sHEL.

The signaling properties of tolerant B cells show characteristic features. While naïve B cells activate ERK, NFAT, JNK, and NF-κB signaling pathways, tolerant B cell are unable to activate JNK and NF-κB signaling.[13] When *ex vivo* isolated B cells from PKCδ$^{-/-}$IgHEL; sHEL mice were stimulated by HEL *in vitro*, these B cells activated NF-κB and JNK, while B cells from PKCδ$^{+/+}$IgHEL; sHEL mice showed decreased activation of JNK and NF-κB. B cells from PKCδ$^{-/-}$ IgHEL; sHEL mice show increased nuclear translocation and DNA-binding activity upon BCR stimulation compared to B cells from PKCδ$^{-/-}$ IgHEL or non-transgenic mice (FIG. 5). Moreover, without stimulation, B cells from PKCδ$^{-/-}$ IgHEL;sHEL mice showed a substantial amount of NF-κB activation, as determined by EMSA. Interestingly the degradation of IκBα is not efficient compare to that in B cells from PKCδ$^{-/-}$IgHEL mice. Comparison of B cells from PKCδ$^{-/-}$ IgHEL and IgHEL; sHEL suggest the difference in the mechanism of NF-κB activation. Although the primary cause for efficient NF-κB activation is unclear, it may be related to the altered pattern of BCR-induced tyrosine phosphorylation (data not shown). The observed signaling changes yield a new signaling pattern characteristic for PKCδ$^{-/-}$ IgHEL cells that develop during the chronic exposure to sHEL self-antigen. Alteration of nuclear signaling in tolerant B cells is a possible reason to cause global change in gene expression.

Overall, our findings indicate that PKCδ is indispensable component of a signaling pathway specific for the induction of tolerance in B cells.

CONCLUSION

Our data demonstrate that PKCβ and PKCδ regulate B cell immunity in a specific and characteristic manner. However, neither PKC isoform plays an essential role in early B cell development. This lack of changes to B cell precursors in the bone marrow may suggest that PKCs have redundant functions in early B cell development. Understanding the role of PKC in early B cell development may require analyzing mice with multiple PKC-deficiencies.

REFERENCES

1. MELLOR, H. & P.J. PARKER. 1998. The extended protein kinase C superfamily. Biochem. J. **332:** 281–292.
2. COUSSENS, L., L. RHEE, P.J. PARKER & A. ULLRICH. 1987. Alternative splicing increases the diversity of the human protein kinase C family. DNA **6:** 389–394.
3. SEYFERT, V.L., S. MCMAHON, W. GLENN, *et al.* 1990. Egr-1 expression in surface Ig-mediated B cell activation: kinetics and association with protein kinase C activation. J. Immunol. **145:** 3647–3653.
4. MISCHAK, H., W. KOLCH, J. GOODNIGHT, *et al.* 1991. Expression of protein kinase C genes in hemopoietic cells is cell-type- and B cell-differentiation stage specific. J. Immunol. **147:** 3981–3987.
5. LEITGES, M., C. SCHMEDT, R. GUINAMARD, *et al.* 1996. Immunodeficiency in protein kinase cb-deficient mice. Science **273:** 788–791.
6. KHAN, W.N. 2001. Regulation of B lymphocyte development and activation by Bruton's tyrosine kinase. Immunol. Res. **23:** 147–156.
7. MORI, M., S.C. MORRIS, T. OREKHOVA, *et al.* 2000. IL-4 promotes the migration of circulating B cells to the spleen and increases splenic B cell survival. J. Immunol. **164:** 5704–5712.
8. KARIN, M. & Y. BEN-NERIAH. 2000. Phosphorylation meets ubiquitination: the control of NF-κB activity. Annu. Rev. Immunol. **18:** 621–663.
9. PETRO, J.B., S.M. RAHMAN, D.W. BALLARD & W.N. KHAN. 2000. Bruton's tyrosine kinase is required for activation of IκB kinase and nuclear factor κB in response to B cell receptor engagement. J. Exp. Med. **191:** 1745–1754.
10. BAJPAI, U.D., K. ZHANG, M. TEUTSCH, *et al.* 2000. Bruton's tyrosine kinase links the B cell receptor to nuclear factor κB activation. J. Exp. Med. **191:** 1735–1744.
11. KANG, S.W., M.I. WAHL, J. CHU, *et al.* 2001. PKCβ modulates antigen receptor signaling via regulation of Btk membrane localization. EMBO J. **20:** 5692–5702.
12. MECKLENBRAUKER, I., K. SAIJO, N.Y. ZHENG, *et al.* 2002. Protein kinase Cδ controls self-antigen-induced B-cell tolerance. Nature **416:** 860–865.
13. HEALY, J.I., & C.C. GOODNOW. 1998. Positive versus negative signaling by lymphocyte antigen receptors. Annu. Rev. Immunol. **16:** 645–670.
14. GOODNOW, C.C., J. CROSBIE, S. ADELSTEIN, *et al.* 1988. Altered immunoglobulin expression and functional silencing of self-reactive B lymphocytes in transgenic mice. Nature **334:** 676–682.

A Modified Digestion-Circularization PCR (DC-PCR) Approach to Detect Hypermutation-Associated DNA Double-Strand Breaks

SARAH K. DICKERSON AND F. NINA PAPAVASILIOU

Laboratory of Lymphocyte Biology, The Rockefeller University, New York, New York 10021, USA

ABSTRACT: Hypermutation of immunoglobulin variable region genes occurs in B cells during an immune response, and is vital for the development of high-affinity antibodies. The molecular mechanism of the hypermutation reaction is unknown, but it seems to correlate with the generation of locus-specific, double-strand breaks. These DNA breaks have been measured by ligation-mediated polymerase chain reaction (LM-PCR), a technique that relies upon the ligation of a small linker to DNA breaks followed by the specific amplification of such breaks using combinations of locus- and linker-specific primers. Here, we describe a modified version of the digestion-circularization PCR (DC-PCR) technique, which can be used to amplify and measure DNA breaks directly.

KEYWORDS: B lymphocyte; immunoglobulin gene; somatic hypermutation; DNA double-strand breaks

Support for the existence of DNA strand lesions in somatic hypermutation (SHM) comes from recent experiments describing the constitutively hypermutating Ramos cell line, and the ability of terminal deoxynucleotidyl transferase (TdT)-transfected Ramos cells to accumulate untemplated nucleotide additions near mutational hotspots within the Ig V region.[1] Since TdT can add untemplated nucleotides to either single- or double-strand breaks,[2] the nature of the breaks remained uncertain. To address this issue, we and others[3,4] have probed genomic DNA from Ramos cells[3] or from germinal center cells from immunized mice[4] for locus-specific, double-strand breaks (DSBs) by ligation-mediated polymerase chain reaction (LM-PCR) (FIG. 1A).[5] Blunt DSBs were readily detectable over the Ramos Ig heavy and light chain V regions (V_H and V_l) (FIG. 1B),[3] but could not be amplified from the heavy chain constant region or from a panel of other genomic loci.[3] In addition, these same breaks could not be amplified with reverse orientation primers (which would detect the "downstream" end of the break, defined as the end lying on the

Address for correspondence: F. Nina Papavasiliou, Laboratory of Lymphocyte Biology, The Rockefeller University, Box 39, 1230 York Avenue, New York, NY 10021.Voice: 212-327-7857; fax: 212-327-7319.

papavasiliou@rockefeller.edu

Ann. N.Y. Acad. Sci. 987: 135–139 (2003).

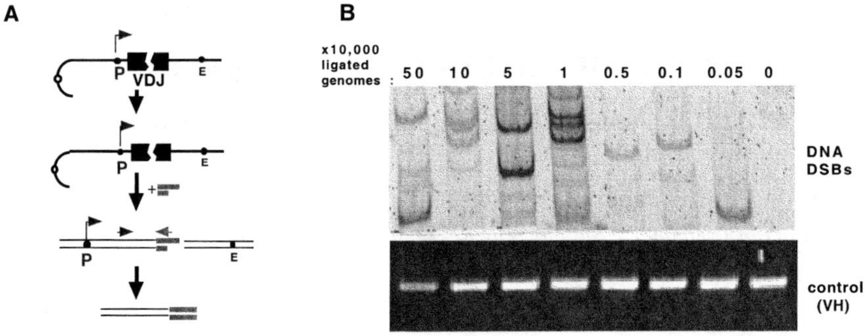

FIGURE 1. LM-PCR amplification of DSBs from the Ramos heavy-chain, variable-region gene. (**A**) Schematic of the principle behind LM-PCR: a double-stranded DNA linker is ligated to DNA ends, and the region proximal to the end is amplified through two rounds of PCR using locus-specific and linker-specific nested primers. (**B**) Dilution analysis for the quantitation of variable-region DSBs using the indicated cell equivalents of linker-ligated DNA (two rounds of PCR: 10 cycles in the first round, and 30 cycles in the second).[3] Amplification products representing DNA DSBs over the Ramos VH are run on a polyacrylamide gel that is directly stained with SybrGreen and visualized on a FluorImager. Generic PCR amplification of the entire VH (30 cycles) serves as a DNA loading control.

piece of DNA separated from the promoter[3]). Hence, DSB formation in somatic hypermutation results in asymmetric products: upstream ends that are usually blunt, and downstream ends that almost always contain a modification that prevents their detection.

An alternative explanation for the asymmetry would be that the process actually creates single-strand gaps.[6] If such gaps were largely confined to one strand of the DNA, and if the linker could ligate to the gaps, we would expect an asymmetric distribution of LM-PCR products, such as we observe. To determine the ligation efficiency of a double-stranded linker to a gap, we generated two substrates (one with a 19-nt gap and the other with a DSB) that could be detected by the same specific PCR primers. In competitive LM-PCR experiments, gaps were detected 100- to 1000-fold less efficiently than DSBs (FIG. 2), indicating that our LM-PCR assay would detect at most 1% of gaps. Since the assay detects a product from about 4–5% of Ig variable region alleles in Ramos cells, gaps would have to exist at a rate of at least 400 per 100 alleles to account for our data. This is improbable, but not impossible. To deal with this issue, we devised a modified digestion-circularization PCR (DC-PCR) assay to look at the same DSBs.

We reasoned that if the upstream ends of the hypermutation-associated DSBs were true blunt DSBs, one should be able to digest the DNA upstream of the V region with a restriction enzyme that produces blunt ends: subsequent ligation then would seal the two blunt ends and produce a circle (FIG. 3A). One can amplify the artificial "joint" by choosing appropriate oligonucleotide primers.

We digested 100 µL of agarose-embedded Ramos DNA with EcoRV, which recognizes a site 3 kb upstream from the Ramos V region (FIG. 3B). Then, 24 hours lat-

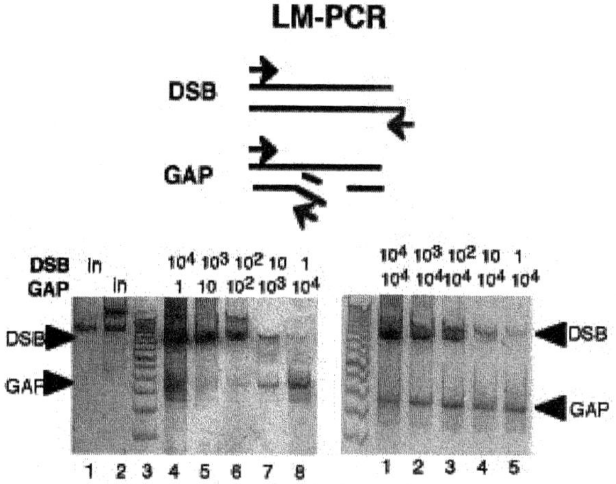

FIGURE 2. Comparison of the ligation efficiency of a double-stranded DNA linker to a gap vs. a DSB. We created a 550-nt DNA fragment with recognition sequences for N.BstNB1, a site-specific endonuclease that catalyzes a single-strand break 4 nts beyond the 3′ side of the recognition sequence, GAGTCNNNN/. N.BstNB1 cleaves twice in the substrate, and as the sites of cleavage are on the same strand and 19 nts apart, the reaction results in the formation of a 19-nt gap, 290 nts from the one end of the 550-nt double-stranded fragment. Formation of the gap can be monitored, as it is accompanied by the loss of a BsmF1 restriction enzyme site that lies within the 19-nt gapped region. In competitive LM-PCR experiments, gaps were detected with 100- (*left*) to 1000-fold (*right*) lower efficiency than DSBs (*left*: compare GAP and DSB signal in lanes 7,8; *right*: compare DSB and GAP signal in lanes 4 and 5). in: input DNA.

er, we diluted the digest to 600 μL and ligated the reaction overnight, followed by PCR amplification (~30 cycles) of the generated joints. We could easily amplify bands only from agarose-embedded Ramos DNA when we would digest with EcoRV prior to ligation. The appearance of the bands was dependent on ligation (FIG. 3A, lanes 2 and 3 vs. lanes 5 and 6), and the bands disappear if, in addition to EcoRV, the genomic DNA was digested with SnaB1, an enzyme that is predicted to cleave the DNA between the EcoRV site and the hypermutation-associated DSBs. Sequencing of the PCR products revealed that the location of these DSBs is not random, but like the LM-PCR amplified breaks, these DSBs also tended to co-localize with mutational hotspots (Ref. 3 and unpublished results).

Because of its dependence on a particular restriction site, this assay is less promiscuous in DSB detection when compared with LM-PCR. In addition, because it does not depend on linker ligation, it only detects true blunt DSBs (and not gaps). We could not amplify specific joints between the downstream end and the end generated by blunt-cutting enzymes located downstream of the V region DSB (FIG. 3B). These data strongly suggest that the SHM-associated lesions we detect are not nicks or gaps but true blunt DSBs, which are asymmetrically detectable. It is possible that a subset of DSBs (perhaps consisting of the few that are detected from both directions) is related to normal breakage and repair processes active in rapidly replicating cells and, as such, are unrelated to SHM. But given the frequency and the location

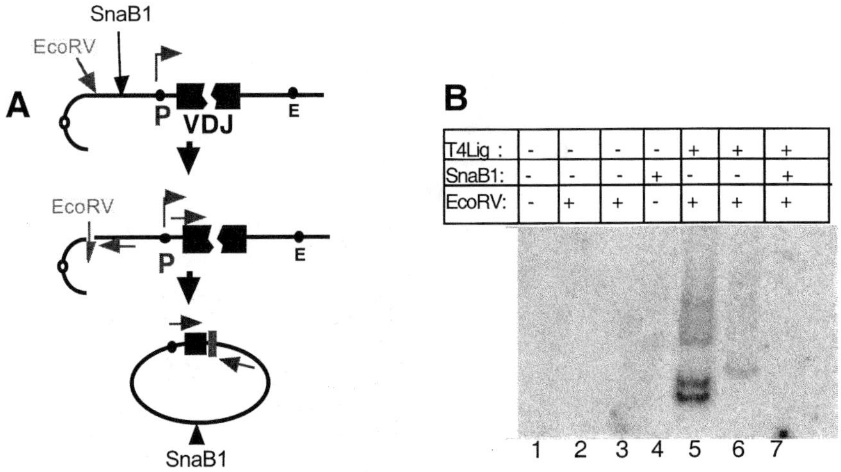

FIGURE 3. DC-PCR amplification of DSBs from the Ramos heavy-chain, variable-region gene. (**A**) Schematic representation of the modified DC-PCR assay. The locus is cut with a locus-specific, blunt restriction enzyme (in this case, EcoRV). In addition, the locus carries a mutation-generated DSB, the upstream end of which is blunt. Self-ligation of the region will lead to the circularization of the EcoRV-generated ends and of the hypermutation-generated ends. Appropriately chosen primers can then amplify the artificial "joint." (**B**) Amplification of variable-region DSBs from 50,000 cell equivalents (*lane 5*) and from 5,000 cell equivalents (*lane 6*) (single round of PCR, 30 cycles). Amplification was with appropriate primers, and products representing DNA DSBs over the Ramos VH are run on a polyacrylamide gel that is directly stained with SybrGreen and visualized on a FluorImager.

of the DSBs over the mutating V regions, as well as the absence of such breaks over other genomic loci, we conclude that in Ramos cells, DSBs over the variable regions are tightly linked to somatic hypermutation.

Thus, there is a body of data which suggests that SHM involves the generation and processing of DNA DSBs. But the identity of the putative nuclease(s), as well as the manner by which it is specifically targeted to immunoglobulin loci, are crucial issues yet to be resolved. Clearly, our understanding of the mechanism of SHM remains rudimentary, but significant advances may come quickly through a better definition of the processes that target and generate the relevant DNA lesions.

REFERENCES

1. SALE, J.E. & M.S. NEUBERGER. 1998. TdT-accessible breaks are scattered over the immunoglobulin V domain in a constitutively hypermutating B cell line. Immunity **9:** 859–869.
2. GRAWUNDER, U. & M. LIEBER. 1997. A complex of RAG-1 and RAG-2 proteins persists on DNA after single-strand cleavage at V(D)J recombination signal sequences. Nucleic Acids Res. **25:** 1375–1382.
3. PAPAVASILIOU, F.N. & D.G. SCHATZ. 2000. Cell-cycle-regulated DNA double-stranded breaks in somatic hypermutation of immunoglobulin genes. Nature **408:** 216–221.

4. BROSS, L. *et al.* 2000. DNA double-strand breaks in immunoglobulin genes undergoing somatic hypermutation. Immunity **13:** 589–597.
5. SCHLISSEL, M. & D. BALTIMORE. 1989. Activation of immunoglobulin kappa gene rearrangement correlates with induction of germline kappa gene transcription. Cell **58:** 1001–1007.
6. KONG, Q. & N. MAIZELS. 2001. DNA breaks in hypermutating immunoglobulin genes: evidence for a break-and-repair pathway of somatic hypermutation. Genetics **158:** 369–378.

Ectopic Germinal Center Formation in Rheumatoid Synovitis

CORNELIA M. WEYAND AND JÖRG J. GORONZY

Departments of Internal Medicine and Immunology, Mayo Clinic, Rochester, Minnesota 55905, USA

ABSTRACT: Synovial inflammation in rheumatoid arthritis is closely related to the formation of ectopic lymphoid microstructures. In synovial tissue from some patients, one finds seemingly diffuse infiltrates; in others, T cells and B cells cluster in aggregates with interdigitating dendritic cells (DCs) but no follicular DCs (FDCs). In a third group, T cell/B cell follicles with germinal center (GC) reactions are generated. Within a given patient, aggregates and GCs are mutually exclusive and stable over time. Because antigen storage capacity, lymphoid density, and three-dimensional topography of GCs optimize immune responses, synovial GCs should play a crucial role in the breakdown of self-tolerance. We have identified factors critical for ectopic GCs, thereby transforming the synovial inflammatory process. Tissues with GCs produced 10- to 20-fold higher amounts of the chemokines CXCL13 and CCL21. CXCL13 derived from three sources, endothelial cells, synovial fibroblasts, and FDC networks. The level of CXCL13 transcripts strongly predicted GCs; however, some tissues had high levels of CXCL13 but lacked GCs. Tissue expression of LT-β emerged as a second key factor. LT-β protein was detected on follicular center and mantle zone B cells. Multivariate regression analysis identified CXCL13 and LT-β as the only cytokines predicting GCs. Remarkably, LT-α did not contribute independently. The contribution of B cells to ectopic lymphoid organogenesis was not limited to LT-β production. Rather, synovial tissue B cells were critical in regulating T cell activation. In adoptive transfer experiments in human synovium-SCID mouse chimeras, activation of synovium-derived CD4 T cells was strictly dependent on T cell/B cell follicles. Depletion of synovial tissue B cells abrogated T cell function, and non-B cell antigen-presenting cells could not maintain T cell stimulation. Unexpectedly, GC function in the rheumatoid lesion was also dependent on CD8 T cells. The majority of T cell receptors derived from CD8 T cells were shared between distinct GCs. Depletion of CD8 T cells disrupted synovial GCs, FDC networks disappeared, and transcription of LT-β, IgG, and Igκ declined. Follicle-sustaining CD8 T cells were located at the edge of or within the mantle zone. Cell–cell communication in the mantle zone, including CD8 T cells, appears to be critical for ectopic GC formation in rheumatoid synovitis.

KEYWORDS: antigen presentation; autoimmunity; lymphoid organization; rheumatoid arthritis

Address for correspondence: Cornelia M. Weyand, M.D., Guggenheim 401, 200 First Street SW, Mayo Clinic, Rochester, Minnesota 55905. Voice: 507-284-1650; fax: 507-284-5045. weyand.cornelia@mayo.edu

Ann. N.Y. Acad. Sci. 987: 140–149 (2003). © 2003 New York Academy of Sciences.

Rheumatoid arthritis (RA) is an autoimmune disease that has crippling potential and also shortens the life expectancy of the affected patient. Pain and disability are caused by chronic destructive inflammation of small and large joints, referred to as rheumatoid synovitis. The premature death of patients with RA also results from abnormal immune responses. Immune and inflammatory pathways have been implicated in precipitating complications of coronary atherosclerosis, particularly the rupture of atherosclerotic plaque. Evidence is accumulating that mechanisms of immune-mediated tissue damage are shared between the synovial and the atherosclerotic lesion, thus predisposing the patient with RA to accelerated coronary artery disease.[1]

Normally, the synovium is a thin membrane of intimal cells overlaying the subintimal, matrix-rich space. The synovial membrane has a dense capillary network and characteristically lacks epithelial features such as a basal lamina or tight junctions. Typical changes associated with inflammation in this structure are hyperplasia of the resident cells and accumulation of lymphocytes and macrophages in the subintimal layer. Most tissues respond to injury by mobilizing fibroblasts that attempt to repair by proliferating and secreting extracellular matrix. A classic example is the process of intimal hyperplasia leading to lumen occlusion in inflammatory vasculopathies.[2] In the setting of synovial inflammation, resident fibroblasts and macrophages respond with hyperproliferation and acquire tissue-destructive capabilities, which ultimately leads to the damage of bone, cartilage, and tendons.

The autoimmune nature of RA was first documented almost 50 years ago when rheumatoid factors were identified as autoantibodies that reacted against the Fc portion of immunoglobulins. Rheumatoid factors remain the most valuable diagnostic tool in assessing patients with inflammatory joint disease. The underlying mechanisms causing breakdown of self-tolerance in selected hosts are incompletely understood, but the emerging paradigm predicts that a multitude of risk determinants cooperate to permit autoreactive and destructive immune responses.

THE SYNOVIAL LESION IN RA: A DIVERSE SET OF LYMPHOID ORGANIZATIONS

The major cellular components of the inflamed synovial lesion are T cells, macrophages, B cells, and dendritic cells (DCs). The recruitment and topographical arrangement of lymphocytes and phagocytic cells in this tissue site result from a nonrandom process. Tissue-infiltrating lymphocytes are often arranged in follicles. The traditional view has been that B cells and CD4 T cells come together in clusters, whereas CD8 T cells occupy the interfollicular areas.[3] To begin to understand the diversity of the synovial lesion and to study pathways that regulate the structural organization of the lymphocytic infiltrate, we have analyzed the morphology and functional activity of a large consecutive series of synovial tissue biopsies harvested from patients with RA. To prevent contamination of the cohort with non-rheumatoid inflammatory arthropathies, we carefully screened the patients and excluded them if their clinical presentation, the autoantibody analysis, and radiographic examination left any doubt on the nature of the underlying disease process. Initial analysis revealed three morphologic patterns.[4] Many of the biopsies had diffuse infiltrates of T cells and B cells in the subintimal layer. In other patients, lymphocytes accumulated in follicular structures, and a small cohort of patients had typical granulomatous reactions in the synovial membrane. Synovial granulomas resembled the rheumatoid nodules that are often found outside of the joint and that are a sign of extraarticular progression of RA.

Analysis of synovial tissue follicles for cellular components and functional activity demonstrated that this category of rheumatoid synovitis could be partitioned into two categories.[5] One category included patients that formed classic germinal centers (GCs). In the other category of patients with synovial follicles, GC reactions were absent. T cells and B cells grouped together in a fixed ratio to form aggregates. However, these aggregates lacked central follicular dendritic cells (FDCs), IgD^+ B cells were dispersed instead of arranged in a mantle zone, and proliferating B cells were infrequent. A structure resembling synovial T cell/B cell aggregates has not yet been identified in secondary lymphoid organs. In a series of 64 consecutive synovial tissue biopsies, diffuse synovitis accounted for 56% of the cases; GCs were detected in 24% of the cases; and 20% of the patients presented with aggregate synovitis.[6]

The lymphoid organization in the synovial membrane has immediate implications for the disease process.[7,8] Profiles of cytokines produced in the tissue are closely linked to the organizational pattern.[4] The characteristic cytokine pattern for diffuse synovitis is that of low levels of IFN-γ, IL-1β, TNF-α, and IL-10. IL-4-specific sequences are also present in such tissues. If the biopsy contains T cell/B cell follicles, IFN-γ, IL-1β, and TNF-α are transcribed at high levels. Follicular tissues also contain high concentrations of the supposedly anti-inflammatory cytokine IL-10, but they essentially lack IL-4 and may, therefore, insufficiently counteract TH1 responses.

SYNOVIAL TISSUE LYMPHOID MICROSTRUCTURES: A CHOICE OF THE HOST

Recognizing diversity in the lymphoid microstructures in rheumatoid synovitis raises a series of questions. The diversity could reflect different developmental stages of the inflammatory process. Diffuse synovitis could simply be the precursor of the more organized types. Aggregate synovitis could possibly be the equivalent of a primary follicle, on its way to becoming a secondary follicle. Several findings indicate that such a model is not valid. First, lymphoid aggregates, GCs, and granulomatous structures generally do not co-occur in the same lesion or even in different joints of the same patient. Sampling distinct joints showed stability of the pattern within a given patient. In addition, by following patients over time, we secured serial synovial tissue biopsies. The topographic arrangement of tissue-infiltrating T cells and B cells remained stable over time, even in samples collected years apart (C.M. Weyand and J.J. Goronzy, unpublished observations). Finally, tissues with lymphoid aggregates were analyzed for the presence of FDCs, a structural hallmark of primary follicles. Aggregates lack FDCs, as documented by the absence of the $CD21_{long}$ isoform, and thus are not primary follicles. These data indicate that the different forms of lymphoid organizations reflect distinct entities of synovial inflammation that are characteristic for individual patients.[6]

CHEMOKINES AND CYTOKINES IN ECTOPIC GERMINAL CENTER FORMATION

The process of lymphoid organogenesis is regulated by a series of cytokines and chemokines.[9–15] It has been proposed that lymphoid neo-organogenesis, occurring at an ectopic site, involves the same pathways.[16,17] Specifically, lymphotoxin (LT)-α, found to be critical in the organogenesis of lymphoid organs, has been suggested to be a key player in inflammation-associated lymphoid organization.[18]

In the search for cytokines and chemokines involved in the ectopic generation of synovial GCs, we have conducted a comparative analysis of 64 synovial tissue biopsies.[6] Similar levels of CCL2 were produced in tissues with diffuse, aggregate, or GC synovitis, indicating that all tissues had equivalent inflammatory activity. Also, DC function appeared to be similar because the concentrations of CCL18 transcripts were indistinguishable in the three synovitis types. However, tissues with GC reactions produced markedly higher levels of the chemokines CXCL13 and CCL21, which are implicated in the recruitment of B cells and T cells, respectively (FIG. 1). It is important to note that aggregate tissues have similar numbers of T cells and B cells as those with GCs.[5] Thus, the process of attracting T cells as well as B cells to the synovial membrane must be different in patients committed to distinct patterns of lymphoid microstructures.

Tissue samples with GC reactions could be distinguished from the other biopsies by the production of LT-α and LT-β (FIG. 2). Both cytokines were significantly higher in GC^+ tissues; LT-α was almost undetectable in diffuse synovitis. LT-β receptor transcription was similar in all cases, suggesting that the target cells for LT-β were equally represented in the different tissues.

To estimate the relative contribution of the different variables in the process of synovial GC formation and to test whether CXCL13, CCL21, LT-α, and LT-β were dependent or independent factors, we have used multivariate logistic regression modeling.[6] Tissue levels of CXCL13 and LT-β emerged as independent predictors of GC formation, suggesting that ectopic production of these two mediators is a parallel and not a consecutive process. Recursive partitioning was used to identify threshold levels of the two independent markers, CXCL13 and LT-β. Modeling of the process of GC formation demonstrated that 69% of the cases with GCs could be correctly predicted if either of these two factors was above the threshold level. In some patients, high production of LT-β or CXCL13 could compensate for the low production of the other factor. However, if both CXCL13 and LT-β were lacking, GCs were not detected. The series also included tissues that fulfilled the cytokine criteria, but GCs were absent, indicating that the presence of CXCL13 and LT-β in the ectopic site is insufficient to guarantee the development of GC reactions. Additional components must be present to drive lymphoid neogenesis in the synovium. Remarkably, LT-α emerged as a dependent variable, suggesting that synovial tissue production of LT-α is a secondary event and not an initiating factor of GC formation. There was a trend for CCL21 to also meet the requirements of being an independent variable and contributing to the events leading to GC assembly. In summary, our regression analysis is consistent with a hierarchical model of mechanistic factors influencing synovial GC formation and with the production of CXCL13 and LT-β being early and likely causative events.

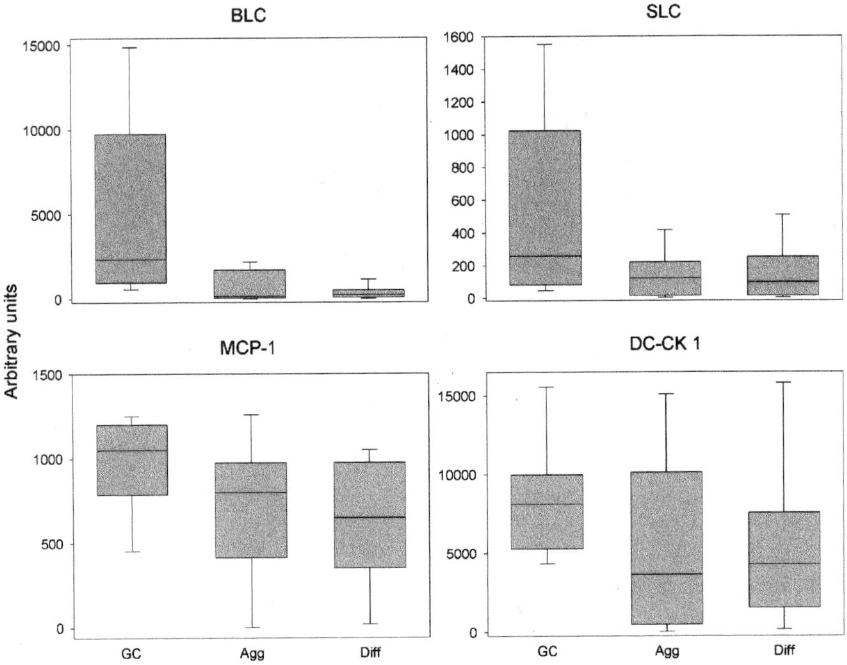

FIGURE 1. The role of chemokine production in ectopic germinal center formation in rheumatoid synovitis. *In situ* chemokine transcription was semiquantified by PCR and oligonucleotide hybridization in synovial tissues with germinal center (GC) formation, tissues with GC-negative lymphoid aggregates (Agg), and tissues without any discrete lymphoid organizations (diffuse synovitis, Diff). Box plots show medians, 25th and 75th percentiles as boxes, and 10th and 90th percentiles as whiskers. GC formation correlated with increased transcription of CXCL13 (BLC) and CCL21 (SLC), but not CCL18 (DC-CK 1) or CCL2 (MCP-1). (From Takemura *et al.*[6] Reprinted with permission from the *Journal of Immunology.*)

The cellular sources of CXCL13 and LT-β were identified by immunohistochemical staining.[6] CXCL13 derived from three distinct cell types. In patients producing this chemokine, anti-CXCL13 antibodies bound select microvessels with endothelial cells staining uniformly. Also, some synovial fibroblasts contained CXCL13 protein. Chemokine-producing synoviocytes were dispersed throughout the tissue. No accumulation of B cells around such fibroblasts was seen. As expected, FDC networks stained positive for CXCL13. In the follicular centers, CXCL13 was also contained in the cytoplasm of selected B cells. We have found that CXCL13 production continued in tissues in which GC and FDC networks have been destroyed (J.J. Goronzy and C.M. Weyand, unpublished observations).

LT-β protein is difficult to detect. The major source of LT-β consisted of B cells. A subpopulation of B cells in the follicular centers reacted with anti-LT-β antibodies. More importantly, LT-β-producing B cells were also found in the mantle zone. Most patients with GC reactions had a small number of LT-β^+ T cells that were localized in the T cell zones surrounding the follicles.

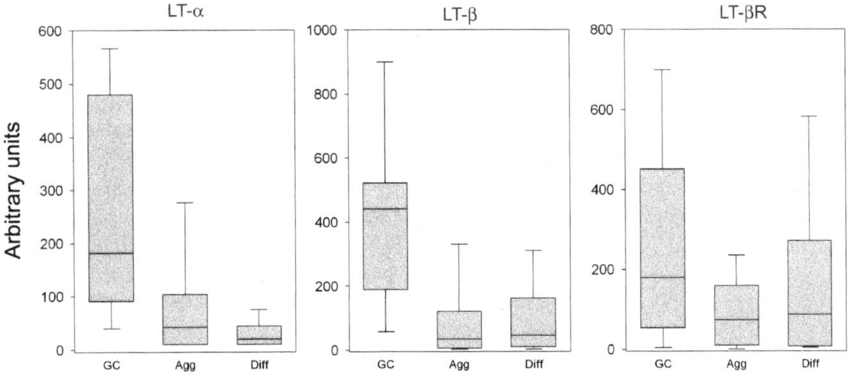

FIGURE 2. Correlation of tissue cytokine patterns and lymphoid microstructures. Tissues from patients with rheumatoid arthritis were stratified as described in FIGURE 1 based on the presence or absence of lymphoid microstructures, and the transcription of cytokines and cytokine receptors was semiquantified. Results are shown as box plots. GC formation was associated with an increased transcription of LT-β and, to a lesser degree, with LT-α; LT-β receptor was equally expressed in all tissues. Using logistic regression modeling, only the difference for LT-β remained significant; LT-α transcription was not an independent variable. (From Takemura *et al.*[6] Reprinted with permission from the *Journal of Immunology.*)

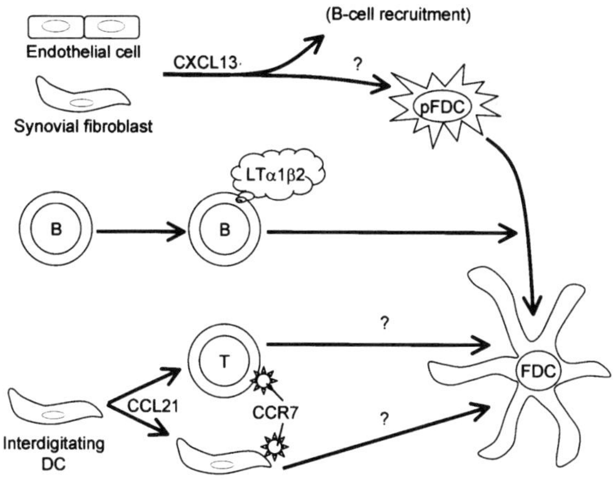

FIGURE 3. A hypothetical model of ectopic germinal center formation in the rheumatoid inflammation.

The emerging model for synovial GC formation is shown in FIGURE 3. GC reactions in the synovium are restricted to tissues containing FDC networks, assigning a checkpoint function to FDC development/maturation. Also, FDC precursors need to be recruited to this tissue site that is normally free of lymphoid structures. At least three independent pathways must contribute to this key event. Endothelial cells and synoviocytes begin to release CXCL13. The recruitment of B cells by this chemokine is unlikely to be the only function it fulfills in rheumatoid synovitis. In tissues with T cell/B cell aggregates without GC formation, CXCL13 is lacking despite the accumulation of B cells. We are proposing that the role of CXCL13 may relate to securing FDC precursors as participants in the inflammatory infiltrate. It is well known that LT-β affects the differentiation and maturation of the FDC network, and this is likely also the role of this chemokine in rheumatoid synovitis. Finally, CCL21, recruiting CCR7-expressing T cells and DCs, appears to influence the process of FDC network generation and GC reaction. Possibly, T cells and DCs responding to the chemokine CCL21 have a direct effect on FDC development and, thus, participate in lymphoid neogenesis. The complexity of the process, requiring three independent pathways to coincide, explains why only a subset of patients progresses to the establishment of tertiary lymphoid tissue in the joint. Yet, patients who form ectopic GCs in the synovial membrane will do so in distinct joints and at different time points in their disease process. This finding suggests host predisposition for each of the independent variables as well as the co-occurrence of these immune pathways.

FUNCTIONAL SIGNIFICANCE OF SYNOVIAL TISSUE GERMINAL CENTERS

The organization of tissue-invading lymphocytes in sophisticated microstructures should have consequences for the functional competence and activity of the inflammatory infiltrate.[19] GCs are known to have optimal antigen storage and the ability to retain antigen on the surface of FDCs for extended periods.[20] With the clustering of T cells, DCs, B cells, and FDCs, cell densities are reached that allow for very effective cell–cell contact and communication. The three-dimensional topography of the GC structure may also amplify communication pathways between cell partners that would not occur if these cells were randomly arranged and dispersed through a tissue site.[21] In essence, a new dimension is added to an immune response by the creation of a complex cellular microstructure. Lymphoid tissues function by optimizing the chance for immune responses. Therefore, the presence of lymphoid organization in peripheral tissue sites should impose an immediate risk for immune recognition and breakdown of self-tolerance.

The function of synovial GCs has been studied in human synovium-SCID mouse chimeras.[22,23] Treatment of chimeras with anti-CD20 antibodies expectedly led to the destruction of B cell follicles.[22] It also resulted in the loss of FDC networks. The most surprising finding was that T cell activation in the synovial lesion was disrupted. IFN-γ production by tissue-invasive T cells decreased in a dose-dependent fashion (FIG. 4). IL-1β, synthesized by synovial tissue macrophages and fibroblasts, declined in parallel with the reduction of IFN-γ. Further support for a critical role of GCs in maintaining T cell stimulation came from experiments in which CD4 T cell clones were adoptively transferred into RA synovium-SCID mouse chimeras. These CD4 T cell clones had been isolated by microdissecting GCs with surrounding T cell zones and identifying those T cell receptor sequences that were shared among spatially separated follicles. Upon the transfer of such CD4 T cell clones into SCID mice implanted with syngeneic or HLA-DR4-matched synovium, production of IFN-γ, IL-1β, and TNF-α in the synovium increased twofold to threefold. However, *in situ* activation of these CD4 T cells was bound to the presence of B cells in GC

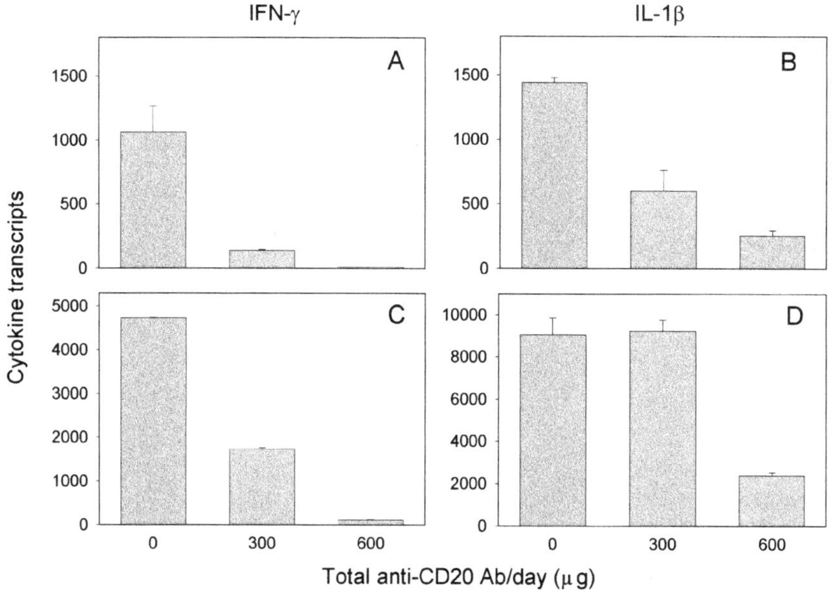

FIGURE 4. Functional importance of germinal center formation for synovial tissue T cell responses. Rheumatoid synovium with GC formation was implanted into SCID mice. Human tissue/SCID mouse chimeras were treated with anti-CD20 antibody (Ab) to deplete human B cells. Synovial tissues were explanted, and T cell responses were assessed by semi-quantifying IFN-γ production. Results are shown as box plots. Depletion of CD20$^+$ B cells diminished IFN-γ transcription and the IFN-γ-dependent IL-1 transcription.[27] (From Takemura *et al.*[22] Reprinted with permission from the *Journal of Immunology.*)

centers. Tissues depleted of B cells and tissues that lacked B cells were unable to support the functional activity of CD4 T cells.

GCs in rheumatoid arthritis have been shown to include immunoglobulins that have undergone mutation, demonstrating that these sites can function in affinity maturation.[24] There is now evidence that synovial GCs may have a broader role by providing unique support for T cell activation in the disease process. Although synovial infiltrates have interdigitating DCs, activated macrophages, and activated fibroblasts, T cell activation in synovial tissues containing GCs can be abrogated by the elimination of B cells.[25] One conclusion is that B cells, themselves, harbor the relevant antigen. Alternatively, the three-dimensional organization of GCs creates conditions that are not present in tissues without this lymphoid assembly. The central role of synovial GCs in driving T cell responses makes these microstructures preferred targets for immunosuppression.[26]

SUMMARY AND CONCLUSION

Rheumatoid synovitis is a process closely associated with lymphoid neogenesis. Distinct patterns of lymphoid microstructures are assembled in the subintimal space

TABLE 1. Morphologic patterns of lymphoid organization in the synovium of patients with rheumatoid arthritis

	Diffuse	Aggregate	Germinal center
T cells	Yes	Yes	Yes
B cells	Variable numbers. Some tissues lack B cells	Yes	Yes
Interdigitating dendritic cells	Yes	Yes	Yes
Follicular dendritic cells	No	No	Yes
IgD^+ mantle zone	No	No	Yes
Proliferating B cells	No	No	Yes
Pattern maintained over time	Yes	Yes	Yes
Equivalent structure in lymphoid organs	?	?	Secondary follicle

of the synovial membrane (TABLE 1). The patterns are specific for each patient and are maintained over time. In a subset of patients, T cells and variable numbers of B cells infiltrate the synovium in a seemingly random process. In other patients, T cells and B cells cluster in aggregates that include interdigitating DCs but lack FDCs. In yet other patients, typical GC reactions emerge. The consistency of the synovitis pattern in individual patients suggests that host determinants regulate the process of lymphoid neogenesis. Several pathways have been identified that contribute to the formation of ectopic GCs. These include the aberrant production of the chemokine CXCL13 by endothelial cells and fibroblasts. Another necessary component of the process is the production of the cytokine LT-β by B cells, including mantle zone B cells. A role of T cells and DCs is suggested by the importance of the chemokine CCL21, which is involved in the recruitment of CCR7-expressing T cells and DCs. These independent variables must coincide to promote the process of GC development. Once formed, these sophisticated lymphoid structures are critical in supporting T cell activation. Depletion of B cells abrogates production of the T cell product, IFN-γ, and macrophage-derived proinflammatory cytokines. The key position of GCs in T cell stimulation suggest that the structure itself is critical in the loss of self-tolerance and the chronic destructive nature of the synovial immune response.

REFERENCES

1. WEYAND, C.M., J.J. GORONZY, G. LIUZZO, *et al.* 2001. T-cell immunity in acute coronary syndromes. Mayo Clin. Proc. **76:** 1011–1020.
2. WEYAND, C.M. & J.J. GORONZY. 1999. Arterial wall injury in giant cell arteritis. Arthritis Rheum. **42:** 844–853.
3. KUROSAKA, M. & M. ZIFF. 1983. Immunoelectron microscopic study of the distribution of T cell subsets in rheumatoid synovium. J. Exp. Med. **158:** 1191–1210.
4. KLIMIUK, P.A., J.J. GORONZY, J. BJORNSSON, *et al.* 1997. Tissue cytokine patterns distinguish variants of rheumatoid synovitis. Am. J. Pathol. **151:** 1311–1319.
5. WAGNER, U.G., P.J. KURTIN, A. WAHNER, *et al.* 1998. The role of CD8+ CD40L+ T cells in the formation of germinal centers in rheumatoid synovitis. J. Immunol. **161:** 6390–6397.

6. TAKEMURA, S., A. BRAUN, C. CROWSON, *et al.* 2001. Lymphoid neogenesis in rheumatoid synovitis. J. Immunol. **167:** 1072–1080.
7. WEYAND, C.M., P.A. KLIMIUK & J.J. GORONZY. 1998. Heterogeneity of rheumatoid arthritis: from phenotypes to genotypes. Springer Semin. Immunopathol. **20:** 5–22.
8. WEYAND, C.M., P.J. KURTIN & J.J. GORONZY. 2001. Ectopic lymphoid organogenesis: a fast track for autoimmunity. Am. J. Pathol. **159:** 787–793.
9. FU, Y.X. & D.D. CHAPLIN. 1999. Development and maturation of secondary lymphoid tissues. Annu. Rev. Immunol. **17:** 399–433.
10. NGO, V.N., H. KORNER, M.D. GUNN, *et al.* 1999. Lymphotoxin alpha/beta and tumor necrosis factor are required for stromal cell expression of homing chemokines in B and T cell areas of the spleen. J. Exp. Med. **189:** 403–412.
11. FU, Y.X., G. HUANG, Y. WANG, *et al.* 1998. B lymphocytes induce the formation of follicular dendritic cell clusters in a lymphotoxin alpha-dependent fashion. J. Exp. Med. **187:** 1009–1018.
12. ENDRES, R., M.B. ALIMZHANOV, T. PLITZ, *et al.* 1999. Mature follicular dendritic cell networks depend on expression of lymphotoxin beta receptor by radioresistant stromal cells and of lymphotoxin beta and tumor necrosis factor by B cells. J. Exp. Med. **189:** 159–168.
13. LUTHER, S.A., H.L. TANG, P.L. HYMAN, *et al.* 2000. Coexpression of the chemokines ELC and SLC by T zone stromal cells and deletion of the ELC gene in the plt/plt mouse. Proc. Natl. Acad. Sci. USA **97:** 12694–12699.
14. FORSTER, R., A. SCHUBEL, D. BREITFELD, *et al.* 1999. CCR7 coordinates the primary immune response by establishing functional microenvironments in secondary lymphoid organs. Cell **99:** 23–33.
15. CYSTER, J.G. 1999. Chemokines and cell migration in secondary lymphoid organs. Science **286:** 2098–2102.
16. KRATZ, A., A. CAMPOS-NETO, M.S. HANSON, *et al.* 1996. Chronic inflammation caused by lymphotoxin is lymphoid neogenesis. J. Exp. Med. **183:** 1461–1472.
17. LUTHER, S.A., T. LOPEZ, W. BAI, *et al.* 2000. BLC expression in pancreatic islets causes B cell recruitment and lymphotoxin-dependent lymphoid neogenesis. Immunity **12:** 471–481.
18. RUDDLE, N.H. 1999. Lymphoid neo-organogenesis: lymphotoxin's role in inflammation and development. Immunol. Res. **19:** 119–125.
19. WEYAND, C.M., A. BRAUN, S. TAKEMURA, *et al.* 2000. Lymphoid microstructures in rheumatoid arthritis. *In* Rheumatoid Arthritis. J.J. Goronzy & C.M. Weyand, Eds.: 168–187. Karger. Basel.
20. ZINKERNAGEL, R.M., S. EHL, P. AICHELE, *et al.* 1997. Antigen localisation regulates immune responses in a dose- and time-dependent fashion: a geographical view of immune reactivity. Immunol. Rev. **156:** 199–209.
21. ZINKERNAGEL, R.M. 2000. Localization dose and time of antigens determine immune reactivity. Semin. Immunol. **12:** 163–171.
22. TAKEMURA, S., P.A. KLIMIUK, A. BRAUN, *et al.* 2001. T cell activation in rheumatoid synovium is B cell dependent. J. Immunol. **167:** 4710–4718.
23. KANG, Y.M., X. ZHANG, U.G. WAGNER, *et al.* 2002. CD8 T cells are required for the formation of ectopic germinal centers in rheumatoid synovitis. J. Exp. Med. **195:** 1325–1336.
24. SCHRODER, A.E., A. GREINER, C. SEYFERT, *et al.* 1996. Differentiation of B cells in the nonlymphoid tissue of the synovial membrane of patients with rheumatoid arthritis. Proc. Natl. Acad. Sci. USA **93:** 221–225.
25. WEYAND, C.M., J.J. GORONZY, S. TAKEMURA, *et al.* 2000. Cell-cell interactions in synovitis: interactions between T cells and B cells in rheumatoid arthritis. Arthritis Res. **2:** 457–463.
26. EDWARDS, J.C. & G. CAMBRIDGE. 2001. Sustained improvement in rheumatoid arthritis following a protocol designed to deplete B lymphocytes. Rheumatology (Oxford) **40:** 205–211.
27. KLIMIUK, P.A., H. YANG, J.J. GORONZY, *et al.* 1999. Production of cytokines and metalloproteinases in rheumatoid synovitis is T cell dependent. Clin. Immunol. **90:** 65–78.

The V(D)J Recombination/DNA Repair Factor Artemis Belongs to the Metallo-β-Lactamase Family and Constitutes a Critical Developmental Checkpoint of the Lymphoid System

DESPINA MOSHOUS,[a] ISABELLE CALLEBAUT,[b] RÉGINA DE CHASSEVAL,[a] CATHERINE POINSIGNON,[a] ISABELLE VILLEY,[a] ALAIN FISCHER,[a] AND JEAN-PIERRE DE VILLARTAY[a]

[a]*Unité Développement Normal et Pathologique du Système Immunitaire, INSERM U429, Hôpital Necker Enfants-Malades, Paris, France*

[b]*Système Moléculaires & Biologie Structurale, LMP, CNRS UMR 7590, Université Paris 6 et Paris 7, Paris, France*

ABSTRACT: V(D)J recombination constitutes a critical checkpoint in the development of the immune system as shown in several animal models as well as severe combined immune deficiency (SCID) condition in humans. We recently cloned the *Artemis* gene, whose mutations are responsible for RS-SCID, a condition characterized by an absence of both B and T lymphocytes and associated with increased sensitivity to ionizing radiations. *Artemis* is ubiquitously expressed and is localized in the nucleus. Artemis belongs to the metallo-β-lactamase superfamily and defines a new group, β-CASP, within this family. β-CASP proteins are β-lactamases acting on nucleic acids. While RS-SCID patients harbor *Artemis* loss-of-function mutations, we identified four patients with a combined immunodeficiency characterized by a low but detectable number of both B and T lymphocytes caused by hypomorphic mutations in the *Artemis* gene. Two of these patients developed aggressive B cell lymphomas, a condition that suggests Artemis may be considered a "caretaker" factor, similarly to the other V(D)J recombination/DNA repair actors.

KEYWORDS: V(D)J recombination; *Artemis* gene; metallo-β-lactamase family

The highly polymorphic antigen receptors on B and T lymphocytes are composed of variable (V), diversity (D), and joining (J) gene segments that undergo somatic rearrangement prior to their expression by a mechanism known as V(D)J recombina-

Address for correspondence: Jean-Pierre de Villartay, INSERM U429, Hôpital Necker Enfants Malades, 149 rue de Sèvres, 75015 Paris, France. Voice: (33) 01 44 49 50 81; fax: (33) 01 42 73 06 40.
devillar@infobiogen.fr

Ann. N.Y. Acad. Sci. 987: 150–157 (2003).

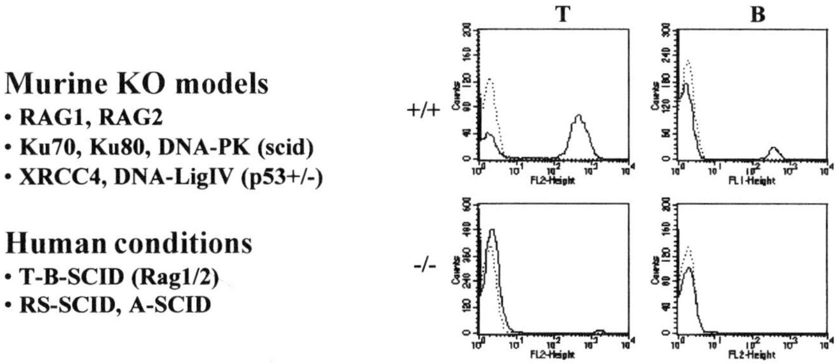

FIGURE 1. V(D)J recombination deficiencies in animal models and in human SCIDs. All these situations are characterized by an absence of T and B lymphocytes, as shown here on FACS profiles, using anti-CD3 and anti-CD19 antibodies, respectively.

tion.[1] V(D)J recombination can be roughly divided into three steps. First, the RAG1 and RAG2 proteins initiate the rearrangement process by introducing a DNA double-strand break at recombination-specific sequences (RSS) that flank all V, D, and J segments.[2,3] The DNA damage is then resolved by the general DNA repair machinery through the DNA non-homologous end-joining (NHEJ) pathway first described in *Saccharomyces cerevisiae* (reviewed in Haber[4]). First, the Ku70/Ku80/DNA-PKcs complex acts as a DNA damage-signaling complex (reviewed in Jackson and Jeggo[5]). Then the double-strand break is repaired by the XRCC4/DNA-ligase IV complex.[6,7] V(D)J recombination represents a critical checkpoint in the development of the immune system. Indeed, all the animal models carrying a defective gene of either one of the known V(D)J recombination factors, either natural (murine or equine SCID) or engineered through homologous recombination, exhibit a profound defect in the lymphoid developmental program owing to an arrest of the B and T cell maturation at early stages (FIG. 1).[8–18]

A fully defective T cell development program is also observed in a number of rare human conditions denominated severe combined immune deficiency (SCID).[19] In about 20% of SCID conditions, the phenotype consists in a virtually complete absence of both circulating T and B lymphocytes, associated with a defect in the V(D)J recombination process, while natural killer (NK) cells are normally present and functional (FIG. 1). Mutations in either the *RAG1* or *RAG2* gene account for a subset of patients with T-B-SCID.[20–22] In some patients (RS-SCID), the defect is not caused by *RAG1* or *RAG2* mutations and is accompanied by an increased sensitivity to ionizing radiations of both bone marrow cells (CFU-GMs) and primary skin fibroblasts,[23] as well as a defect in V(D)J recombination in fibroblasts.[24] We recently cloned the *Artemis* gene and demonstrated that it was indeed responsible for the RS-SCID condition (Fig. 2).[25] *Artemis* is expressed in all tissues[25] and is localized in the nucleus (FIG. 3), as expected for a protein involved in V(D)J recombination and DNA repair.

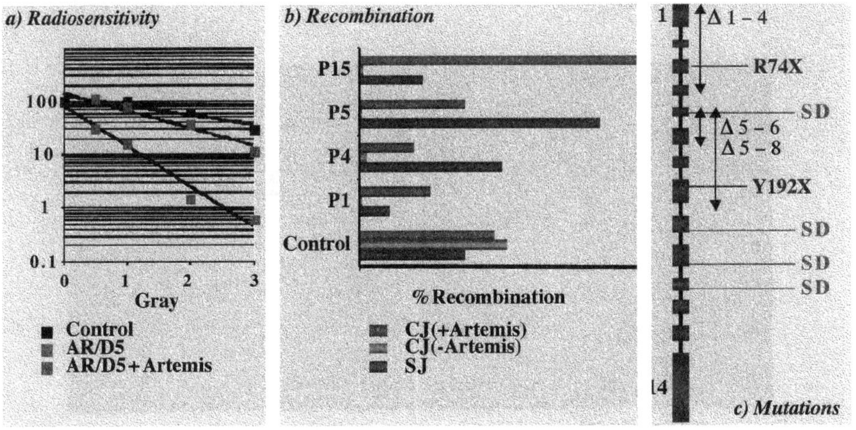

FIGURE 2. Artemis deficiency leads to RS-SCID. (**a**) Complementation of the increased sensitivity to ionizing radiation by wt Artemis in AR/D5, an RS-SCID fibroblast line. (**b**) Complementation of the V(D)J recombination defect by wt Artemis in RS-SCID fibroblasts. (**c**) Spectrum of *Artemis* gene mutations in RS-SCIDs.

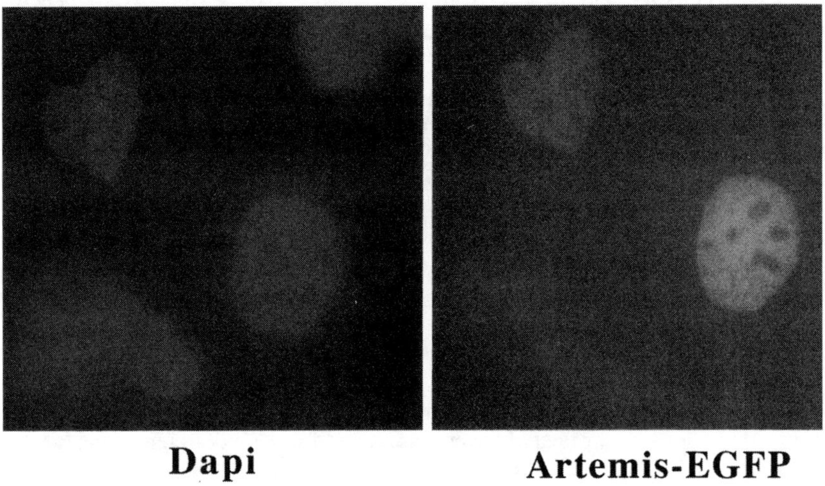

FIGURE 3. Nuclear localization of Artemis. An Artemis/GFP fusion protein was introduced in fibroblast and analyzed by fluorescence microscopy. Nuclei were stained with DAPI.

ARTEMIS DEFINES β-CASP, A NEW GROUP WITHIN THE METALLO-β-LACTAMASE FAMILY

Protein databases search revealed that Artemis does not have a global homologue in other species, although we cloned a cDNA encoding the murine Artemis, which presents 78% identity with its human counterpart at the protein level. A detailed sequence analysis of the Artemis protein nonetheless indicated that it belongs to the

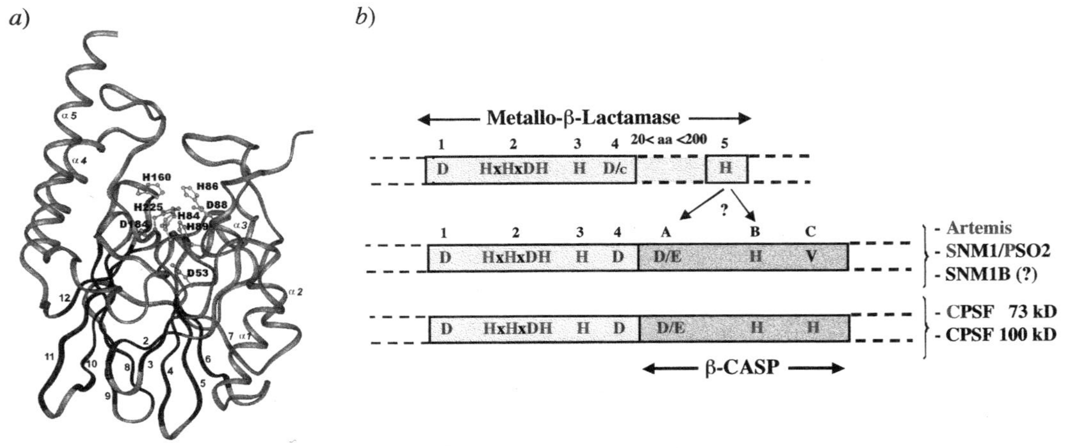

FIGURE 4. The metallo-β-lactamase/β-CASP domain. (**a**) Three-dimensional structure of the metallo-β-lactamase domain of *Strenotrophomonas maltophilia* (1SML). (**b**) Sequence comparison among metallo-β-lactamase proteins highlighting a new domain, β-CASP, present in CPSF, Artemis, SNM1, and PSO2.

metallo-β-lactamase superfamily.[25] Metallo-β-lactamases are enzymes that were first described in bacteria, where they are responsible for cleavage of the β-lactam ring of certain antibiotics. Metallo-β-lactamase fold consists of a four-layered β-sandwich with two mixed β-sheets flanked by α-helices, with the metal-binding sites located at one edge of the β-sandwich (FIG. 4a).[26] Five sequence motifs mostly comprising histidine and aspartic acids constitute critical residues of the catalytic site.[27,28] Motifs 1 to 4 are clearly conserved in Artemis (FIG. 4b), while motif 5 was not easily recognizable. However, a sequence downstream of motif 4 presented significant homology with a wide variety of proteins, including the DNA repair factors SNM1 and PSO2 as well as the RNA cleavage and polyadenylation-specific factor (CPSF). Based on these homologies, we named this new domain β-CASP (FIG. 4b). The β-CASP domain, which is observed in all living organisms, is always associated with the metallo-β-lactamase domain. Several highly conserved residues in the β-CASP domain are likely to play a key role in the structure and/or function of this family within the metallo-β-lactamase superfamily.

WHICH FUNCTION FOR ARTEMIS?

Several distinct enzymatic activities constitute the V(D)J recombinase. The RAG1 and RAG2 proteins initiate the reaction, and the five known factors of the NHEJ pathway are responsible for signaling and repairing the DNA damage. To date, one important activity has somehow remained elusive: the hairpin-opening activity. Indeed, the RAG1/2-generated double-strand break is left as a hairpin-sealed coding end that obviously needs to be processed before it can be repaired. Several candidates have been suggested for this activity. The RAD50/MRE11/NBS1 complex, which is known to participate in the NHEJ pathway,[29,30] was found at the site of the rearranging T cell receptor (TCR) genes, arguing for its possible involvement in the DNA repair phase of the V(D)J recombination.[31] Moreover, MRE11 was found to have hairpin-opening activity *in vitro*, although only in the non-physiologic Mn^{2+} conditions. The finding of normal coding joint formation in NBS patients does not strongly support this candidate. The RAG1/2 complex has also been proposed to carry the hairpin-opening activity. Indeed, this complex remains on the broken ends following the initial DNA double-strand break and is capable of hairpin activity *in vitro*.[32–35] One major drawback of this hypothesis is that hairpins at coding ends accumulate in the murine SCID where RAG1/2 proteins are present and normal. Although one may argue that DNA-PKcs are necessary to activate RAG1/2 for this activity, there is no experimental data that support this hypothesis. We proposed that Artemis could be the factor responsible for hairpin opening for several reasons.[25] First, it is unlikely that Artemis participates directly in the rejoining phase, as one would expect a much stronger clinical phenotype if this were to be the case. After all, DNA ligase IV and XRCC4 knockout mice are lethal to embryos, and NBS patients have evident extra-immunological manifestations ("bird-like face, growth retardation, cancer predisposition), as is the case for defects in the other players of the NHEJ pathway (review in Roth and Gellert[36]). Moreover, the rejoining of linearized DNA constructs, introduced in RS-SCID fibroblasts, or the V(D)J signal joint formation is normal in Artemis-deficient patients. In fact, human RS-SCID patients and SCID mice are the only two situations where V(D)J recombination-associated DNA repair defect affects only the formation of the V(D)J coding joints. In all the other situations analyzed, the signal joints are absent in the context of a defective V(D)J recombination. Given the accumulation of hairpin structures in SCID mice and the likely hydrolase activity of Artemis through its metallo-β-lactamase domain, it was tempting to propose that Artemis is indeed the hairpin opener. It is exactly what Michael Lieber and colleagues demonstrated in a very elegant paper.[37] Artemis forms a complex with (and is phosphorylated by) DNA-PKcs, which activates its endonuclease activity, leading to the opening of RAG1/2-generated hairpin structures.

HYPOMORPHIC MUTATIONS OF *ARTEMIS* ARE RESPONSIBLE FOR ATYPICAL RS-SCIDS

Beside RS-SCID, several other diseases are characterized by an immunodeficiency associated with increased sensitivity to ionizing radiation, such as ataxia telangectasia (AT) and Nijmegen breakage syndrome (NBS).[38] In our survey of such conditions, we came across four patients who presented a combined immune deficiency characterized by a severe B and T lymphocytopenia and a profound hypogammaglobulinemia. Mutations leading to truncation of Artemis in the C-terminus domain were responsible for these conditions. The low T and B cell count, the partial V(D)J recombination activity in fibroblast of some of these patients, and the incomplete capacity of these truncated Artemis forms to complement the ionizing radiation sensitivity of a fully Artemis-deficient cell line attested to the hypomorphic character of these new mutations. Hypomorphic mutations in NHEJ factors have already been described in the case of DNA ligase IV,[39] for which complete loss-of-function mutations are not expected, as DNA ligase IV knockout proved to be lethal to embryos in mice.[17,18]

CAN ARTEMIS BE CONSIDERED A "CARETAKER"?

Among the four above-mentioned patients with hypomorphic mutation of *Artemis*, two developed aggressive, Epstein-Barr virus (EBV)-associated, disseminated B cell lymphomas. This high susceptibility to lymphomas is also seen in murine models, where the defect of an NHEJ factor (Ku80, DNA-PKcs, XRCC4, or DNA ligase IV) is crossed onto the $P53^{-/-}$ background (review in Ferguson and Alt[40]). In fact, these animal models recapitulate the Kinzler and Vogelstein model of tumorigenesis whereby the concomitant defects in both "caretaker" and "gatekeeper" factors are required for the development of a tumor.[41] In these animal models, the NHEJ factors are considered as the caretakers, and their targeted inactivation indeed leads to genetic instability. The parallel between these animal models and the situation of the patients harboring hypomorphic mutation of *Artemis* strongly suggests that Artemis may as well be considered a "caretaker" factor. The development of appropriate model of Artemis deficiency should help accredit this hypothesis.

ACKNOWLEDGMENTS

We thank Christophe Pannetier, Françoise Le Deist, Marina Cavazzana-Calvo, Serge Romana, Elizabeth Macintyre, Danielle Canioni, Nicole Brousse, and Jean-Laurent Casanova for their various contributions to this work. This work was supported by a grant from the Institut National de la Santé et de la Recherche Médicale (INSERM), as well as grants from the Commissariat à l'Energie Atomique (CEA, LRC-7), the Association de Recherche sur le Cancer (ARC), and the Louis Jeantet Foundation.

REFERENCES

1. TONEGAWA, S. 1983. Somatic generation of antibody diversity. Nature **302:** 575–581.
2. SCHATZ, D.G. *et al.* 1989. The V(D)J recombination activating gene, RAG-1. Cell **59:** 1035–1048.
3. OETTINGER, M.A., D.G. SCHATZ, C. GORKA & D. BALTIMORE. 1990. RAG-1 and RAG-2, adjacent genes that synergistically activate V(D)J recombination. Science **248:** 1517–1523.
4. HABER, J.E. 2000. Partners and pathways repairing a double-strand break. Trends Genet. **16:** 259–264.
5. JACKSON, S.P. & P.A. JEGGO. 1995. DNA double-strand break repair and V(D)J recombination: involvement of DNA-PK. Trends Biochem. Sci. **20:** 412–415.
6. LI, Z.Y. *et al.* 1995. The Xrcc4 gene encodes a novel protein involved in DNA double-strand break repair and V(D)J recombination. Cell **83:** 1079–1089.
7. ROBINS, P. & T. LINDAHL. 1996. DNA ligase IV from HeLa cell nuclei. J. Biol. Chem. **271:** 24257–24261.
8. MOMBAERTS, P. *et al.* 1992. RAG-1 deficient mice have no mature B and T lymphocytes. Cell **68:** 869–877.
9. SHINKAI, Y. *et al.* 1992. RAG-2 deficient mice lack mature lymphocytes owing to inability to initiate V(D)J rearrangement. Cell **68:** 855–867.
10. ZHU, C. *et al.* 1996. Ku86-deficient mice exhibit severe combined immunodeficiency and defective processing of V(D)J recombination intermediates. Cell **86:** 379–389.
11. NUSSENZWEIG, A. *et al.* 1996. Requirement for Ku80 in growth and immunoglobulin V(D)J recombination. Nature **382:** 551–555.
12. JHAPPAN, C. *et al.* 1997. DNA-PKcs: a T-cell tumour suppressor encoded at the mouse SCID locus. Nat. Genet. **17:** 483–486.
13. GAO, Y. *et al.* 1998. A targeted DNA-PKcs-null mutation reveals DNA-PK-independent functions for KU in V(D)J recombination. Immunity **9:** 367–376.
14. TACCIOLI, G.E. *et al.* 1998. Targeted disruption of the catalytic subunit of the DNA-PK gene in mice confers severe combined immunodeficiency and radiosensitivity. Immunity **9:** 355–366.
15. SHIN, E.K. *et al.* 1997. A kinase-negative mutation of DNA-PKcs in equine SCID results in defective coding and signal joint formation. J. Immunol. **158:** 3565–3569.
16. FRANK, K.M. *et al.* 1998. Late embryonic lethality and impaired V(D)J recombination in mice lacking DNA ligase IV. Nature **396:** 173–177.
17. GAO, Y. *et al.* 1998. A critical role for DNA end-joining proteins in both lymphogenesis and neurogenesis. Cell **95:** 891–902.
18. BARNES, D.E. *et al.* 1998. Targeted disruption of the gene encoding DNA ligase IV leads to lethality in embryonic mice. Curr. Biol. **31:** 1395–1398.
19. FISCHER, A. *et al.* 1997. Naturally occurring primary deficiencies of the immune system. Annu. Rev. Immunol. **15:** 93–124.
20. SCHWARZ, K. *et al.* 1996. RAG mutations in human B cell-negative SCID. Science **274:** 97–99.
21. CORNEO, B. *et al.* 2000. 3D clustering of human RAG2 gene mutations in severe combined immune deficiency (SCID). J. Biol. Chem. **275:** 12672–12675.

22. VILLA, A. *et al.* 2001. V(D)J recombination defects in lymphocytes due to RAG mutations: severe immunodeficiency with a spectrum of clinical presentations. Blood **97:** 81–88.
23. CAVAZZANA-CALVO, M. *et al.* 1993. Increased radiosensitivity of granulocyte macrophage colony-forming units and skin fibroblasts in human autosomal recessive severe combined immunodeficiency. J. Clin. Invest. **91:** 1214–1218.
24. NICOLAS, N. *et al.* 1998. A human SCID condition with increased sensitivity to ionizing radiations and impaired V(D)J rearrangements defines a new DNA recombination/repair deficiency. J. Exp. Med. **188:** 627–634.
25. MOSHOUS, D. *et al.* 2001. Artemis, a novel DNA double-strand break repair/V(D)J recombination protein, is mutated in human severe combined immune deficiency. Cell **105:** 177–186.
26. WANG, Z. *et al.* 1999. Metallo-beta-lactamase: structure and mechanism. Curr. Opin. Chem. Biol. **3:** 614–622.
27. ARAVIND, L. 1997. An evolutionary classification of the metallo-β-lactamase fold. *In Silico Biology* **1:** 69–91.
28. DAIYASU, H. *et al.* 2001. Expansion of the zinc metallo-hydrolase family of the beta-lactamase fold. FEBS Lett. **503:** 1–6.
29. CARNEY, J.P. *et al.* 1998. The hMre11/hRad50 protein complex and Nijmegen breakage syndrome: linkage of double-strand break repair to the cellular DNA damage response. Cell **93:** 477–486.
30. VARON, R. *et al.* 1998. Nibrin, a novel DNA double-strand break repair protein, is mutated in Nijmegen breakage syndrome. Cell **93:** 467–476.
31. CHEN, H.T. *et al.* 2000. Response to RAG-mediated V(D)J cleavage by NBS1 and gamma-H2AX. Science **290:** 1962–1965.
32. YARNELL SCHULTZ, H. *et al.* 2001. Joining-deficient RAG1 mutants block V(D)J recombination in vivo and hairpin opening in vitro. Mol. Cell **7:** 65–75.
33. QIU, J.X. *et al.* 2001. Separation-of-function mutants reveal critical roles for RAG2 in both the cleavage and joining steps of V(D)J recombination. Mol. Cell **7:** 77–87.
34. SHOCKETT, P.E. & D.G. SCHATZ. 1999. DNA hairpin opening mediated by the RAG1 and RAG2 proteins. Mol. Cell. Biol. **19:** 4159–4166.
35. BESMER, E. *et al.* 1998. Hairpin coding end opening is mediated by RAG1 and RAG2 proteins. Mol. Cell **2:** 817–828.
36. ROTH, D.B. & M. GELLERT. 2000. New guardians of the genome. Nature **404:** 823–825.
37. MA, Y. *et al.* 2002. Hairpin opening and overhang processing by an Artemis/DNA-dependant protein kinase complex in nonhomologous end joining and V(D)J recombination. Cell **108:** 781–794.
38. SHILOH, Y. 1997. Ataxia-telangiectasia and the Nijmegen breakage syndrome: related disorders but genes apart. Annu. Rev. Genet. **31:** 635–662.
39. O'DRISCOLL, M. *et al.* 2001. DNA ligase IV mutations identified in patients exhibiting developmental delay and immunodeficiency. Mol. Cell **8:** 1175–1185.
40. FERGUSON, D.O. & F.W. ALT. 2001. DNA double strand break repair and chromosomal translocation: lessons from animal models. Oncogene **20:** 5572–5579.
41. KINZLER, K.W. & B. Vogelstein. 1997. Cancer-susceptibility genes: gatekeepers and caretakers. Nature **386:** 761–763.

Hypermutation in Human B Cells *in Vivo* and *in Vitro*

SANDRA WELLER,[a] AHMAD FAILI,[a] SAID AOUFOUCHI,
QUENTIN GUÉRANGER, MORITZ BRAUN, CLAUDE-AGNÈS REYNAUD,[b]
AND JEAN-CLAUDE WEILL[b]

INSERM U373, Faculté de Médecine Necker-Enfants Malades, Paris, France

ABSTRACT: We develop our previous observation that a subpopulation of circulating memory IgM+IgD+CD27+ B cells belongs to a separate pathway of differentiation in humans. This subpopulation, which represents 5–25% of peripheral B cells, is also present in spleen in the same proportion and displays a marginal-zone-like B cell phenotype. In addition, we describe a short-time *in vitro* induction model for somatic hypermutation by using the BL2 Burkitt's lymphoma cell line stimulated by a combination of antibodies directed against different surface receptors. This short-time assay allows us to show that mutations are stably introduced in one DNA strand of the BL2 VH gene in the G1 phase of the cell cycle.

KEYWORDS: hypermutation; B cells; DNA breaks

A HUMAN B CELL SUBPOPULATION

According to Klein *et al.*,[1] human circulating B cells have been classified into two different compartments, naive B cells, which are $IgM^+IgD^+CD27^-$ and represent 60% of the total B cell count, and memory B cells, which are divided into three categories according to their suface Ig phenotype—switched $IgM^-IgD^-CD27^+$ (15%), IgM-only $CD27^+$ (10%) and $IgM^+IgD^+CD27^+$ B cells (15%). Memory B cells are considered as post-germinal center B cells because they carry a mutated Ig receptor.

In our own analysis, while we have observed naive $CD27^-$, switched and $IgM^+IgD^+CD27^+$ B cells in normal individuals, IgM-only $CD27^+$ B cells did not appear consistently as a distinct population (FIG. 1), with values reaching only a few percent of the IgD^-CD27^+ subset. Only in specific cases, for example, in some young children or in some patients with autoimmune diseases, was a clear IgM-only $CD27^+$ B cell subpopulation detectable.[2]

To find out whether memory IgM^+ B cells, with or without surface IgD, are *bona fide* post-germinal center B cells, or whether they could belong to another differen-

[a]S.W. and A.F. contributed equally to this work.

[b]C-A.R. and J-C.W. share equal senior authorship.

Address for correspondence: Jean-Claude Weill, INSERM U373, Faculté de Médecine Necker-Enfants Malades, 156 rue de Vaugirard, 75730 Paris Cedex 15, France. Voice: 00 33 1 40 61 53 80; fax: 00 33 1 40 61 55 90.

weill@necker.fr

Ann. N.Y. Acad. Sci. 987: 158–165 (2003). © 2003 New York Academy of Sciences.

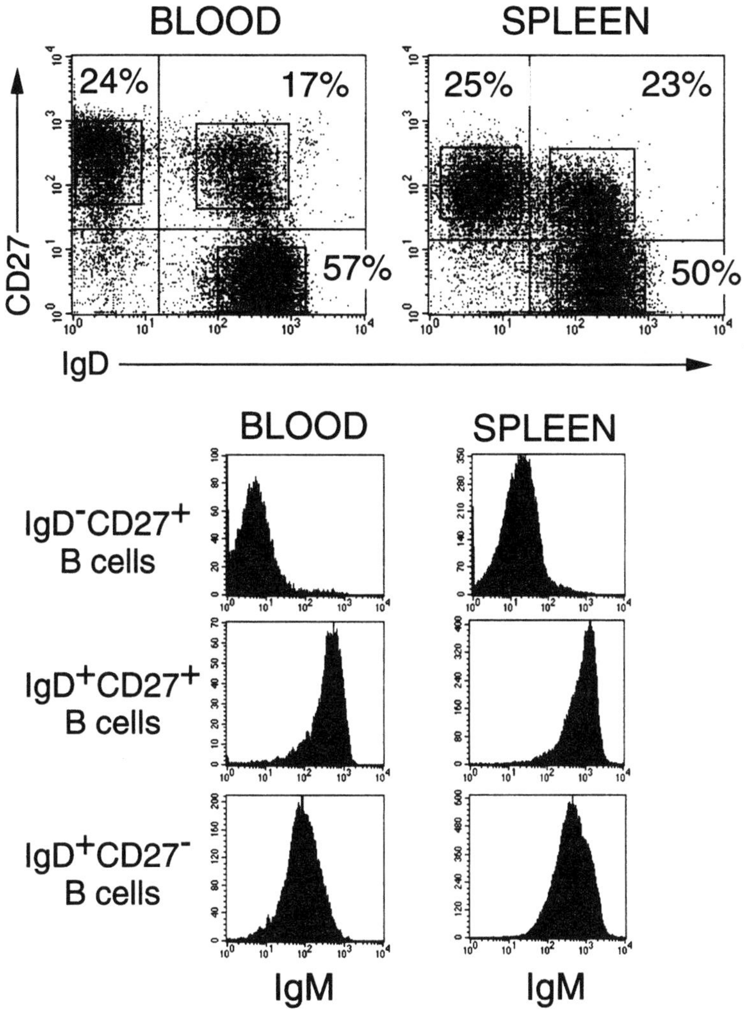

FIGURE 1. Comparison of B cell subpopulations from blood and spleen. CD19-gated B cells are analyzed for their IgD and CD27 phenotype. The three compartments defined are further analyzed for IgM expression. Similar IgM fluorescence intensity is observed for the corresponding splenic and peripheral subsets: $CD27^-$ cells are IgM-intermediate, $CD27^+IgD^+$ are IgM-high, and $CD27^+IgD^-$ cells are mostly IgM-negative. The following antibodies are used: CD19-APC, IgD-FITC, CD27-PE, IgM-biotin revealed by either streptavidin-tricolor (blood) or streptavidin-Red613 (spleen).

tiation pathway, we studied hyper-IgM patients who carry an invalidating mutation either of the CD40 ligand (CD40L) or of the CD40 gene, and who therefore do not possess any germinal centers.

CD40L-deficient patients (HIGM1 syndrome) display a population of circulating $IgM^+IgD^+CD27^+$ memory B cells while they do not possess any switched and IgM-only $CD27^+$ B cells.[3,4] We have observed that these $IgM^+IgD^+CD27^+$ peripheral B cells carry a mutated Ig receptor at their surface, which implies that these cells can hypermutate in the absence of germinal centers and cognate T–B interaction.[4] Recently, the analysis of a CD40-deficient patient (HIGM3 syndrome) showed a similar profile, with 5% of $IgM^+IgD^+CD27^+$ B cells harboring mutated Ig genes,[5] in the absence of an IgM-only memory subset.[2]

According to these results, we have proposed a new scheme of B cell differentiation in humans in which $IgM^+IgD^+CD27^+$ B cells develop separately from the other memory B cells since they diversify their Ig receptors outside the germinal center microenvironment, and we hypothesized that this B cell subset could be involved in T-independent responses[4] (FIG. 2).

In the mouse, splenic marginal zone (MZ) B cells have been shown to be involved in T-independent responses against blood-borne antigens such as polysaccharidic determinants from encapsulated bacteria. We therefore analyzed spleens from individuals at different ages, which have undergone a splenectomy for non-immunological reasons (such as spherocytosis) to find out whether the $IgM^+IgD^+CD27^+$ subset

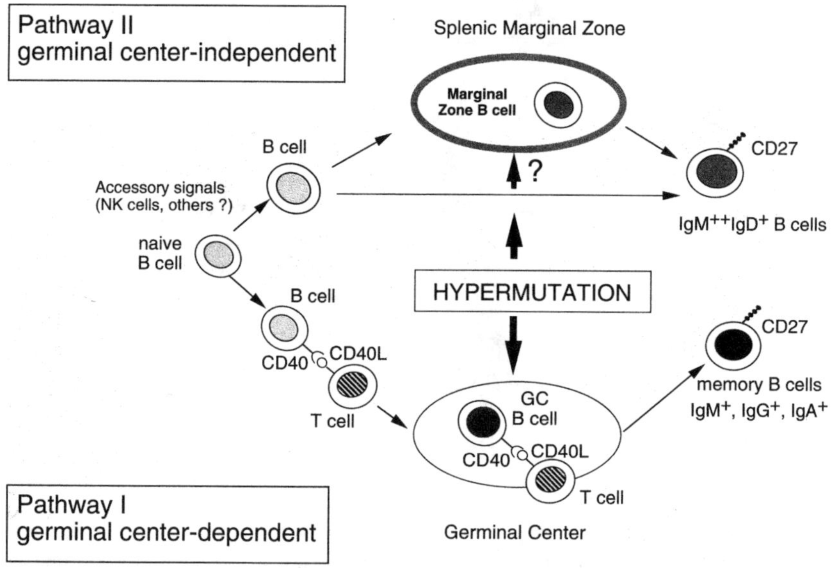

FIGURE 2. Proposed scheme for two pathways of B cell differentiation in humans. In pathway I, B cells are engaged in T-dependent immune responses, hypermutation occurring in germinal centers. In pathway II, B cells hypermutate their Ig receptor prior to or after antigen stimulation (or both), the immune response occurring in the splenic marginal zone.

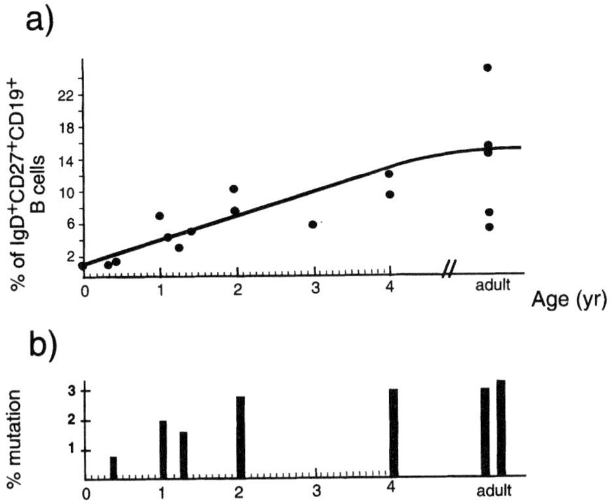

FIGURE 3. Ontogenic appearance of circulating IgM+IgD+CD27+ B cells. The proportion of IgM+IgD+CD27+ cells among total circulating B cells (**a**) and the mutation frequency observed for their rearranged V3-23 genes (**b**) are determined at various ages in normal individuals.

is present in this tissue. Approximately 20–30% of total splenic B cells belong to the IgM+IgD+CD27+ compartment. Like peripheral B cells, their precise Ig phenotype is IgM++IgD+ (FIG. 1). Moreover, these cells are also CD21+, CD1c+, and predominantly CD23−. These markers are the hallmark of the marginal zone B cell population.[6] Strikingly, peripheral IgM+IgD+CD27+ B cells are also CD21+, CD23−, and CD1c+. Moreover, similar mutation frequencies are observed in Ig V genes from peripheral and splenic B cells with the same surface phenotype, the frequency in the IgM+IgD+CD27+ subset being strikingly lower than in switched memory B cells.[2]

These results imply that (1) peripheral IgM+IgD+CD27+ B cells are the circulating counterpart of the splenic MZ B cells and (2) splenic and peripheral MZ-like B cells mutate their Ig receptor in the absence of cognate T–B interactions.

Some species such as sheep have evolved hypermutation as an antigen-independent mode of generation of their pre-immune repertoire in gut-associated lymphoid tissues.[7] One would therefore like to know whether MZ-like B cells mutate their Ig receptor in response to T-independent antigens in the splenic MZ, or whether this diversification is part of a program of development to generate a large immune repertoire with various affinities and specificities, or both.

When the IgM+IgD+27+ peripheral subpopulation is studied during ontogeny in normal individuals, these cells appear in the first months of life and their number gradually increases until 10–15 years, at which stage they reach the adult value of 5–25% of the total peripheral B cell count. Strikingly enough, this population is well developed and rather well mutated at 1 year, an age at which young children do not yet respond to bacterial polysaccharidic antigens (FIG. 3).[2] The fact that circulating MZ B cells have already mutated their Ig receptor at 1 year, before they can respond to T-independent antigens, the splenic MZ being probably not yet mature at this stage, suggests that this diversification may not require antigen stimulation.

Many questions remain:

1. Where and when does this diversification take place during development, and does antigen stimulation induce a new round of hypermutation?
2. Is there only one wave of production of these cells during fetal or early newborn life in humans, as recently shown for mouse MZ B cells?[8,9]
3. Is the splenic MZ the unique site where $IgM^+IgD^+CD27^+$ B cells respond to antigens?
4. What exactly is the role of the circulating MZ-like B cells?

AN *IN VITRO* MODEL OF HYPERMUTATION

Burkitt's lymphoma cell lines have recently been used to study Ig hypermutation *in vitro*. In some of them, such as Ramos, hypermutation is constitutive,[10] while in others, such as CL01 and BL2, it is inducible after aggregation of their surface Ig and co-culture with normal peripheral T cells, or with a normal or transformed T helper clone.[11–13]

To substitute for the signals provided by T cells in this system, we have stimulated BL2 with anti-IgM in association with antibodies against other surface receptors. We found that ligation of CD19 and CD21, together with a strong cross-linking of the IgM receptor, was sufficient to induce hypermutation of the BL2 VH gene in the absence of any T cell help.[14]

Our investigation[14] produced the following results:

1. Hypermutation induced with the system of three antibodies may arise as early as 1 h 30 after incubation at 37 °C.
2. When cells separated in the different phases of the cell cycle (G1, S, G2/M) are induced for hypermutation, mutations are stably introduced in the VH gene in the G1 phase of the cell cycle. While they can also be induced in G2/M, their frequency drops after 24 h, suggesting that they are repaired for most of them (FIG. 4).
3. Mutations arise in one DNA strand and are fixed on the other strand by replication, as shown by analysis of single cells taken either at 1 h 30 or after 24–48 h after one or two divisions in isolated wells (FIG. 5).
4. Induction of mutations in the short-time assay is a post-transcriptional event since mutations occur in the presence of actinomycin D.
5. The short-time induction is dependent of AID (activation-induced cytidine deaminase[15]), since inactivation by homologous recombination of the AID gene on both alleles of the BL2 completely inhibits the process.
6. No specific single- or double-stranded breaks can be detected in the targeted VH gene during the induction process in the G1 phase.

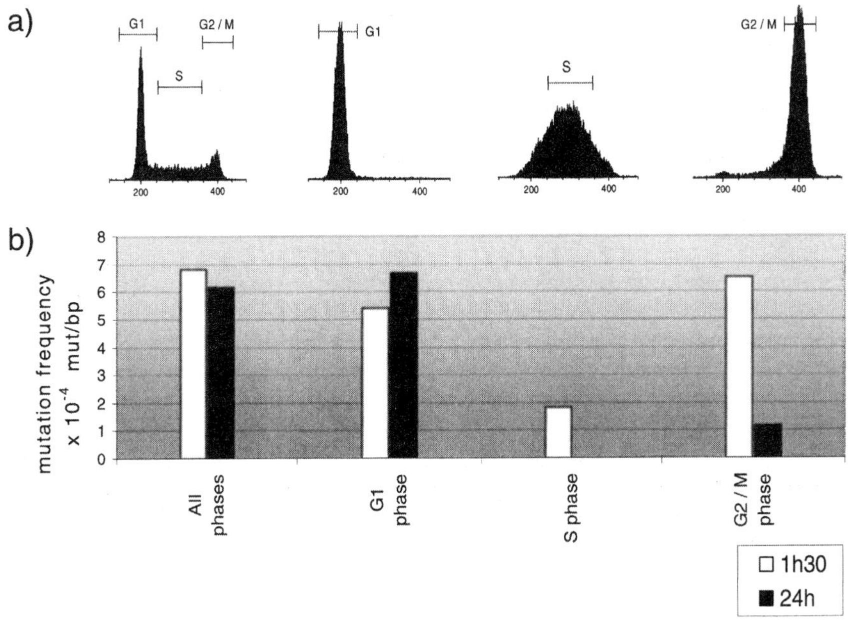

FIGURE 4. Induction of somatic mutation in the BL2 cell line in the different phases of the cell cycle. The BL2 cell line is purified by elutriation in the different phases of the cell cycle. The profile obtained by propidium iodide staining is shown (**a**). Mutations are induced by the assay with three antibodies, and the frequency of these mutations are determined after 1 h 30 and 24 h of stimulation (**b**). Mutations are stably introduced in G1, but they are largely repaired in G2/M after 24 h.

Our data are in contradiction with previous reports. Effectively, double-stranded breaks have been detected by two groups at specific V gene hotspots during hypermutation.[16,17] For one of them, these breaks are observed in G2/M, leading the authors to propose an error-prone copying of the sister chromatid as the molecular basis of hypermutation.[16] Our results obtained with the BL2 cell line, however, imply that hypermutation can occur in G1 on a single strand of DNA in the absence of another VH gene copy. They also show that AID is indispensable, thus validating the use of Burkitt's cell lines to study hypermutation.

There remain many unanswered questions:

1. How is the process directed specifically to the V gene region?
2. How does the entry into DNA occur? By single- or double-stranded breaks, or by the action of another enzymatic reaction, such as a DNA glycosylase?
3. What error-prone DNA polymerases are involved? Although several of the translesion DNA polymerases recently characterized have been proposed as part of the mutasome,[18] their precise role remains to be firmly established.

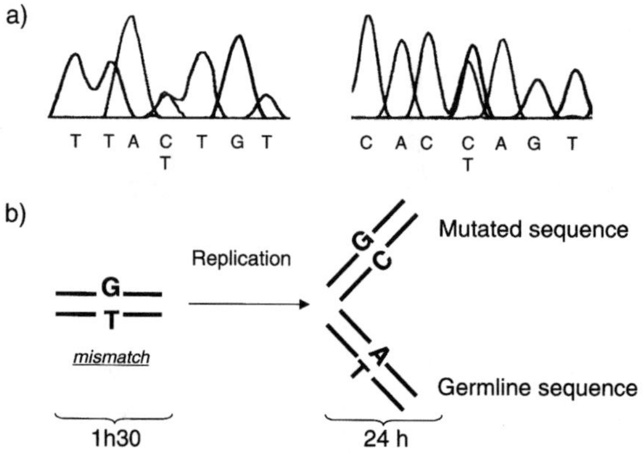

FIGURE 5. Mode of introduction of mutations in the VH gene of the BL2 cell line in the short-time assay. The BL2 cell line is purified in G1 by elutriation and induced for hypermutation. (**a**) The VH gene is analyzed by PCR in single cells, taken either 1 h 30 after stimulation, or 24 h after stimulation after one division in isolated wells (*left* and *right panels*, respectively). (**b**) The mutation appears as a double peak in the electropherogram, which corresponds to the introduction of mutation in one of the two DNA strands, followed by fixation of the mutation by replication in one of the two daughter cells.

ACKNOWLEDGMENTS

S.W. is supported by a fellowship of the Société de Secours des Amis des Sciences and M.B. by the Deutsche Forschungsgemeinschaft.

REFERENCES

1. KLEIN, U., K. RAJEWSKY & R. KUPPERS. 1998. Human immunoglobulin (Ig)M+IgD+ peripheral blood B cells expressing the CD27 cell surface antigen carry somatically mutated variable region genes: CD27 as a general marker for somatically mutated (memory) B cells. J. Exp. Med. **188:** 1679–1689.
2. WELLER, S. *et al.* 2003. In preparation.
3. AGEMATSU, K. *et al.* 1998. Absence of IgD-CD27(+) memory B cell population in X-linked hyper-IgM syndrome. J. Clin. Invest. **102:** 853–860.
4. WELLER, S. *et al.* 2001. CD40-CD40L independent Ig gene hypermutation suggests a second B cell diversification pathway in humans. Proc. Natl. Acad. Sci. USA **98:** 1166–1170.
5. FERRARI, S. *et al.* 2001. Mutations of CD40 gene cause an autosomal recessive form of immunodeficiency with hyper IgM. Proc. Natl. Acad. Sci. USA **98:** 12614–12619.
6. OLIVER, A.M. *et al.* 1997. Marginal zone B cells exhibit unique activation, proliferative and immunoglobulin secretory responses. Eur. J. Immunol. **27:** 2366–2374.
7. REYNAUD, C.-A. *et al.* 1995. Hypermutation generating the sheep immunoglobulin repertoire is an antigen-independent process. Cell **80:** 115–125.
8. HAO, Z. & K. RAJEWSKY. 2001. Homeostasis of peripheral B cells in the absence of cell influx from the bone marrow. J. Exp. Med. **194:** 1151–1163.
9. CARVALHO, T.L. *et al.* 2001. Arrested B lymphopoiesis and persistence of activated B cells in adult interleukin $7^{-/-}$ mice. J. Exp. Med. **194:** 1141–1150.

10. SALE, J.E. & M.S. NEUBERGER. 1998. TdT-accessible breaks are scattered over the immunoglobulin V domain in a constitutively hypermutating B cell line. Immunity **9:** 859–869.
11. DENEPOUX, S. *et al.* 1997. Induction of somatic mutation in a human B cell line in vitro. Immunity **6:** 35–46.
12. ZAN, H. *et al.* 1999. Induction of Ig somatic hypermutation and class switching in a human monoclonal IgM^+ IgD^+ B cell line in vitro: definition of the requirements and modalities of hypermutation. J. Immunol. **5:** 985–990.
13. POLTORATSKY, V. *et al.* 2001. Expression of error-prone polymerases in BL2 cells activated for somatic hypermutation. Proc. Natl. Acad. Sci. USA **98:** 7976–7981.
14. FAILI, A. *et al.* 2002. AID-dependent somatic hypermutation occurs as a DNA single-strand event in the BL2 cell line. Nat. Immunol. **3:** 815–821.
15. MURAMATSU, M. *et al.* 2000. Class switch recombination and hypermutation require activation-induced cytidine deaminase (AID), a potential RNA editing enzyme. Cell **102:** 553–563.
16. PAPAVASILIOU, F.N. & D.G. SCHATZ. 2000. Cell-cycle-regulated DNA double-stranded breaks in somatic hypermutation of immunoglobulin genes. Nature **408:** 216–221.
17. BROSS, L. *et al.* 2000. DNA double-strand breaks in immunoglobulin genes undergoing somatic hypermutation. Immunology **13:** 589–597.
18. DIAZ, M. & P. CASALI. 2002. Somatic immunoglobulin hypermutation. Curr. Opin. Immunol. **14:** 235–240.

Gene Expression Dynamics during Germinal Center Transit in B Cells

ULF KLEIN,[a] YUHAI TU,[b] GUSTAVO A. STOLOVITZKY,[b] JEFFREY L. KELLER,[c] JOSEPH HADDAD, JR.,[c] VLADAN MILJKOVIC,[a] GIORGIO CATTORETTI,[a] ANDREA CALIFANO,[d] AND RICCARDO DALLA-FAVERA[a]

[a]*Institute for Cancer Genetics, Columbia University, Departments of Pathology and Genetics and Development, New York, New York 10032, USA*

[b]*IBM T.J. Watson Research Center, Yorktown Heights, New York 10598, USA*

[c]*Division of Pediatric Otolaryngology, Columbia-Presbyterian Medical Center, Babies and Children's Hospital, New York, New York 10032, USA*

[d]*First Genetic Trust, Inc., Lyndhurst, New Jersey 07071, USA*

ABSTRACT: The germinal center (GC) reaction in T cell dependent antibody responses is crucial for the generation of B cell memory and plays a critical role in B cell lymphomagenesis. To gain insight into the physiology of this reaction, we identified the transcriptional changes that occur in B cells during the GC-transit (naïve B cells → CD77$^+$ centroblasts (CBs) → CD77$^-$ centrocytes (CCs) → memory B cells) by DNA microarray experiments and the subsequent data analysis employing unsupervised and supervised hierarchical clustering. The naïve B cell is characterized by a nonproliferative, anti-apoptotic phenotype and the expression of various chemokine and cytokine receptors. The transition from naïve B cells to CBs is associated with (1) the up-regulation of genes associated with cellular proliferation, DNA-repair, and chromatin remodeling; (2) the acquisition of a pro-apoptotic phenotype; (3) the down-regulation of cytokine, chemokine, and adhesion receptors expressed in the naïve cells; and (4) the expression of a distinct adhesion repertoire. The CB and the CC revealed surprisingly few gene expression differences, suggesting that the CC is heterogeneous in its cellular composition. The CB/CC to memory B cell transition shows a general reversion to the profile characteristic for the naïve B cells, with the exception of the up-regulation of several surface receptors, including CD27, CD80, and IL-2Rβ, and the simultaneous expression of both anti- and pro-apoptotic genes. These gene expression profiles of the normal B cell subpopulations are being used to identify the signals occurring during GC development, the cellular derivation of various types of B cell malignancies, and the genes deregulated in GC-derived tumors.

KEYWORDS: germinal center; DNA microarray; gene expression; hierarchical clustering

Address for correspondence: Ulf Klein, Institute for Cancer Genetics, Columbia University, 1150 St. Nicholas Avenue, New York, NY 10032. Voice: 212-851-5270; fax: 212-851-5256. uk30@columbia.edu

Ann. N.Y. Acad. Sci. 987: 166–172 (2003). © 2003 New York Academy of Sciences.

Antibody affinity maturation and B cell memory derive from genetic and phenotypic changes of antigen-activated B cells during T cell-dependent immune responses. Antigen stimulation drives a "naïve" B cell into a specific microenvironment in the lymphoid organs, the germinal center (GC). The B cell begins to proliferate and forms a large clone of cells that, due to the somatic hypermutation process acting at the rearranged IgV gene, differ in their fine specificity for the immunizing antigen. Mutants expressing high-affinity antibody are positively selected and programmed to either secrete large amounts of antibody (plasma cells) or to have the capacity to memorize the cognate antigen and quickly respond to repeated exposure (memory B cells). The GC reaction is also thought to play a critical role in the genesis of B cell tumors: occasional errors in the control of DNA replication and (probably even more important) in the processes of somatic hypermutation and Ig isotype switching may put the GC B cell at a high risk to obtain oncogenic chromosome translocations or somatic mutations in the coding or regulatory region of proto-oncogenes.

We determined the global gene expression changes of a B cell during GC transit with the following aims: (1) to improve our understanding of the developmental processes a B cell undergoes in T-dependent immune responses, and (2) to obtain a gene expression map of the normal B cell subsets, that is, a map that would provide a reference point for the study of the various B cell malignancies regarding their cellular derivation or the identification of tumor-specific genes.

Outgoing from the phenotypic and functional characterization of the major human B cell subpopulations,[1,2] namely, naïve B cells, GC centroblasts (CB) and centrocytes (CC), and memory B cells, isolation strategies based on magnetic cell separation (MACS-system, Miltenyi Biotech) were developed to obtain large cell numbers at a high purity. Naïve B cells (IgD^+, $CD27^-$, $CD38^{low}$, and unmutated V-region genes) were isolated by depleting GC B cells (CD10, CD27), memory B cells (CD27), plasma cells (CD27), T cells (CD3), and macrophages (CD14), followed by a positive enrichment of IgD^+ cells. CBs and CCs can be flow cytometrically distinguished by the differential expression of CD77, a neutral glycolipid (CBs are $CD38^{high}$, $CD77^+$, and CCs are $CD38^{high}$, $CD77^-$). CBs were isolated in a single step by staining for $CD77^+$ GC B cells. CCs were obtained by first depleting tonsillar mononuclear cells of CBs (CD77), naïve, memory, and plasma cells (CD39), and T cells (CD3), and subsequent enrichment for CD10. Memory B cells ($CD27^+$, $CD38^{low}$, and mutated V-region genes) were purified by depletion of GC B cells (CD10, CD38), plasma cells (CD38), T cells (CD3), and macrophages (CD14), followed by a positive enrichment for $CD27^+$ cells. Each cell population was purified from tonsillar mononuclear cells of five donors. Somatic hypermutation analysis of rearranged antibody genes amplified from two samples of each of the subsets confirmed—consistent with previous findings (see Refs. 1 and 3)—that the naïve B cells expressed unmutated IgV genes, whereas the majority of the transcripts derived from the CBs, CCs, and memory cells showed somatic mutations.

Total RNA was isolated from each of the purified B cell samples (five samples each of naïve B cells, CBs, CCs, and memory B cells), and the corresponding labeled cRNAs were generated and hybridized to Affymetrix U95A arrays representative of around 12,000 genes, as previously described.[4] Using the Genes@Work software platform, which is based on the SPLASH pattern discovery algorithm,[5] the 20 data sets were analyzed by unsupervised hierarchical clustering to determine whether the phenotypically defined B cell subsets clustered into separate groups also with regard

TABLE 1. Genes known to be differentially expressed between the B cell subsets as identified by unsupervised clustering

Cluster (A | B)	Known genes in cluster	Gene expression differential (A | B)
CB;CC | N;M	A-myb	24.1
	CD38	16.0
	Ki67	7.4
	CD10	7.2
	RGS13	5.1
	BCL7A	4.4
	PCNA	4.2
	BCL–6	3.9
	TTG–2	2.5
	CD39	–8.8
	CD44	–8.0
	BCL–2	–7.7
	CD62	–4.6
	CDw32	–4.1
M | N	CD27	14.1
	CD95/FAS	9.8
	CD80	5.2
	CD86	2.5
	CD23	–4.1
CC | CB	RAG-1	10.8
	TdT	7.1
	14.1 sLC	4.8

NOTE: Gene expression differential calculated in terms of multiples. For example, in the first row, the 24.1 means that a 24.1-fold greater expression was found in the A cluster than in the B cluster. CB: centroblast; CC: centrocyte; M: memory B cell; N: naïve B cell.

to their global gene expression. The most dominant gene expression pattern divides the 20 data sets into two groups composed of the 10 CB and CC samples on the one hand and the 10 naïve and memory samples on the other. A second round of clustering correctly divided the naïve and memory B cell samples, and the CB and CC samples, into subgroups. The corresponding patterns included genes known to be differentially expressed among GC and non-GC B cells, naïve and memory B cells, and CBs and CCs (TABLE 1). Stainings of cytospins generated from the CB and CC fraction for BCL-6 confirmed that the vast majority of GC B cells were BCL-6$^+$ (results not shown). Taken together, the above results validated the identity of the four purified B cell subpopulations.

To identify the gene expression changes of B cells during the GC response, we used the supervised clustering approach of the Genes@Work software platform (see above). Supervised clustering allows the identification of differentially expressed genes between two cell types defined *a priori* according to a given criterion, that is, cell phenotype (see Refs. 4 and 5). A crucial requirement for the identification of dif-

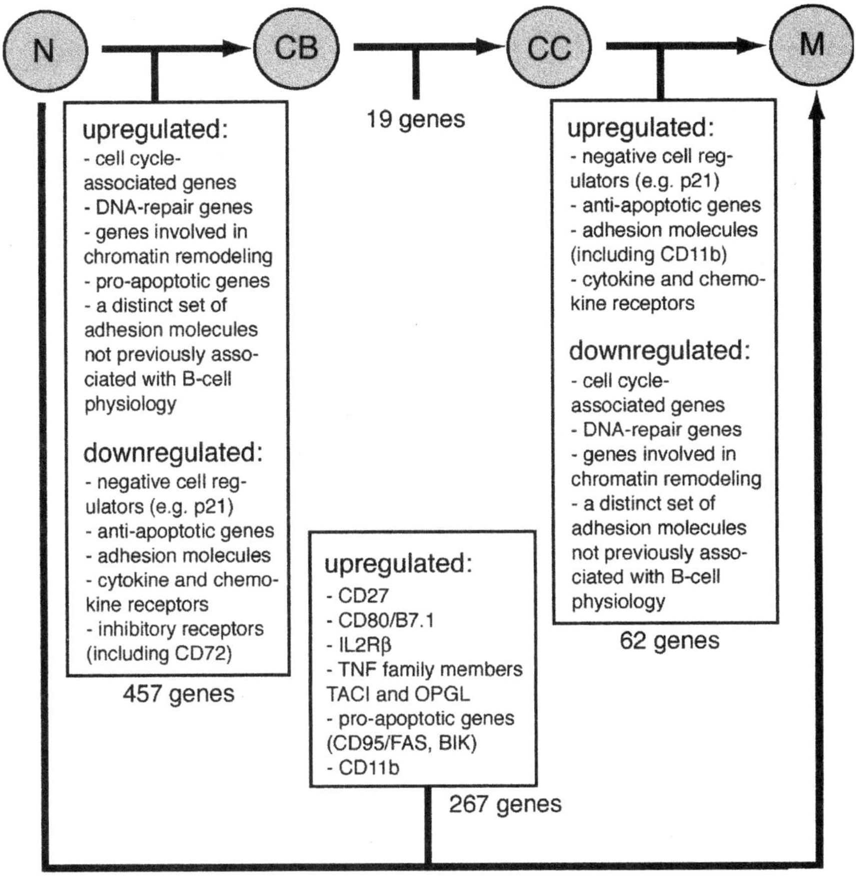

FIGURE 1. Changes in gene expression during the GC transit of B cell subpopulations identified by DNA microarray analysis and supervised hierarchical clustering. Shown are the transitions of the B cell subsets in the proposed developmental order, together with the transition of the pre-and post-GC B cells. The number of genes changing in the individual transitions is indicated. For experimental approach, see text.

ferentially expressed genes was the consistent expression of a given gene across all five samples of a given subset. The changes between each pair of B cell subpopulations (transitions) are shown in the proposed developmental order,[1] together with the pre- to post-GC B cell transition (FIG. 1).

The N → CB transition was found to involve 457 genes (FIG. 1). Consistent with the fact that the CBs constitute a highly proliferating population, genes involved in DNA replication (the A, B, E, and F-type cyclins and several CDC genes, G_1/S and G_2/M-transition genes, and mitosis-associated and -controlling genes) were found to be up-regulated in the N → CB transition. Accordingly, negative cell cycle regulators, including p21/WAF-1 and the D-type cyclins, were down-regulated in this transition. Genes associated with DNA-repair functions, as well as genes involved in chromatin-remodeling, such as histones and high mobility group proteins, were up-regulated in the CBs. Consistent with the observation that GC B cells are prone to undergo programmed cell death, pro-apoptotic genes were up-regulated, while anti-apoptotic genes where down-regulated in the N → CB transition. Among the latter was a gene named TOSO, which has recently been shown to prevent FAS-mediated apoptosis in T cells.[6] RNA encoding several cytokine and chemokine receptors was strongly down-regulated in the CB. Regarding genes implicated in adhesion, several known adhesion molecules were found to be down-regulated in the N → CB transition; on the contrary, most of the genes up-regulated in the CB have so far not been associated with adhesion functions in lymphocytes. Finally, mRNAs of five molecules with known inhibitory functions in B cell receptor (CD72, FcγRII, SHP-1, and LAIR-1) or T cell receptor signaling (SLAP) were slightly down-regulated in the N → CB transition.

The CB → CC transition involved only 19 genes (FIG. 1). This finding might either indicate that the $CD77^+$ and $CD77^-$ GC B cell subsets are very similar in their gene expression profile, or that the $CD77^-$ CC fraction is rather heterogeneous in its cellular composition—for example, it likely contains the immediate precursors of plasma or memory cells. The genes found to characterize the $CD77^-$ CC fraction analyzed here included not only the previously described TdT, RAG-1, and surrogate light chain (LC) genes,[7] but also the transcription factor LEF-1, which is required for the pro-B cell to the pre-B cell stage during early B cell development.[8] Stainings of cytospins generated from the tonsillar subsets confirmed the CC-specific TdT expression and revealed around 1% of the TdT^+ cells in this fraction (results not shown). Recent studies in the mouse showed that TdT^+/RAG^+/surrogate LC^+ B cells represent immature B cells that do not up-regulate the RAG genes in the periphery. In human tissues, histological analysis suggests that the vast majority of those cells actually reside outside the GC (G. Cattoretti, unpublished results). This would imply that the CC fractions purified by our group and the CC fractions purified by others (see Refs. 1 and 7) contained TdT^+/RAG^+/surrogate LC^+ non-GC B cells with a phenotype identical to the one that defines the CC by flow cytometrical analysis.

The CC → M transition was found to involve 267 genes (FIG. 1). Overall, genes involved in proliferation, DNA metabolism, adhesion, as well as cytokine and chemokine receptors regained expression levels comparable to those found in naïve B cells. Message encoding the adhesion molecule CD11b/Mac-1 was specifically up-regulated in the CC → M transition. This observation, together with the recent identification of CD11b on a subset of human peripheral blood $CD27^+$ B cells,[9] suggests that this molecule may characterize a memory B cell subset in the human, similar to the one described in the mouse.[10] Also up-regulated in the CC → M transition was mRNA for the recently identified TNF family member TACI.

The N → M transition involved 62 genes (FIG. 1). Compared to the massive gene expression changes at the entry and the exit of the GC reaction, this relatively small number appears surprising given that a B cell has been reprogrammed during the GC reaction to respond to its specific antigen with a different kinetic. Several independent *in vitro* activation assays clearly showed that a memory B cell, in contrast to the naïve B cell, reacts quickly to activating stimuli and secretes high amounts of Ig. To bring these seemingly contradictory observations together, one may assume that a B cell that has completed the GC reaction returns to a phenotype largely resembling that of a naïve B cell, and that only few key molecules are required to execute the characteristic cellular response of the memory B cell. First among those molecules appears to be the CD27-antigen, which (upon co-ligation with its ligand CD70 expressed on T cells) has been implicated in the differentiation of a memory B cell into an antibody-secreting cell.[2] mRNA encoding four other cell surface molecules was elevated in the memory B cell fraction. Those were the TNF family members TACI and OPGL, which have been implicated in the control of immune responses to CD80/B7.1 and IL-2Rβ, respectively. Staining of tonsillar sections for IL-2Rβ confirmed the expression of this molecule on the surface of subepithelial memory B cells as well as on memory cells located in the tonsillar marginal zone equivalent (results not shown). Moreover, while the vast majority of GC B cells did not stain for IL-2Rβ, a subset of B cells localized in the light zone, the GC area where B cells are thought to be selected and instructed to differentiate into post-GC cells, was IL-2Rβ$^+$. These findings suggest that IL-2Rβ is up-regulated on a subset of GC B cells that differentiate into memory B cells. Whether IL-2Rβ is expressed on all memory B cells or defines a memory B cell subset, and what its functional significance might be, remains to be addressed. Finally, memory cells, like naïve B cells, expressed the anti-apoptotic Bcl-2 and TOSO (see above). Unlike the naïve cells, however, memory cells (or a subset thereof) are CD95/FAS$^+$ and show up-regulation of the pro-apoptotic Bcl-2 interacting killer. This observation may hint at the possibility that memory B cells are inherently susceptible to apoptotic stimuli transmitted by immune helper cells, for example, in a situation where an aberrantly activated memory B cell needs to be deleted.

The gene expression profiles established here were used to track signals known to be crucial for B cell development in the normal B cell subsets[11] and to identify the non-malignant counterpart of B cell derived malignancies.[4] In a first analysis, we observed that the common gene expression profile of B cell chronic lymphocytic leukemia, a CD5$^+$ malignancy that was long thought to be derived from the CD5$^+$ B cell lineage, is more closely related to memory B cells than to those derived from CD5$^+$ B cells, naïve B cells, and GC B cells.[4]

ACKNOWLEDGMENTS

U.K. was recipient of a fellowship granted by the Human Frontiers Science Program.

REFERENCES

1. PASCUAL, V., Y.-J. LIU, A. MAGALSKI, *et al.* 1994. Analysis of somatic mutation in five B-cell subsets of human tonsil. J. Exp. Med. **180:** 329–339.
2. AGEMATSU, K., S. HOKIBARA, H. NAGUMO, *et al.* 2000. CD27: a memory B-cell marker. Immunol. Today **21:** 204–206.
3. KLEIN, U., T. GOOSSENS, M. FISCHER, *et al.* 1998. Somatic hypermutation in normal and transformed human B cells. Immunol. Rev. **162:** 261–280.
4. KLEIN, U., Y. TU, G. STOLOVITZKY, *et al.* 2001. Gene expression profiling of B-cell chronic lymphocytic leukemia reveals a homogeneous phenotype related to memory B cells. J. Exp. Med. **194:** 1625–1638.
5. CALIFANO, A., G. STOLOVITZKY & Y. TU. 2000. Analysis of gene expression microarrays for phenotype classification. Proc. Int. Conf. Intell. Syst. Mol. Biol. **8:** 75–85.

6. HITOSHI, Y., J. LORENS, S.I. KITADA, *et al.* 1998. TOSO, a cell surface, specific regulator of FAS-induced apoptosis in T cells. Immunity **8:** 461–471.
7. MEFFRE, E., F. PAPAVASILIOU, P. COHEN, *et al.* 1998. Antigen receptor engagement turns off the V(D)J recombination machinery in human tonsil B cells. J. Exp. Med. **188:** 765–772.
8. REYA, T., M. O'RIORDAN, R. OKAMURA, *et al.* 2000. Wnt signaling regulates B lymphocyte proliferation through a LEF-1-dependent mechanism. Immunity **13:** 15–24.
9. BAR-OR, A., E.M. OLIVEIRA, D.E. ANDERSON, *et al.* 2001. Immunological memory: contribution of memory B cells expressing co-stimulatory molecules in the resting state. J. Immunol. **167:** 5669–5677.
10. MCHEYZER-WILLIAMS, L.J., M. COOL & M.G. MCHEYZER-WILLIAMS. 2000. Antigen-specific B-cell memory: expression and replenishment of a novel B220(–) memory B-cell compartment. J. Exp. Med. **191:** 1149–1166.
11. BASSO, K., U. KLEIN, H. NIU, *et al.* 2003. Tracking CD40 signaling during normal germinal center development by gene expression profiling. Ann. N.Y. Acad. Sci. **987:** this volume.

Somatic Hypermutation and B Cell Receptor Selection in Normal and Transformed Human B Cells

RALF KÜPPERS

Institute for Genetics and Department of Internal Medicine I, University of Cologne, Cologne, Germany

ABSTRACT: From the beginning to the end, the life of B cells is dominated by selection of the cells for expression of an appropriate antigen receptor. However, recent studies revealed that there are several diseases in the human where B cells lost their dependence on a B cell receptor (BCR). In classic Hodgkin's lymphoma, the lymphoma cells presumably derive from "crippled" germinal center (GC) B cells that acquired unfavorable somatic Ig gene mutations, which often render originally functional immunoglobulin (Ig) genes nonfunctional. A peculiar situation is observed among Epstein-Barr virus (EBV)-infected B cells in angioimmunoblastic lymphadenopathy with dysproteinaemia (AILD)-type T cell lymphoma, where somatic hypermutation uncoupled from any selection for functionality of the BCR is observed in expanding clones. Clones of EBV-harboring B cells that show ongoing hypermutation during proliferation and are Ig-deficient in at least a fraction of cases were recently also identified in post-transplant lymphoproliferative disorders. Hence, transformed B cells may, in particular settings, escape the normal selectional forces to express a BCR, and EBV may cause dramatic changes in B cell differentiation programs. Somatic hypermutation may be involved in lymphomagenesis by several means. Some chromosomal translocations into Ig loci likely involve DNA-strand breaks associated with hypermutation. Moreover, by aberrant targeting of the CD95 gene, GC B cells and lymphomas developing from them may become resistant to elimination by CD95 ligand-expressing T cells. Finally, aberrant hypermutation of multiple proto-oncogenes appears to be a major factor in diffuse large cell lymphoma pathogenesis.

KEYWORDS: angioimmunoblastic lymphadenopathy; Epstein-Barr virus; germinal center; Hodgkin's lymphoma; post-transplant lymphoproliferative disease; somatic hypermutation

Address for correspondence: Ralf Küppers, University of Cologne, Department of Internal Medicine I, LFI E4 R706, Joseph-Stelzmannstr. 9, D-50931 Cologne, Germany. Voice: ++49 221 478 4490; fax: ++49 221 478 6383.
ralf.kuppers@uni-koeln.de

Ann. N.Y. Acad. Sci. 987: 173–179 (2003). © 2003 New York Academy of Sciences.

ROLE OF THE B CELL RECEPTOR IN B CELL DEVELOPMENT AND SURVIVAL

The main task of B cells in humoral immune responses is to search for infectious agents and other foreign antigens that enter the body and to mediate their elimination by producing large amounts of specific antibody. To be able to recognize a practically unlimited diversity of antigens, each newly generated B cell is equipped with a distinct B cell receptor (BCR). This enormous diversity of antibody specificities is generated during B cell development in the bone marrow by a somatic DNA recombination process that joins gene segments coding for the antigen-binding variable (V) domains of the antibody heavy and light chains.[1] In the course of T cell-dependent immune responses, antibody molecules are further diversified when antigen-activated B lymphocytes establish germinal centers (GCs). In these structures, the process of somatic hypermutation is activated that introduces mutations at a very high rate into the immunoglobulin (Ig) V region genes.[2,3]

As the main role of a B cell is to search for and produce specific antibodies against foreign antigens, it is not surprising that the life of B cells is dominated by selection for the appropriate cell surface receptor.[1] During B cell development in the bone marrow, B cell precursors are first selected for productive heavy-chain gene rearrangements. If a heavy chain is produced and expressed as a pre-BCR, light-chain gene rearrangements are performed, and the cells are selected for expression of surface Ig.[1] Selection for an appropriate BCR is also a critical step in the GC reaction. Among GC B cells producing antibody variants generated by somatic hypermutation, cells carrying antigen receptors with improved affinity for the respective antigen are positively selected.[4] As the hypermutation process is largely random, most GC B cells will acquire disadvantagous mutations and be eliminated within the GG by apoptosis.

Studies in the mouse indicate that even resting mature B cells are controlled for and depend on expression of surface Ig.[5] Hence, at all stages of their development, B cells are selected for expression of an appropriate antigen receptor. The finding that most B cell lymphomas express a BCR suggests that this dependency on a BCR also extends to malignant B cells.[6]

ORIGIN OF HODGKIN AND REED/STERNBERG CELLS IN CLASSIC HODGKIN'S LYMPHOMA FROM CRIPPLED GC B CELLS

Hodgkin's lymphoma (HL) is unusual among human lymphomas because the malignant cells, that is, the typical Hodgkin and Reed/Sternberg (HRS) cells, account for only a small minority of cells in the tumor tissue. In most cases of classic HL, the HRS cells derive from B cells.[7–11] Although HRS cells usually lack expression of typical B cell surface markers, their origin from mature B cells is evident from the finding that they carry rearranged Ig heavy- and light-chain genes.[7–9] Moreover, the presence of somatic mutations in the rearranged V genes in nearly all cases established that these cells participated in a GC reaction. Surprisingly, obviously crippling mutations (that is, nonsense mutations and deletions causing loss of the correct reading frame) that rendered originally functional V gene rearrangements nonfunctional were identified in a quarter of the cases.[7–9,12] As GC B cells that acquire such mutations are normally eliminated within the GC, it is likely that the HRS cells in these cases derived from pre-apoptotic GC B cells that were rescued from apoptosis by some transforming event.[12,13] Moreover, as most disadvantagous somatic mutations cannot be easily identified (for example, replacement mutations that reduce the affinity to the antigen), we speculated that HRS cells as a rule derive from crippled GC B cells.[7,12,13]

LOSS OF THE B CELL IDENTITY OF HRS CELLS: A STRATEGY TO ESCAPE SELECTION FOR A FUNCTIONAL BCR?

One of the reasons why the origin of the HRS cells in classic HL was unclear for a long time is their unusual immunophenotype. This immunophenotype does not correspond to any normal lymphoid cell. HRS cells often show co-expression of markers typical for distinct cell lineages. Typical B cells markers, such as CD19, CD20, CD79a, and CD79b, are expressed only in a small fraction of cases.[14] Moreover, HRS cells in classic HL lack expression of Ig (note that most crippling mutations, for example, unfavorable replacement mutations, do not intrinsically cause a down-regulation of Ig transcription).[9]

To study the down-regulation of B cell lineage markers in HRS cells systematically, we recently performed gene expression profiling studies with HL cell lines. In a SAGE (serial analysis of gene expression) analysis, the gene expression profile of the HRS cell line L1236 was compared with a profile from human GC B cells. In another approach, high-density oligonucleotide microarrays were used to compare four HL lines (L428, L1236, KMH2, and HDLM2) with samples from normal human naive, GC, and memory B cells. Both studies revealed a global down-regulation of the B cell lineage gene expression program.[15] The down-regulated genes included B cell-specific genes (such as A-myb and RP105), lymphocyte-specific genes (such as CD52 and Pu-B), and also genes expressed by various hematopoietic cell types, but with important roles in B cell physiology (such as c-src and Lyn). Hence, there is not only down-regulation of some B cell markers, but a fundamental defect in the maintenance of the B cell differentiation program in HRS cells. As other B cell lymphomas usually retain most or at least many features of their normal counterparts, reflecting a close association between transforming events and the differentiation program, one may speculate that the nature of at least one transforming event in HL pathogenesis differs fundamentally from those in other B cell lymphomas. Perhaps a master regulator of cell fate decisions is affected in HRS cells, causing the unusual phenotype of these cells.

Intriguingly, the global down-regulation of the B cell-specific gene expression program in HRS cells offers an explanation for the survival of Ig-deficient B cells in classic HL. As the stringent selection of B cells for expression of a BCR likely depends on the typical gene expression program of these cells, HRS cells may survive the loss of BCR expression by losing their B cell identity.

GENERATION AND SURVIVAL OF CRIPPLED EBV-INFECTED B CELLS IN T CELL LYMPHOMA OF THE AILD TYPE

Angioimmunoblastic lymphadenopathy with dysproteinaemia (AILD) is one of the most frequent T cell lymphomas. Often, large numbers of EBV-infected B cells are found in the AILD lymphoma tissue.[16,17] We isolated single EBV-infected cells from six EBV-rich cases of AILD and analyzed their rearranged Ig genes to determine their clonal composition and differentiation stage.[18] In all cases, clonal expansions of varying numbers and sizes were observed among the EBV-infected B cells. Most clones were characterized by mutated Ig V genes, and intraclonal sequence diversity was observed in all large clones.[18] Surprisingly, many V region genes in clones with ongoing hypermutation carried crippling mutations. In several clones, hypermutation continued after acquisition of crippling mutations early in the clonal expansion. This indicates that in these EBV-infected B cells, somatic hypermutation occurred without any selection for functionality and even in the absence of BCR expression. Hence, this analysis shows that crippled B cells can develop and persist also in diseases other than classic HL. However, the two situations differ markedly. In HL, the HRS cell precursors likely represent pre-apoptotic GC B cells that turned *off* somatic hypermutation when they underwent malignant transformation, as there is no intraclonal sequence diversity in HRS cells of classic HL.[7,9] In AILD, somatic hypermutation was presumably turned *on* when the EBV-infected B cells started to proliferate, and this mutation activity continues even after acquisition of destructive mutations.[18] The reason for the sustained hypermutation activity in EBV-infected B cells in AILD is unknown, but may be related to influences of the cellular microenvironment, which is characterized by dense networks of follicular dendritic cells and the CD4 T cell tumor clone.

CRIPPLED EBV-INFECTED B CELLS IN POST-TRANSPLANTLYMPHOMAS

Post-transplant lymphoproliferative diseases frequently developing in immunosuppressed transplant patients usually represent proliferations of EBV-infected B cells.[19] Besides polyclonal or oligoclonal expansions of these cells, malignant B cell lymphomas may also develop.[19] In a recent analysis of five cases of post-transplant lymphoproliferative disease by micromanipulation and single-cell PCR for rearranged V genes, two cases were encountered that showed ongoing mutation in large EBV-positive B cell clones in the absence of follicular dendritic cells and T helper cells.[20] One of these cases was a monomorphic centroblastic lymphoma, which showed ongoing mutation after acquisition of crippling Ig gene mutations, similar to what we observed previously in EBV-infected B cells in AILD. The other case was peculiar as it harbored two large clonal expansions. One clone was represented by typical HRS cells and likely represented the malignant tumor clone, as the histology of the case was typical for classic HL. However, among smaller EBV-infected B cells, a second large unrelated (tumor?) B cell clone was identified. This clone showed ongoing mutation during clonal expansion, while hypermutation was silenced in the HRS cell clone. Thus, expanded clones of EBV-infected B cells may retain or acquire features of GC B cells (that is, somatic hypermutation) in an unphysiological setting.

THE ROLE OF SOMATIC HYPERMUTATION IN B CELL LYMPHOMAGENESIS

Reciprocal chromosomal translocations involving Ig loci are a hallmark of many B cell lymphomas.[21,22] Such translocations usually result in the dysregulated expression of oncogenes that become juxtaposed to the Ig enhancers. The molecular structure of the translocation breakpoints suggests that most of them are generated as byproducts of either V gene recombination, class-switching, or somatic hypermutation.[23] The frequent involvement of these processes in the generation of translocations is likely attributable to the association of each of these processes to DNA-strand breaks.

Somatic hypermutation may be involved in lymphomagenesis also by targeting non-Ig genes. The bcl-6 gene is mutated in about one-third of normal human GC and memory B cells, and consequently also in many GC B cell-derived lymphomas.[24–26] It remains to be determined whether some bcl-6 mutations cause dysregulated expression of the gene that may promote lymphoma development. The CD95 gene is mutated in a small fraction of GC B cells[27] and a considerable fraction of GC B cell-derived lymphomas, many of which are associated with autoimmune diseases.[28,29] GC B cells with inactivating CD95 gene mutations can be selected from human tonsils, suggesting that resistance to CD95-mediated apoptosis can be acquired by normal GC B cells.[27] As CD95 is a death receptor inducing apoptosis upon cross-linking, CD95-inactivating mutations may prevent elimination of GC B cells and lymphoma cells developing from them by CD95 ligand-expressing T cells.

Somatic hypermutation plays a particularly important role in the pathogenesis of diffuse large cell lymphomas. In these lymphomas, four proto-oncogenes (c-myc, Pax-5, Rho/TTF, Pim-1) were found mutated in a large fraction of cases.[30] The mutation pattern suggests that they resulted from (aberrant) somatic hypermutation. The particular genes were mutated at significant frequency only in diffuse large cell lymphomas, but not in other B cell lymphomas or normal GC B cells, pointing to a diffuse large cell lymphoma-specific process.[30] The aberrant hypermutation activity in these lymphoma cells or their precursors may cause lymphomagenesis by changing gene regulation or coding sequence of these proto-oncogenes. Moreover, the four hypermutable genes are susceptible to chromosomal translocations in the same regions that are targeted by hypermutation, suggesting that the generation of these translocations is linked to the generation of DNA-strand breaks during hypermutation in these genes.[30]

ACKNOWLEDGMENTS

This work was supported by the Deutsche Forschungsgemeinschaft (grant SFB502) and a Heisenberg award to the author.

REFERENCES

1. RAJEWSKY, K. 1996. Clonal selection and learning in the antibody system. Nature **381:** 751–758.
2. KÜPPERS, R. *et al.* 1993. Tracing B cell development in human germinal centres by molecular analysis of single cells picked from histological sections. EMBO J. **12:** 4955–4967.
3. NEUBERGER, M.S. & C. MILSTEIN. 1995. Somatic hypermutation. Curr. Opin. Immunol. **7:** 248–254.
4. WEISS, U., R. ZOEBELEIN & K. RAJEWSKY. 1992. Accumulation of somatic mutants in the B cell compartment after primary immunization with a T cell-dependent antigen. Eur. J. Immunol. **22:** 511–517.
5. LAM, K.P., R. KÜHN & K. RAJEWSKY. 1997. In vivo ablation of surface immunoglobulin on mature B cells by inducible gene targeting results in rapid cell death. Cell **90:** 1073–1083.
6. KÜPPERS, R. *et al.* 1999. Cellular origin of human B-cell lymphomas. N. Engl. J. Med. **341:** 1520–1529.
7. KANZLER, H. *et al.* 1996. Hodgkin and Reed-Sternberg cells in Hodgkin's disease represent the outgrowth of a dominant tumor clone derived from (crippled) germinal center B cells. J. Exp. Med. **184:** 1495–1505.
8. KÜPPERS, R. *et al.* 1994. Hodgkin disease: Hodgkin and Reed-Sternberg cells picked from histological sections show clonal immunoglobulin gene rearrangements and appear to be derived from B cells at various stages of development. Proc. Natl. Acad. Sci. USA **91:** 10962–10966.
9. MARAFIOTI, T. *et al.* 2000. Hodgkin and Reed-Sternberg cells represent an expansion of a single clone originating from a germinal center B-cell with functional immunoglobulin gene rearrangements but defective immunoglobulin transcription. Blood **95:** 1443–1450.
10. MÜSCHEN, M. *et al.* 2000. Rare occurrence of classical Hodgkin's disease as a T cell lymphoma. J. Exp. Med. **191:** 387–394.
11. SEITZ, V. *et al.* 2000. Detection of clonal T-cell receptor gamma-chain gene rearrangements in Reed-Sternberg cells of classic Hodgkin disease. Blood **95:** 3020–3024.
12. KÜPPERS, R. 2002. Molecular biology of Hodgkin's lymphoma. Adv. Cancer. Res. **44:** 277–312.
13. KÜPPERS, R. & K. RAJEWSKY. 1998. The origin of Hodgkin and Reed/Sternberg cells in Hodgkin's disease. Annu. Rev. Immunol. **16:** 471–493.
14. WATANABE, K. *et al.* 2000. Varied B-cell immunophenotypes of Hodgkin/Reed-Sternberg cells in classic Hodgkin's disease. Histopathology **36:** 353–361.
15. SCHWERING, I., A. BRÄUNINGER, U. KLEIN, *et al.* 2003. Loss of the B-lineage-specific gene expression program in Hodgkin and Reed-Sternberg cells of Hodgkin lymphoma. Blood **101:** 1505–1512.
16. ANAGNOSTOPOULOS, I. *et al.* 1992. Heterogeneous Epstein-Barr virus infection patterns in peripheral T-cell lymphoma of angioimmunoblastic lymphadenopathy type. Blood **80:** 1804–1812.
17. WEISS, L.M. *et al.* 1992. Detection and localization of Epstein-Barr viral genomes in angioimmunoblastic lymphadenopathy and angioimmunoblastic lymphadenopathy-like lymphoma. Blood **79:** 1789–1795.
18. BRÄUNINGER, A. *et al.* 2001. Survival and clonal expansion of mutating "forbidden" (immunoglobulin receptor-deficient) Epstein-Barr virus-infected B cells in angioimmunoblastic T cell lymphoma. J. Exp. Med. **194:** 927–940.
19. SWERDLOW, S.H. 1992. Post-transplant lymphoproliferative disorders: a morphologic, phenotypic and genotypic spectrum of disease. Histopathology **20:** 373–385.
20. BRÄUNINGER, A., T. SPIEKER, A.S. BAUR, *et al.* 2003. Submitted for publication.
21. DALLA-FAVERA, R. & G. GAIDANO. 2001. Molecular biology of lymphomas. *In* Cancer: Principles and Practice of Oncology. V.T. De Vita, S. Hellman & S.A. Rosenberg, Eds.: 2215–2235. 6th edit. Lippincott Williams & Wilkins. Philadelphia, PA.
22. SIEBERT, R. *et al.* 2001. Molecular features of B-cell lymphoma. Curr. Opin. Oncol. **13:** 316–324.

23. KÜPPERS, R. & R. DALLA-FAVERA. 2001. Mechanisms of chromosomal translocations in B cell lymphomas. Oncogene **20:** 5580–5594.
24. MIGLIAZZA, A. *et al.* 1995. Frequent somatic hypermutation of the 5′ noncoding region of the BCL6 gene in B-cell lymphoma. Proc. Natl. Acad. Sci. USA **92:** 12520–12524.
25. PASQUALUCCI, L. *et al.* 1998. BCL-6 mutations in normal germinal center B cells: evidence of somatic hypermutation acting outside Ig loci. Proc. Natl. Acad. Sci. USA **95:** 11816–11821.
26. SHEN, H.M. *et al.* 1998. Mutation of BCL-6 gene in normal B cells by the process of somatic hypermutation of Ig genes. Science **280:** 1750–1752.
27. MÜSCHEN, M. *et al.* 2000. Somatic mutation of the CD95 gene in human B cells as a side-effect of the germinal center reaction. J. Exp. Med. **192:** 1833–1840.
28. GRONBAEK, K. *et al.* 1998. Somatic Fas mutations in non-Hodgkin's lymphoma: association with extranodal disease and autoimmunity. Blood **92:** 3018–3024.
29. MÜSCHEN, M. *et al.* 2002. The origin of CD95 gene mutations in B cell lymphoma. Trends Immunol. **23:** 75–80.
30. PASQUALUCCI, L. *et al.* 2001. Hypermutation of multiple proto-oncogenes in B-cell diffuse large-cell lymphomas. Nature **412:** 341–346.

Dendritic Cells

Controllers of the Immune System and a New Promise for Immunotherapy

JACQUES BANCHEREAU,[a] SOPHIE PACZESNY,[a] PATRICK BLANCO,[a] LYNDA BENNETT,[a,b] VIRGINIA PASCUAL,[a,b] JOSEPH FAY,[a] AND A. KAROLINA PALUCKA[a]

[a]*Baylor Institute for Immunology Research, Dallas, Texas, USA*

[b]*University of Texas Southwestern Medical Center, Dallas, Texas, USA*

ABSTRACT: The immune system is controlled by dendritic cells (DCs). Just as lymphocytes comprise different subsets, DCs comprise several subsets that differentially control lymphocyte function. In humans, the myeloid pathway includes Langerhans cells (LCs) and interstitial DCs (intDCs). While both subsets produce IL-12, only intDCs make IL-10 and induce B cell differentiation. Another pathway includes plasmacytoid DCs, which promptly secrete large amounts of IFN-α/β upon viral encounter. Thus, insights into *in vivo* DC functions are important to understand the launching and modulation of immunity.

KEYWORDS: dendritic cells; Langerhans cells; interstitial DCs; T cells; immunotherapy

INTRODUCTION

Dendritic cells (DCs) induce and regulate immune responses. The discovery of *in vitro* culture systems yielding large amounts of mouse and human[1,2] DCs accelerated studies on their biology and led us to understand their complexity. Indeed, DCs have different functions at different stages of maturation. While immature (non-activated) DCs that capture self-antigens (e.g., apoptotic cells) induce tolerance in the steady state,[3,4] mature antigen-loaded DCs induce antigen-specific immunity.[5,6] An additional level of complexity comes with the existence of several DC subsets with different functions[7,8] (FIG. 1). In humans, two major DC pathways are described. A myeloid pathway, with two subsets, including Langerhans cells (LCs), found in stratified epithelia such as skin and interstitial DCs (intDCs), found in all other tissues. While both subsets can produce IL-12, only intDCs can make IL-10 and induce naïve B cell differentiation.[9] The other main pathway includes the plasmacytoid DCs (pDCs), which secrete, within a few hours following viral encounter, large amounts of type I interferon,[10,11] a family of cytokines with antiviral and im-

Address for correspondence: A. Karolina Palucka, Baylor Insitute for Immunology Research, 3434 Live Oak, Dallas, TX 75204. Voice: 214-820-7450; fax: 214-820-4813.
karolinp@baylorhealth.edu

Ann. N.Y. Acad. Sci. 987: 180–187 (2003). © 2003 New York Academy of Sciences.

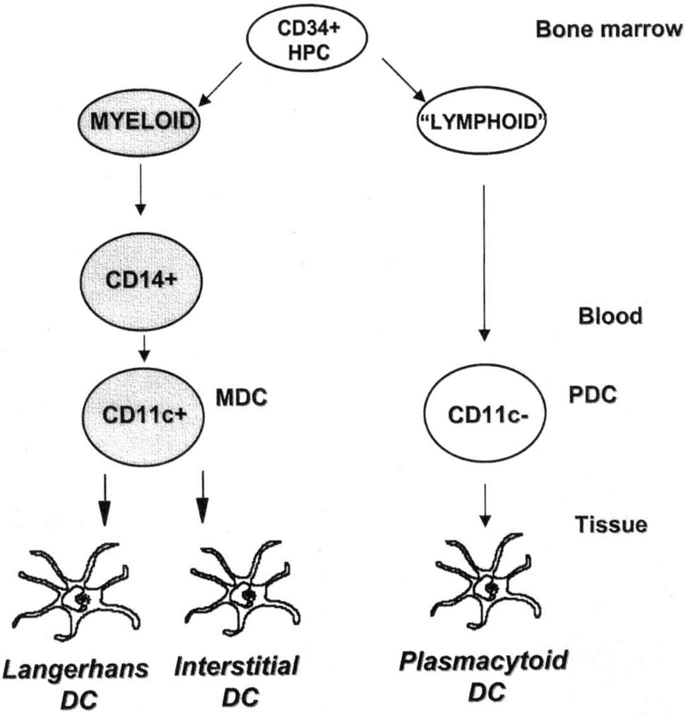

FIGURE 1. Subsets of human dendritic cells. Blood DCs, mobilized by FLT3 ligand, contain both CD11c$^+$ myeloid DC and CD11c$^-$ plasmacytoid DC. Most clinical studies to date have been carried out with DC made by culturing monocytes with GM-CSF and IL-4. These preparations contain cells that resemble intDCs and are devoid of LCs. These DCs are immature and require exogenous factors (CD40 ligand or macrophage cytokines) for maturation. DCs can also be generated by culturing CD34$^+$ HPC with GM-CSF and TNF-α that allows to obtain two DC subsets: LCs and intDCs. Adding IL-4 to CD34 cultures with GM-CSF/TNF skews differentiation toward intDCs. A distinct subset of HPC, CD34$^+$CD45RA$^-$ gives rise *in vitro* to plasmacytoid DCs upon culture with FLT3 ligand.

munomodulatory properties. Furthermore, depending upon the maturation signal that they receive, pDCs modulate T cell differentiation into either IFN-γ or IL-4 producing T cells[11] or into regulatory T cells.[12] Thus, distinct human DC subsets differentially control lymphocytes *in vitro.*

DENDRITIC CELLS AS VACCINES TO ENHANCE IMMUNITY

The *in vitro* generation of large numbers of human DCs prompted their use in immunotherapy approaches, particularly in cancer. Studies in mice have shown that injection of DCs loaded with tumor-associated antigens (TAAs) leads to anti-tumor immune responses resulting in tumor rejection.[13] Early trials in humans have shown the safety of TAA-loaded DCs as well as some clinical and immune responses. Recent studies concentrated on establishing maximal immune responses to control antigens and tumor antigens. Many issues remain to be addressed before DC therapy becomes an integral part of active immunotherapy, that is, modulation of the immune system for therapeutic purposes.[13] These include the choice of the DC subset to be administered and the way to generate it. The most popular way is to culture blood monocytes with the cytokines that permit generation of interstitial-like DC, that is, GM-CSF and IL-4, which yield a uniform population of immature DC resembling intDCs. This contrasts with hematopoietic stem cells that, when cultured with GM-CSF and TNF, yield preparations that are composed of two distinct DC subsets, LC and intDC. While GM-CSF/IL-4–induced DCs require additional maturation factors, $CD34^-$ DCs do not because they are generated in the presence of TNF-α, a DC-activation factor. Two precursor subpopulations emerge under these culture conditions, $CD1a^+$ and $CD14^+$, which can differentiate into LC and intDC, respectively.[9,14] Both precursor subsets mature at day 12–14 into DC with typical morphology, phenotype (CD80, CD83, CD86, CD58, high HLA class II) and function. $CD1a^+$ precursors give rise to cells with Langerhans cell characteristics (Birbeck granules, Lag antigen and E-cadherin). In contrast, the $CD14^+$ precursors mature into $CD1a^+$ DC lacking Birbeck granules, E-cadherin and Lag antigen, but expressing CD2, CD9, CD68 and the coagulation factor XIIIa described in dermal dendritic cells. These DC subsets differ with regard to (1) antigen capture—intDC are more efficient in Dextran uptake and in binding immune complexes than LC, (2) lysosomal activity—intDC express higher levels of enzymatic activity, (3) capacity to induce naïve B cell differentiation—intDC are uniquely able to induce the priming of naïve B cells and their differentiation into IgM secreting plasma cells through the secretion of IL-12,[15,16] and (4) while both subsets express IL-12 upon CD40 ligation, only intDC express IL-10. While the unique function of LCs remains to be established, there are some reports suggesting that CD34-DCs with LCs component are more efficient in CD8 T cell priming than monocyte-derived DCs.[17,18] In the circumstances such as the induction of tumor-specific CTLs, CD34-DCs could thus be advantageous.

We have vaccinated 18 HLA A^*0201^+ patients with stage IV melanoma with CD34-HPC-derived DCs pulsed with six Ags: influenza matrix peptide (Flu-MP), KLH, and peptides derived from the four melanoma Ags: MART-1/Melan A, gp100, tyrosinase, and MAGE-3.[19] The DCs were loaded with KLH, Flu matrix peptide and HLA-A2 binding peptides derived from four melanoma antigens MAGE-3, MART-1/MELAN A, GP-100, and tyrosinase and administered biweekly over six weeks (four s.c. injections). Vaccination of patients with advanced melanoma with antigen-pulsed CD34-DCs has proven to be well tolerated and results in enhanced immunity to both "non-self" (viral peptide and KLH protein) and "self" (melanoma peptides) antigens.[19] It was observed that the level of immune responses in the blood correlated with early outcome at the tumor sites, thus providing further stimulus for the idea that the measurement of immune responses in the blood helps evaluate vaccine potency. Indeed, antigen-pulsed $CD34^+$ HPC-derived DCs induced primary and recall immune responses detectable directly in the blood. An immune response to control antigens (KLH, Flu-MP) was observed in 16 of 18 patients. An enhanced immune response to one or more melanoma antigens (MelAg) was seen in these same 16 patients, including 10 patients who responded to > 2 MelAg (FIG. 2). The MelAg specific T cells elicited after DC vaccine are functional, and detectable in effector T cell

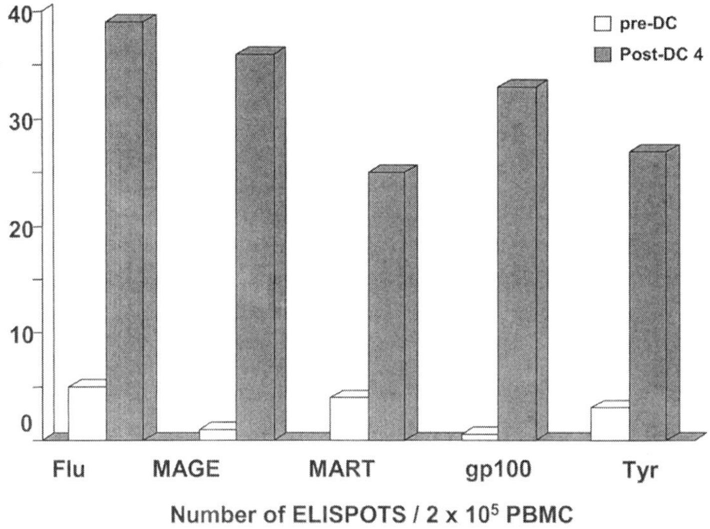

FIGURE 2. Immune responses to CD34-derived DC vaccine in advanced melanoma in one best-case scenario. Circulating viral and melanoma-specific effector cells. PBMC obtained at baseline and after four DC vaccinations are cultured overnight with each of the four melanoma peptides used for vaccination or with Flu-MP (10 μM). The specific T cell response in each of the evaluated patients is expressed as number of IFN-γ ELISPOTS/2 × 105 PBMC. The values obtained with control gag peptide or in the wells with no peptide are subtracted (on average 1 spot/2 ×105 PBMC, range 0–8) .

assays without the need for prior *ex vivo* expansion. They are also capable of proliferation and effector function after short-term (1 week) co-culture with antigen bearing DCs, without the need for exogenous cytokines or multiple restimulations with antigen. The two patients who failed to respond to both control and tumor antigens experienced rapid tumor progression. Of 17 patients with evaluable disease, 6 of 7 patients with immunity to 2 or less MelAg had progressive disease at 10 weeks after study entry, in contrast to tumor progression in only 1 of 10 patients with immunity to > 2 MelAg. Regression of > 1 tumor metastasis was observed in 7 of these patients. The overall immunity to melanoma antigens following DC vaccination was associated with clinical outcome ($P = 0.015$). Thus, CD34-DCs represent an efficient vector for immunization.

Another important parameter to establish is the dose and frequency of DC administration. Unlike traditional chemotherapy, the highest dose may not yield the best clinical response. Likewise, too frequent administration may result in activation-induced cell death, resulting in elimination of T cells that are able to kill cancer cells. It is believed that optimal anti-tumor effects will be obtained with many vaccinations possibly over a lifelong schedule. Our results show that Ags-pulsed CD34-DCs lead to rapid induction of $CD4^+$ and $CD8^+$ T cell immunity. Indeed, a single DC vaccination was sufficient for induction of KLH-specific responses in 6 patients and Flu-MP–specific responses in 8 patients. A single DC vaccine was sufficient to induce tumor-specific effectors to ≥ 1 melanoma antigen in 5 patients. None of these 5 patients showed early disease progression. Only 1 of 6 patients with rapid KLH-response experienced early disease progression. Rapid and slow Flu-MP responders did not differ with regard to disease progression. Thus, the patient's ability to rapidly respond to CD4 epitopes or to melanoma Ags could be an early indicator of clinical outcome after DC-based vaccination.

As with many classical vaccines, DC vaccines will likely be administered through the skin. Antigen loading is the subject of considerable research is the antigen loading. At present, DCs are mostly loaded with peptides from defined (tumor) antigens that bind to MHC class I and II antigens. This presents numerous limitations such as (1) the restriction to a given MHC type; (2) the limited number of TAAs, which restricts vaccination therapy to mostly melanoma, for which many TAAs have been identified; and (3) the limited repertoire of elicited immune effectors which may not allow eradication of the multiple tumor variants.

TARGETING DENDRITIC CELLS TO INDUCE TOLERANCE

Controlling an immune response is equally complex as is its launching. Indeed, upon entry of the pathogen, for instance, the influenza virus, DCs will capture and present microbe-derived antigens to immune effectors to induce immunity. However, DCs will present not only virus antigens but also self-antigens. Thus, a mechanism must be in place not only to allow the elimination of virally infected cells, but also to prevent the immune system attack on the components of self. Two mechanisms were created—central and peripheral tolerance—both of which are controlled and maintained by DCs. Central tolerance occurs in thymus where newly generated T cells with a receptor that recognizes components exposed by mature thymic DCs are deleted. However, many self-antigens may not access the thymus while other are expressed later in life. Upon activation, these autoreactive cells may lead to autoimmunity. Hence, the need for peripheral tolerance, which occurs in lymphoid organs by induction of T cell anergy, that is, unresponsiveness, rather than deletion. The development of peripheral tolerance involves immature DCs.[4] These cells sitting within tissues capture the remains of cells that die in the process of physiological tissue turnover. Because there is no inflammation accompanying this process, the DCs remain immature and migrate toward the draining lymph nodes. These immature DCs, which lack co-stimulatory molecules, present the tissue antigens to autoreactive T cells, which in the absence of co-stimulation, enter into a state of anergy. Hence, increased availability of mature DCs may result in autoimmunity. This concept is illustrated by our recent studies in patients suffering from systemic lupus erythematosus. Blood monocytes from active SLE patients were found to act as DCs inasmuch as they are able to induce the proliferation of naïve allogeneic T cells (FIG. 3).[20] Further analysis demonstrated that the serum of active patients was able to induce the differentiation of healthy monocytes into cells with properties of DCs. The critical factor involved in DC maturation was found to be interferon-α[20] though the increased levels of FLT3 ligand[21] may contribute to the activation of monocytes. pDCs were considered as a possible source of the INF-α because they represent the most efficient producers of IFN-α in the blood. Curiously, the numbers of pDCs ($CD123^{+}CD11c^{-}$) were found to be ~70% reduced in SLE blood. These $CD123^{+}$

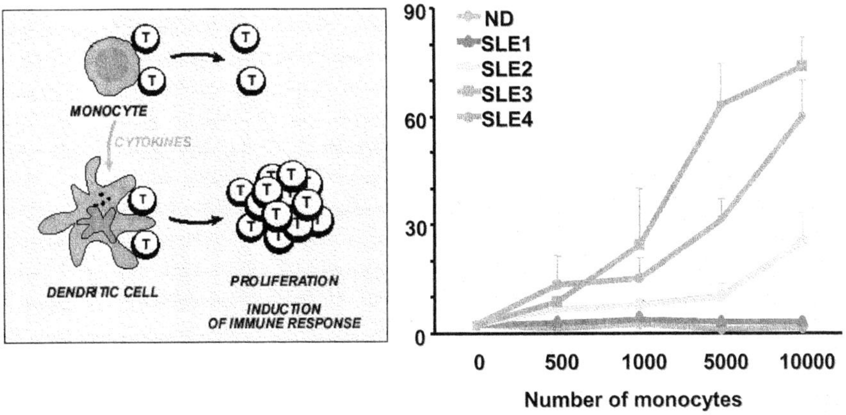

FIGURE 3. Monocytes freshly isolated from blood of patients with SLE with high serum levels of IFN-α are able to stimulate MLR. Indicated numbers of CD14$^+$ monocytes isolated from blood of SLE patients and normal donors (>95% pure) are cultured for 5 days with 105 allogeneic CD4$^+$ T cells. Small letters indicate different patients.

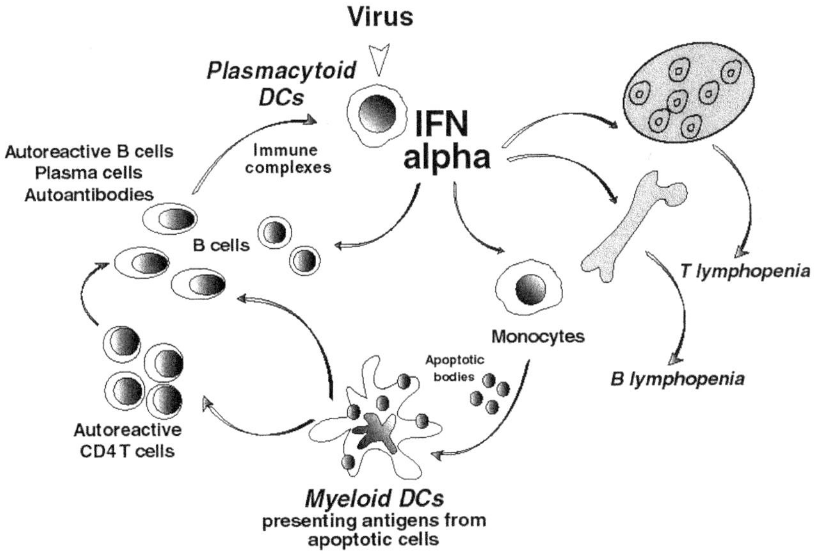

FIGURE 4. The central role of DC subsets and IFN-α in the pathogenic loop of SLE. Plasmacytoid DCs, triggered by viruses, release high levels of IFN-α which induces monocytes to differentiate into DCs able to capture circulating apoptotic cells. Upon subsequent IFN-α–induced maturation, these DCs present self-antigens to autoreactive CD4 T cells. Those DCs also directly support B cell proliferation and differentiation, and such a "menage à trois" generates high numbers of plasma cells producing autoantibodies. Autoantibodies bind circulating nucleosomes forming immune complexes that sustain IFN-α production by pDCs. High levels of IFN-α in SLE could also explain the T and B lymphopenia through the direct suppressive effect on thymus and bone marrow, respectively.

SLE pDCs produce as much IFN-α as healthy donor pDCs in response to viral triggering.[20] However, in contrast to healthy donors, SLE blood also contains $CD123^-$ cells that produce IFN-α.[20]

Thus, we now view SLE as a disease where IFN-induced DCs phagocytose cell nuclei that circulate in the blood of these patients and its components are presented to autoreactive T cells and B cells, leading to the generation of anti-nuclear antibodies including anti-double-stranded DNA antibodies, the hallmark of SLE. These autoantibodies form immune complexes that deposit in the kidneys, or the vessel walls eventually leading to kidney failures or vasculitis, life-threatening complications of the disease. Thus, unabated myeloid DC induction and alterations in the balance between DC subsets contribute to autoimmunity (FIG. 4). Yet, only a fraction of patients display circulating IFN-α, thereby raising the question of whether this reflects disease heterogeneity. Because IFN-α may represent a major therapeutic target in SLE, as TNF does in arthritis, we approached this question by analyzing the genes expressed by patient leukocytes using oligonucleotide microarray technology. We found that blood mononuclear cells from all active SLE patients overexpress IFN-regulated genes.[22] This signature is extinguished by treatment with high-dose intravenous steroids, further pointing to IFN as a specific target for therapeutic intervention. Thus, understanding the pathophysiology of the disease may allow the generation of new drugs that will target the cause of the disease (excessive IFN-α production leading to unabated DC activation) rather than attempt to treat the late symptoms. In the present case, one would like to find ways to inhibit the excessive interferon production or to prevent its effects.

CONCLUSION

What we learn from studying autoimmunity will help us induce strong tumor-specific immunity. What we learn from *ex vivo*–generated DC vaccines will permit us to develop "intelligent missiles." These generic vaccines equipped with (tumor) antigens, chaperons, DC activation molecules, and specific ligands will target distinct DC subsets to induce the type of immunity desired to counteract the pathologic insult.

REFERENCES

1. ROMANI, N., S. GRUNER, D. BRANG, *et al.* 1994. Proliferating dendritic cell progenitors in human blood. J. Exp. Med. **180:** 83–93.
2. CAUX, C., C. DEZUTTER-DAMBUYANT, D. SCHMITT & J. BANCHEREAU. 1992. GM-CSF and TNF-alpha cooperate in the generation of dendritic Langerhans cells. Nature **360:** 258–261.
3. HUANG, F.P., N. PLATT, M. WYKES, *et al.* 2000. A discrete subpopulation of dendritic cells transports apoptotic intestinal epithelial cells to T cell areas of mesenteric lymph nodes. J. Exp. Med. **191:** 435–444.
4. STEINMAN, R.M., S. TURLEY, I. MELLMAN & K. INABA. 2000. The induction of tolerance by dendritic cells that have captured apoptotic cells. J. Exp. Med. **191:** 411–416.
5. FINKELMAN, F.D., A. LEES, R. BIRNBAUM, *et al.* 1996. Dendritic cells can present antigen in vivo in a tolerogenic or immunogenic fashion. J. Immunol. **157:** 1406–1414.
6. HEATH, W.R. & F.R. CARBONE. 2001. Cross-presentation, dendritic cells, tolerance and immunity. Annu. Rev. Immunol. **19:** 47–64.

7. Banchereau, J., F. Briere, C. Caux, *et al.* 2000. Immunobiology of dendritic cells. Annu. Rev. Immunol. **18:** 767–812.
8. Liu, Y.J. 2001. Dendritic cell subsets and lineages, and their functions in innate and adaptive immunity. Cell **106:** 259–262.
9. Caux, C., C. Massacrier, B. Vanbervliet, *et al.* 1997. CD34+ hematopoietic progenitors from human cord blood differentiate along two independent dendritic cell pathways in response to granulocyte-macrophage colony-stimulating factor plus tumor necrosis factor alpha: II. Functional analysis. Blood **90:** 1458–1470.
10. Siegal, F.P., N. Kadowaki, M. Shodell, *et al.* 1999. The nature of the principal type 1 interferon-producing cells in human blood [In process citation]. Science **284:** 1835–1837.
11. Kadowaki, N., S. Antonenko, J.Y. Lau & Y.J. Liu. 2000. Natural interferon alpha/beta-producing cells link innate and adaptive immunity. J. Exp. Med. **192:** 219–226.
12. Gilliet, M. & Y.J. Liu. 2002. Generation of human CD8 T regulatory cells by CD40 ligand-activated plasmacytoid dendritic cells. J. Exp. Med. **195:** 695–704.
13. Banchereau, J., B. Schuler-Thurner, A.K. Palucka & G. Schuler. 2001. Dendritic cells as vectors for therapy. Cell **106:** 271–274.
14. Caux, C., B. Vanbervliet, C. Massacrier, *et al.* 1996. CD34+ hematopoietic progenitors from human cord blood differentiate along two independent dendritic cell pathways in response to GM-CSF+TNF alpha. J. Exp. Med. **184:** 695–706.
15. Dubois, B., B. Vanbervliet, J. Fayette, *et al.* 1997. Dendritic cells enhance growth and differentiation of CD40-activated B lymphocytes [see comments]. J. Exp. Med. **185:** 941–951.
16. Dubois, B., C. Massacrier, B. Vanbervliet, *et al.* 1998. Critical role of IL-12 in dendritic cell-induced differentiation of naive B lymphocytes. J. Immunol. **161:** 2223–2231.
17. Mortarini, R., A. Anichini, M. Di Nicola, *et al.* 1997. Autologous dendritic cells derived from CD34+ progenitors and from monocytes are not functionally equivalent antigen-presenting cells in the induction of melan-A/Mart-1(27-35)-specific CTLs from peripheral blood lymphocytes of melanoma patients with low frequency of CTL precursors. Cancer Res. **57:** 5534–5541.
18. Ferlazzo, G., J. Klein, X. Paliard, *et al.* 2000. Dendritic cells generated from CD34+ progenitor cells with flt3 ligand, c-kit ligand, GM-CSF, IL-4, and TNF-alpha are functional antigen-presenting cells resembling mature monocyte-derived dendritic cells. J. Immunother. **23:** 48–58.
19. Banchereau, J., A.K. Palucka, M. Dhodapkar, *et al.* 2001. Immune and clinical responses in patients with metastatic melanoma to CD34(+) progenitor-derived dendritic cell vaccine. Cancer Res. **61:** 6451–6458.
20. Blanco, P., A.K. Palucka, M. Gill, *et al.* 2001. Induction of dendritic cell differentiation by IFN-alpha in systemic lupus erythematosus. Science **294:** 1540–1543.
21. Gill, M.A., P. Blanco, E. Arce, *et al.* 2002. Blood dendritic cells and DC-poietins in systemic lupus erythematosus. Hum. Immunol. **63:** 1172–1180.
22. Bennett, L., A.K. Palucka, E. Arce, *et al.* 2002. Interferon and granulopoiesis signatures in systemic lupus erythematosus blood. J. Exp. Med. In press.

Co-Stimulatory Blockade in the Treatment of Murine Systemic Lupus Erythematosus (SLE)

ANNE DAVIDSON, XIAOBO WANG, MASAHIKO MIHARA,[a] MEERA RAMANUJAM, WEIQING HUANG, LENA SCHIFFER, AND JAYASHREE SINHA

Departments of Medicine and Microbiology and Immunology, Albert Einstein College of Medicine, Bronx, New York 10461, USA

ABSTRACT: Although the life span of patients with systemic lupus erythematosus (SLE) has improved considerably over the last several decades, the toxicities of chronic immunosuppressive therapy are major causes of morbidity and mortality. Safer and more effective therapies for SLE are clearly needed. SLE is characterized by excessive activation of both B and T lymphocytes. Activation of these cells requires both antigen engagement and co-stimulatory signals from interacting lymphocytes (Carreno, B.M. & M. Collins, 2002, Annu. Rev. Immunol. 20: 29–53; Grewal, I.S. & R.A. Flavell, 1998, Annu. Rev. Immunol. 16: 111–135). Thus, blockade of co-stimulatory signals offers a new therapeutic approach to SLE. Our short-term goal has been to understand the effect of co-stimulatory blocking reagents on the development, selection, and activation of pathogenic anti-dsDNA antibody producing B cells in mice genetically predetermined to develop SLE and showing signs of either early or advanced disease activity. Our long-term goal is to use the knowledge we gain to design therapeutic regimens for humans that avoid the complications of long-term immunosuppression. As new co-stimulatory molecules are discovered, studying their mechanism of action in animal models and their clinical utility in human autoimmune disease should lead both to a new understanding of disease pathogenesis and also to safer and more effective therapies.

KEYWORDS: SLE; co-stimulatory blockade; B cells

INTRODUCTION

For the last several years, our laboratory has been studying the effects of CTLA4Ig and anti-CD40L in lupus-prone NZB/W F1 mice (Refs. 1 and 2; Wang, in press; Schiffer, submitted). The mechanism of action of these molecules in non-autoimmune mice has been well studied. Anti-CD40L blocks both T and B cell responses to hapten/carrier when given prior to immunization.[3] CD40 ligation is required for expansion rather than priming of the initial T cell immune response, and at limiting antigen dose, it predominantly affects Th2 responses.[4] Three to 12 days following antigen administration, anti-CD40L continues to inhibit B cell but not T

Address for correspondence: Anne Davidson, Albert Einstein College of Medicine, 1300 Morris Park Avenue, U505, Bronx, NY 10461. Voice: 718-430-4107; fax: 718-430-8789. davidson@aecom.yu.edu

[a]Current address: Chugai Pharmaceuticals, Gotemba-shi, Shizuoka 412, Japan.

Ann. N.Y. Acad. Sci. 987: 188–198 (2003). © 2003 New York Academy of Sciences.

cell immune responses.[3] Anti-CD40L has no effect on plasma cell survival or selection, but it attenuates the secondary B cell response to antigen mediated by reactivation of memory cells.[5] Finally, CD40 has multiple roles in the function of other cell types, and anti-CD40L can therefore also inhibit the inflammatory effector component of the immune response.[6]

CTLA4Ig blocks T cell priming and has a more profound effect on Th1 cells than on Th2 cells at low antigen doses.[7] Our own studies using a transplant model have shown that CTLA4Ig blocks both primary and secondary humoral responses.[8] It should be noted that CTLA4Ig therapy is more complex than anti-CD40L therapy because CTLA4Ig also blocks the negative regulatory effects of CTLA4 in T cells,[9,10] and this may modulate its beneficial effects in the treatment of autoimmunity, especially when used as a single agent.

The use of combination therapy with CTLA4Ig and anti-CD40L in autoimmune disease models is based on observations in animal transplantation models where short-term combination therapy resulted in long-term graft survival free of cell infiltration or atherosclerotic damage.[11] In a model of peptide-induced autoimmune oophoritis, either CTLA4Ig or anti-CD40L given at the time of immunization with the inducing peptide interfered with effector function but did not prevent expansion of pathogenic T cells. In contrast, when both agents were given together, expansion of pathogenic T cells was blocked. This result did not appear to be due to permanent deletion of pathogenic T cells or induction of regulatory cells because a second dose of immunizing peptide given a month later without concomitant co-stimulatory blockade induced full-blown autoimmunity.[12]

These experimental situations differ significantly from spontaneous autoimmune diseases such as systemic lupus erythematosus (SLE) where the nature and timing of self-antigen stimulation are not known. Furthermore, SLE is not a purely T cell–mediated disease. B cells have important functions in SLE, both as cells that produce the autoantibodies that initiate tissue damage and also as antigen-presenting cells that can amplify disease through a variety of mechanisms.[13]

We and others have shown that long-term administration of CTLA4Ig to NZB/NZW F1 mice can prevent the onset of SLE, but that nephritis appears within 2–8 weeks of stopping treatment.[1,14] In contrast, a 2-week course of anti-CD40L given early in life caused a longer delay in the onset of disease.[15] Daikh and Wofsy were the first to show that a 2-week course of combination CTLA4Ig and anti-CD40L given to pre-nephritic NZB/W F1 mice resulted in marked delay in disease onset lasting many months after therapy was given.[16] Finally, Daikh and Wofsy showed that nephritis in NZB/W F1 mice could be reversed by combination CTLA4Ig/cyclophosphamide but not by either agent alone.[17] Since the mechanism for these effects was unexplained, our lab began to study the mechanism of action of co-stimulatory blockade in the spontaneous NZB/W F1 model of SLE with an emphasis on the effects on autoreactive B cells.

POSSIBLE EFFECTS OF CO-STIMULATORY BLOCKADE ON B CELLS

Regulatory mechanisms for preventing the expansion of pathogenic autoreactive B cells depend on the affinity of the B cell receptor (BCR) for autoantigen, the presence or absence of T cell help, and the microenvironment. The current working model is that at the time that B cells are susceptible to regulation, a high degree of BCR cross-linking by antigen will result in B cell deletion, a moderate degree of cross-linking will result in the induction of anergy, and a low degree of cross-linking will not affect the B cell (ignorance). This outcome can be modified by the amount of co-stimulation received.[18]

In both murine and human SLE, there is spontaneous production of high levels of polyclonal IgM antibodies including dsDNA-binding antibodies. This B cell hyperreactivity is independent of T cell help.[19] However both CD40 and B7.2 engagement have been reported to rescue naïve B cells from apoptosis.[18,20] Therefore, CTLA4Ig and anti-CD40L could act synergistically to block anti-apoptotic signals and facilitate apoptosis of autoreactive B cells during early B cell activation.

Following B cell activation, some B cells migrate to the germinal centers where they undergo class switching, somatic mutation, and terminal differentiation to plasma cells and memory cells.[21] This process is clearly dependent on cellular collaborations that involve B7/CD28 and CD40/CD40L interactions. B cells that acquire autoreactivity via somatic mutation are usually regulated in the germinal centers by receptor editing or apoptosis.[22] Co-stimulatory blockade could block all aspects of germinal center autoreactive B cell maturation and alter thresholds for B cell signaling leading to receptor editing or apoptosis instead of proliferation.

Memory B cells express B7, which makes them highly efficient antigen-presenting cells that are able to present autoantigens endocytosed via their BCR.[23] This antigen-presentation function of B cells can be blocked by co-stimulatory blockade. Memory B cells do not secrete antibodies, but when re-exposed to antigen and T cell help, they will rapidly proliferate and differentiate into high-affinity antibody-secreting cells. Both B7/CD28 and CD40/CD40L interactions are required for this to occur. In a manner similar to those B cells undergoing early activation, memory B cells undergoing reactivation become sensitive to apoptotic signals and they may therefore be deleted if they do not receive co-stimulatory signals.[24]

In contrast to naïve or memory B cells, plasma cells appear not to require either T cells or antigen restimulation to survive and secrete antibody.[25] Plasma cells are able to survive for long periods in the spleen and bone marrow and are thought to be responsible for maintaining long-term antibody synthesis. Thus, co-stimulatory blockade may have no effect on these cells or on serum antibody levels.

To examine this ordered sequence of B cell development we have developed a series of assays that allow us to analyze:

(1) whether anti-double-stranded DNA (anti-dsDNA) antibody producing B cells are anergized or deleted

- *(a)* by performing ELISpot assays to quantitate the frequency of IgG- and IgM-secreting B cells and IgG and IgM anti-dsDNA antibody producing B cells from spleen and bone marrow;
- *(b)* by generating monoclonal anti-DNA antibodies from spleen cells using lipopolysaccharide stimulation to evaluate the presence of anergic B cells and characterizing the monoclonal antibodies produced.

(2) whether B cells have received T cell help

- *(a)* by analyzing immunoglobulin class switching and clonal expansion;
- *(b)* by rapid molecular analysis of an autoreactive immunoglobulin V region gene, V_HBW-16, which is used as a surrogate marker for analysis of the frequency and pattern of somatic hypermutation;

(3) whether any effects seen in SLE nephritis are mediated by inhibition of the effector phase of the immune response by performing histologic and immunohistochemical studies of kidneys.

We have analyzed the effect of treatment with continuous CTLA4Ig or anti-CD40L given as single agents or in short-term two-week combination with a repeat treatment course at the time of onset of fixed proteinuria of 300 mg/dL (Refs. 1 and 2; Wang, in press). Therapy was begun at the age of 19–21 weeks, when mice had not yet developed high titer anti-DNA antibodies, or at the age of 26 weeks, when these antibodies had appeared. All three protocols resulted in a delay of onset of SLE in NZB/NZW F1 mice when administered at 19–21 weeks; however, treatment was much less effective when given to mice with high-titer anti-DNA antibodies or established nephritis. Anti-CD40L was effective at 26 weeks if given in higher and more frequent doses and resulted in decreases in anti-dsDNA antibody titers and immunologic effects similar to those seen with earlier treatment. The effects of treatment on proteinuria and death are shown in FIGURE 1.

The three treatment protocols had a number of similar immunologic effects. None of the protocols altered the expansion of a precursor pool of immature IgM-producing autoreactive B cells in NZB/NZWF1 mice, nor did they alter the abnormally high serum IgM levels found in these mice; they did, however, block the proliferation of IgM anti-dsDNA antibody producing B cells that occurs in these mice with age. Combination therapy did result in a lower number of B cells in the spleens of treated mice but did not alter selection of autoreactive B cells into the naïve pool.[2] This finding may reflect the need for other important B cell co-stimulators that in-

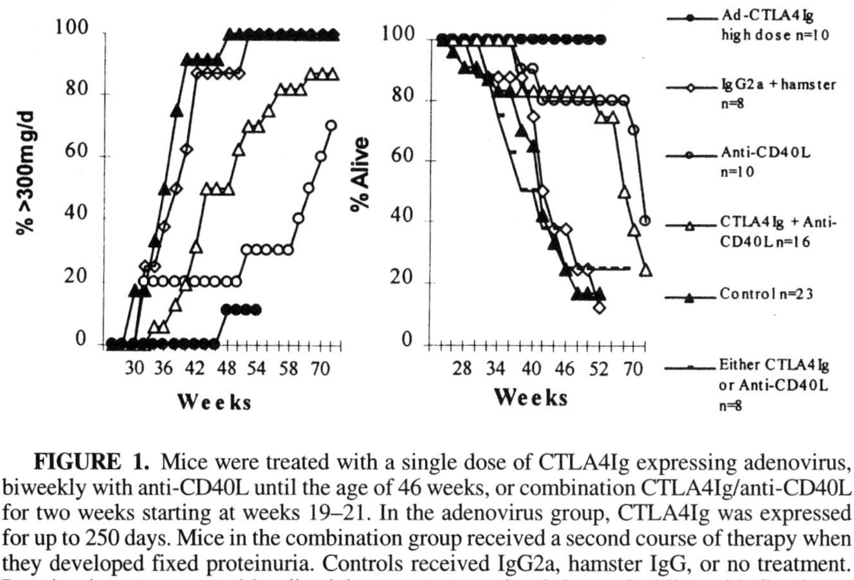

FIGURE 1. Mice were treated with a single dose of CTLA4Ig expressing adenovirus, biweekly with anti-CD40L until the age of 46 weeks, or combination CTLA4Ig/anti-CD40L for two weeks starting at weeks 19–21. In the adenovirus group, CTLA4Ig was expressed for up to 250 days. Mice in the combination group received a second course of therapy when they developed fixed proteinuria. Controls received IgG2a, hamster IgG, or no treatment. Proteinuria was measured by dipstick every two weeks (*left panel*). Mice with fixed proteinuria >300 mg/dL were assessed as positive. Survival curves are shown on the right.

TABLE 1. Pooled mutation analysis of the V_HBW-16 gene in treated mice in the prevention groups

	Weeks 22–25	Weeks 34–42	Treated	*P* Value
Number of genes sequenced	30	74	132	
Total number of mutations	24	309	281	
Average mutations per sequence	1.3	4.18	2.13	<0.0001[a]
Percent CDR2 mutations of a neutral or negatively charged amino acid to Arg or Lys	0	17.8	5.1	0.016[b]

[a]Treated vs. young, and treated vs. controls (Fisher's exact test).
[b]Treated vs. controls (Fisher's exact test).

fluence B cell selection including the newly discovered molecule BAFF that is abnormally expressed in NZB/W F1 mice (see below).

Few high-affinity IgG anti-dsDNA antibody producing B cells were recruited into the autoimmune process during treatment. No IgG class-switch sterile transcripts were detected by RT-PCR from spleen cells during any of the three treatment protocols. Despite the block in class switching, circulating serum IgG levels remained stable in all treated mice for up to 52 weeks, suggesting that the long-lived plasma cell compartment is relatively unaffected by treatment. This was confirmed by the finding that the frequency of IgG-secreting B cells in the bone marrow was not affected by long-term CTLA4Ig treatment.[1]

The block in class switching suggested that B cells either did not receive or did not respond to T cell help. This was confirmed by the finding of a decreased frequency of somatic mutations in the V_HBW-16 heavy-chain gene. V_HBW-16 is strongly associated with pathogenic anti-dsDNA antibody activity both in NZB/NZWF1 and MRL/lpr mice. Despite its presence in the germ line of several normal strains, V_HBW-16 is not expressed in these strains unless they are experimentally induced to mount an anti-dsDNA antibody response, indicating that antibodies using this gene are regulated in the peripheral B cell compartment of normal mice. High-affinity anti-dsDNA antibody activity in NZB/NZW F1 mice is associated with the presence of particular amino acids, particularly arginine, in the CDR regions of some VH genes, including V_HBW-16.[26] The presence of arginine in CDR3 of V_HBW-16 was no different in treated mice from controls, suggesting that selection of this gene into the naïve repertoire was unaffected by treatment. In CDR2, however, fewer arginine and lysine mutations were found in all three groups of treated mice than in untreated mice (TABLE 1). This suggests that there might be an effect of co-stimulatory blockade on selection of those sequences that do undergo mutation.

There were also some instructive differences between the mechanisms of action of each treatment protocol. Continuous CTLA4Ig treatment resulted in a marked decrease in T cell activation as assessed by $CD4^+$ T cell numbers and expression of the CD69 cell surface marker, but it did not prevent the abnormal transition of T cells from the naïve to the memory pool. This could be explained by the presence of other T cell co-stimulators that are more important in maintaining the T cell response after priming has occurred. Once treatment was stopped, disease recurred rapidly in CTLA4Ig-treated mice.[1] This was probably due to the accumulation of memory T cells that appeared to be able to stimulate B cells from untreated mice *in vitro* (Schiffer and Davidson, unpublished data). In contrast, both anti-CD40L treatment and combination treatment prevented the abnormal accumulation of memory T cells, and this effect lasted for many weeks after cessation of treatment. Once abnormal numbers of memory T cells were present, this phenotype was not reversed by therapy even in responding mice.[2]

In contrast to CTLA4Ig treatment, there was a marked delay in disease onset following cessation of anti-CD40L treatment. Class switching recovered after 6–8 weeks, but there was a period of unresponsiveness of autoreactive B cells that lasted a variable time following cessation of treatment. During this time, large numbers of IgG anti-dsDNA reactive cells could be detected in spleens by ELISpot, but they were not sufficiently activated to yield spontaneous hybridomas or to generate a serum anti-dsDNA response (Wang, in press). In mice treated with combination therapy at 19–21 weeks, effects on class switching and somatic mutation lasted at least 16–20 weeks after treatment, suggesting a long-lived effect on T cell help. We were unable to detect anergic B cells in these mice 16–20 weeks after treatment.[2]

While continuous co-stimulatory blockade with either agent was immunosuppressive of the humoral response to the hapten oxazolone, combination therapy treated mice regained their ability to respond to oxazolone within six weeks after completing treatment.[2] At this time, their anti-oxazolone antibodies, unlike those in control mice, were not cross-reactive with dsDNA, suggesting that this treatment resulted in restoration of some aspects of peripheral B cell tolerance. When a second course of combination therapy was given at the time of onset of fixed proteinuria, 70% of these mice responded, compared with less than 30% of controls, suggesting a difference in the pathogenicity of the T and/or B cells that emerged after a first course of treatment.[2]

Results were strikingly different in mice with established nephritis. Mice with fixed proteinuria of >300 mg/dL were treated with 2 weeks of combination therapy and a single dose of cyclophosphamide. In these mice, remissions (proteinuria < 30 mg/dL) occurred within of 2–3 weeks, and 60% of mice that relapsed responded to a second treatment course. By ELISpot and hybridoma analysis, the frequency of activated IgG anti-dsDNA antibody producing B cells decreased for only 2–3 weeks while remissions were sustained for up to 20 weeks. Light microscopy of the kidneys of mice in remission revealed much less inflammation than in controls; however immunofluorescence revealed IgG and C3 staining of glomeruli was no different than controls. Most of the infiltrating cells in control mice could be identified as CD19, CD11c, or CD4-positive cells, but these cells were absent in mice in remission. Preliminary data suggests that remission induction using this combination protocol mediates its effect by influencing the effector response of lymphoid cells to renal inflammatory chemokines, therefore leading to less B and T cell infiltration and nephritis.[27]

In sum, these findings indicate that continuous co-stimulatory blockade with CTLA4Ig and anti-CD40L has profound effects both on T and B cell activation in NZB/NZWF1 mice when used in early disease. Anti-CD40L has more profound effects on T cells than does CTLA4Ig. This may be related to the blockade of negative regulatory functions of CTLA4 by CTLA4Ig. Both therapies are immunosuppressive for the foreign antigen oxazolone. Co-stimulatory blockade with these agents appears to predominantly affect the T cell dependent steps of B cell activation, particularly class switching, somatic mutation, B cell selection, and transition to the long-lived B cell compartments. This in turn may slow the activation of autoreactive T cells by autoreactive B cells,[28] the antigen-presenting effects of memory B cells,[13] and the pro-inflammatory effects of DNA/anti-dsDNA immune complexes.[29] Once activation of autoreactive T cells has occurred in these mice and memory cells are present, neither CTLA4Ig nor anti-CD40L is effective in altering the outcome of disease when used as a single agent in moderate dose. However, two weeks of combination therapy together with a single dose of cyclophosphamide can reverse nephritis in NZB/NZWF1 mice, and this effect persists for many weeks afterwards. This appears to be mediated by a mechanism related to modulation of the effector response to tissue immune complex deposition.

On the basis of these experimental findings, we believe that strong consideration should be given to the use of short-term combination co-stimulatory blockade for human autoimmune disease. The effects of these agents in early disease appear to differ from those in late disease. Thus, we suggest that these agents could be used in combination with conventional therapy to induce remission in early disease or after ablative therapies have restored a more "naïve" repertoire of cells in late disease. It may be possible to repeat therapy for disease flares. Single agents may be effective for maintenance therapy but are likely to have immunosuppressive properties. As new co-stimulatory molecules are discovered, a variety of combination regimens should be tested for efficacy and safety.

FUTURE DIRECTIONS

Since we first started working with CTLA4Ig and anti-CD40L, a large number of new co-stimulatory molecules related to the CD28/B7 family and the TNF/TNFR family have been discovered and characterized (FIG. 2). We are currently working on three of these molecules. The first is ICOS, a CD28-like co-stimulatory molecule that is expressed on activated and memory T cells and preferentially induces IL-10 that is known to exacerbate disease in NZB/NZW F1 mice and in humans through its effect on B cell proliferation and immunoglobulin. ICOS is highly expressed on T cells in the apical zone of germinal centers where somatic mutation occurs. ICOS also appears to predominate over CD28 as a co-stimulator of the secondary B cell response. The ICOS ligand, B7RP-1, is predominantly on B cells; thus, involvement of ICOS/B7RP-1 may be of major importance in co-stimulating the antigen presentation by B cells that contributes to T cell expansion and activation in mice with SLE secretion.[9,30,31]

The second molecule we are examining is PD-1, an inhibitory receptor like CTLA4 that down-regulates T cell receptor mediated lymphocyte proliferation and cytokine secretion. PD-1 is expressed on activated T and B cells, and its ligation leads to suppression of T and B cell proliferation and decreased T cell cytokine secretion. PD-1 ligation is optimally inhibitory in the presence of suboptimal antiCD28 co-stimulation, suggesting that PD-1 agonism will synergize with CD28 antagonism.[9,32,33]

Finally, we have developed a blocker of the CD40L-like co-stimulatory molecule, BAFF (also known as BLyS, TALL-1, THANK, and zTNF). The BAFF receptors TACI, BCMA, and BAFF-R are expressed on resting B cells.[34,35] Engagement of these receptors, but particularly of BAFF-R, is required both for maturation of naïve B cells[36] and for germinal center development.[35,37] Mice deficient in either BAFF or BAFF-R lack all B cell subsets past the T1 stage except B1 cells.[38] BAFF also contributes to germinal center development and may act to prolong the life span of plasma cells[39] High levels of soluble BAFF are found in the serum of NZB/W F1 mice and humans with active SLE.[40] BAFF-transgenic mice have increased B cell numbers, increased serum immunoglobulin levels, and develop anti-DNA antibodies and nephritis.[41] These data suggest that BAFF may contribute to SLE pathogenesis.

Our approach to blockade of ICOS and BAFF and agonism of PD-1 has been to develop fully murine fusion proteins similar to CTLA4Ig. Our rationale, based on similar approaches to the treatment of infectious disease and cancer, is to find combination treatments that synergize by targeting more than one cell activation pathway. All three of our fusion proteins are biologically active in normal mice either *in vitro* or *in vivo*, and we have begun to test them in combination with each other and with CTLA4Ig and anti-CD40L. Our initial experiments in NZB/W F1 mice have shown synergy between ICOS-Ig and CTLA4Ig (but not ICOS-Ig and anti-CD40L) and between PD-1LIg and CTLA4Ig, suggesting that transmitting a negative signal into the cell through PD-1 may compensate for lack of negative CTLA4 signaling.

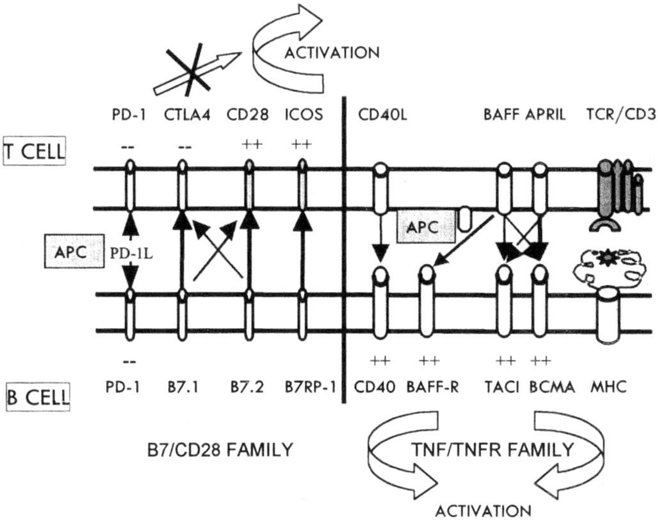

FIGURE 2. Cartoon showing co-stimulatory molecules of current interest to our laboratory. Positive signals are delivered by CD40, BAFF receptors, and CD28/ICOS. PD-1 delivers a negative signal to both T and B cells. Blocking of both CD40 and CD28 is synergistic, compared with blocking either pathway alone. Other synergies are being identified.

Finally, TACI-Ig in short-term combination with CTLA4Ig is as effective a preventive treatment in NZB/W F1 mice as CTLA4Ig/anti-CD40L. While these experiments are still in progress, understanding the mechanism of action of each combination regimen will be essential in bringing such regimens to human clinical trial. Furthermore, since the mechanism of action of co-stimulatory blockade in induction of remission is different to that in prevention studies, these results cannot be used to predict the effects of treatment on established disease. This question can only be answered experimentally.

Many other co-stimulatory pathways important in B and T cell activation, particularly those belonging to the TNF/TNFR family, are potential targets for therapeutic intervention in autoimmunity. These include the Ox40/Ox40L pair that mediates both B and T cell co-stimulation,[42] the CD27/CD70 pair that is involved in terminal B cell differentiation,[43] the T cell co-stimulator LIGHT (whose over-expression causes SLE in mice),[44] and surprisingly, stimulation through the T cell co-stimulator 4-1BB.[45] The coming years should bring exciting new developments in this area in the treatment of human disease.

REFERENCES

1. MIHARA, M. *et al.* 2000. CTLA4Ig inhibits T cell-dependent B-cell maturation in murine systemic lupus erythematosus. J. Clin. Invest. **106:** 91–101.
2. WANG, X. *et al.* 2002. Mechanism of action of combined short term CTLA4Ig and anti-CD40L in murine SLE. J. Immunol. **168:** 2046–2053.
3. HAN, S. *et al.* 1995. Cellular interaction in germinal centers. Roles of CD40 ligand and B7-2 in established germinal centers. J. Immunol. **155:** 556–567.
4. HOWLAND, K.C. *et al.* 2000. The roles of CD28 and CD40 ligand in T cell activation and tolerance. J. Immunol. **164:** 4465–4470.
5. FOY, T.M. *et al.* 1994. gp39-CD40 interactions are essential for germinal center formation and the development of B cell memory. J. Exp. Med. **180:** 157–163.
6. THIENEL, U., J. LOIKE & M.J. YELLIN. 1999. CD154 (CD40L) induces human endothelial cell chemokine production and migration of leukocyte subsets. Cell. Immunol. **198:** 87–95.
7. SCHWEITZER, A.N. & A.H. SHARPE. 1998. Studies using antigen-presenting cells lacking expression of both B7-1 (CD80) and B7-2 (CD86) show distinct requirements for B7 molecules during priming versus restimulation of Th2 but not Th1 cytokine production. J. Immunol. **161:** 2762–2771.
8. REDDY, B. *et al.* 2001. The effect of CD28/B7 blockade on alloreactive T and B cells after liver cell transplantation. Transplantation **71:** 801–811.
9. CARRENO, B.M. & M. COLLINS. 2002. The B7 family of ligands and its receptors: new pathways for costimulation and inhibition of immune responses. Annu. Rev. Immunol. **20:** 29–53.
10. RILEY, J.L. *et al.* 2001. ICOS costimulation requires IL-2 and can be prevented by CTLA-4 engagement. J. Immunol. **166:** 4943–4948.
11. KIRK, A.D. *et al.* 1997. CTLA4-Ig and anti-CD40 ligand prevent renal allograft rejection in primates. Proc. Natl. Acad. Sci. USA **94:** 8789–8794.
12. GRIGGS, N.D. *et al.* 1996. The relative contribution of the CD28 and gp39 costimulatory pathways in the clonal expansion and pathogenic acquisition of self-reactive T cells. J. Exp. Med. **183:** 801–810.
13. LIPSKY, P.E. 2001. Systemic lupus erythematosus: an autoimmune disease of B cell hyperactivity. Nat. Immunol. **2:** 764–766.
14. FINCK, B.K., P.S. LINSLEY & D. WOFSY. 1994. Treatment of murine lupus with CTLA4Ig. Science **265:** 1225–1227.

15. MOHAN, C. *et al.* 1995. Interaction between CD40 and its ligand gp39 in the development of murine lupus nephritis. J. Immunol. **154:** 1470–1480.
16. DAIKH, D.I. *et al.* 1997. Long-term inhibition of murine lupus by brief simultaneous blockade of the B7/CD28 and CD40/gp39 costimulation pathways. J. Immunol. **159:** 3104–3108.
17. DAIKH, D.I. & D. WOFSY. 2001. Cutting edge: reversal of murine lupus nephritis with CTLA4Ig and cyclophosphamide. J. Immunol. **166:** 2913–2916.
18. SATER, R.A., P.C. SANDEL & J.G. MONROE. 1998. B cell receptor-induced apoptosis in primary transitional murine B cells: signaling requirements and modulation by T cell help. Int. Immunol. **10:** 1673–1682.
19. REININGER, L. *et al.* 1996. Intrinsic B cell defects in NZB and NZW mice contribute to systemic lupus erythematosus in (NZB × NZW)F1 mice. J. Exp. Med. **184:** 853–861.
20. RATHMELL, J.C. *et al.* 1998. Repression of B7.2 on self-reactive B cells is essential to prevent proliferation and allow Fas-mediated deletion by CD4(+) T cells. J. Exp. Med. **188:** 651–659.
21. TARLINTON, D. 1998. Germinal centers: form and function. Curr. Opin. Immunol. **10:** 245–251.
22. HANDE, S., E. NOTIDIS & T. MANSER. 1998. Bcl-2 obstructs negative selection of autoreactive, hypermutated antibody V regions during memory B cell development. Immunity **8:** 189–198.
23. MAMULA, M.J. 1998. Epitope spreading: the role of self peptides and autoantigen processing by B lymphocytes. Immunol. Rev. **164:** 231–239.
24. MANSER, T. *et al.* 1998. The roles of antibody variable region hypermutation and selection in the development of the memory B-cell compartment. Immunol. Rev. **162:** 183–196.
25. SLIFKA, M.K. & R. AHMED. 1998. Long-lived plasma cells: a mechanism for maintaining persistent antibody production. Curr. Opin. Immunol. **10:** 252–258.
26. ASH-LERNER, A. *et al.* 1997. Expression of an anti-DNA-associated VH gene in immunized and autoimmune mice. J. Immunol. **159:** 1508–1519.
27. SINHA, J. *et al.* 2001. Short term costimulatory blockade combined with cyclophosphamide does not abrogate glomerular deposition of antibodies but attenuates the kidney inflammatory response. [Abstract.] Arthritis Rheum. **44:** S397.
28. CHAN, O.T., M.P. MADAIO & M.J. SHLOMCHIK. 1999. The central and multiple roles of B cells in lupus pathogenesis. Immunol. Rev. **169:** 107–121.
29. MIYATA, M. *et al.* 2001. CpG-DNA derived from sera in systemic lupus erythematosus enhances ICAM-1 expression on endothelial cells. Ann. Rheum. Dis. **60:** 685–689.
30. HUTLOFF, A. *et al.* ICOS is an inducible T-cell co-stimulator structurally and functionally related to CD28. Nature **397:** 263–266.
31. DONG, C. *et al.* 2001. ICOS co-stimulatory receptor is essential for T-cell activation and function. Nature **409:** 97–101.
32. FREEMAN, G.J. *et al.* 2000. Engagement of the PD-1 immunoinhibitory receptor by a novel B7 family member leads to negative regulation of lymphocyte activation. J. Exp. Med. **192:** 1027–1034.
33. NISHIMURA, H. & T. HONJO. 2001. PD-1: an inhibitory immunoreceptor involved in peripheral tolerance. Trends Immunol. **22:** 265–268.
34. YU, G. *et al.* 2000. APRIL and TALL-I and receptors BCMA and TACI: system for regulating humoral immunity. Nat. Immunol. **1:** 252–256.
35. THOMPSON, J.S. *et al.* 2001. BAFF-R, a newly identified TNF receptor that specifically interacts with BAFF. Science **293:** 2108–2111.
36. BATTEN, M. *et al.* 2000. BAFF mediates survival of peripheral immature B lymphocytes. J. Exp. Med. **192:** 1453–1466.
37. SCHNEIDER, P. *et al.* 1999. BAFF, a novel ligand of the tumor necrosis factor family, stimulates B cell growth. J. Exp. Med. **189:** 1747–1756.
38. ROLINK, A.G. & F. Melchers. 2002. BAFFled B cells survive and thrive: roles of BAFF in B-cell development. Curr. Opin. Immunol. **14:** 266–275.
39. DO, R.K. *et al.* 2000. Attenuation of apoptosis underlies B lymphocyte stimulator enhancement of humoral immune response. J. Exp. Med. **192:** 953–964.

40. ZHANG, J. *et al.* 2001. Cutting edge: a role for B lymphocyte stimulator in systemic lupus erythematosus. J. Immunol. **166:** 6–10.
41. MACKAY, F. *et al.* 1999. Mice transgenic for BAFF develop lymphocytic disorders along with autoimmune manifestations. J. Exp. Med. **190:** 1697–1710.
42. DE SMEDT, T. *et al.* 2002. Ox40 costimulation enhances the development of T cell responses induced by dendritic cells in vivo. J. Immunol. **168:** 661–670.
43. JACQUOT, S. 2000. CD27/CD70 interactions regulate T dependent B cell differentiation. Immunol. Res. **21:** 23–30.
44. WANG, J. *et al.* 2001. The regulation of T cell homeostasis and autoimmunity by T cell-derived LIGHT. J. Clin. Invest. **108:** 1771–1780.
45. SUN, Y. *et al.* 2002. Administration of agonistic anti-4-1BB monoclonal antibody leads to the amelioration of experimental autoimmune encephalomyelitis. J. Immunol. **168:** 1457–1465.

The Transfer of Immunity from Mother to Child

LARS Å. HANSON,[a] MARINA KOROTKOVA,[a] SAMUEL LUNDIN,[b] LILJANA HÅVERSEN,[c] SVEN-ARNE SILFVERDAL,[d] INGER MATTSBY-BALTZER,[c] BIRGITTA STRANDVIK,[e] AND ESBJÖRN TELEMO[a]

Departments of Clinical Immunology,[a] Medical Microbiology and Immunology,[b] Clinical Bacteriology,[c] and Pediatrics,[e] Göteborg University, Göteborg, Sweden

Department of Pediatrics,[d] Örebro Medical Centre Hospital, Örebro, Sweden

ABSTRACT: The newborn's immune system grows fast from a small size at birth by exposure primarily to the intestinal microflora normally obtained from the mother at and after birth. While building up its immune system, the infant is supported by the transplacental IgG antibodies, which also contain anti-idiotypic antibodies, possibly also actively priming the offspring. The second mode of transfer of immunity occurs via the milk. Numerous major protective components, including secretory IgA (SIgA) antibodies and lactoferrin, are present.

The breastfed infant is better protected against numerous common infections than the non-breastfed. Breastfeeding also seems to actively stimulate the infant's immune system by anti-idiotypes, uptake of milk lymphocytes, cytokines, etc. Therefore, the breastfed child continues to be better protected against various infections for some years. Vaccine responses are also often enhanced in breastfed infants. Long-lasting protection against certain immunological diseases such as allergies and celiac disease is also noted.

KEYWORDS: anti-idiotypic; secretory IgA; lactoferrin; Bramwell receptor; immune system; breastfeeding

INTRODUCTION

The immune system develops in fetal life and is qualitatively quite complete at delivery, although certain cytokines are produced only at low levels. Also, many cells such as phagocytes and dendritic cells are not yet adequate in number and function.[1] The lymphocyte population is very limited, and the immune system of newborn mice is reported to be only a few percent of that of an adult.[2]

The major impetus for the expansion of the lymphoid population is the exposure to the microbial flora colonizing the gut from birth on. This is obvious from studies of germ-free animals, which quickly expand their limited immune system after in-

Address for correspondence: Lars Å. Hanson, Department of Clinical Immunology, Göteborg University, Guldheosgatan 10, SE-41346 Göteborg, Sweden. Voice: 46-31-3424916; fax: 46-31-3424621.

lars.a.hanson@immuno.gu.se

Ann. N.Y. Acad. Sci. 987: 199–206 (2003). © 2003 New York Academy of Sciences.

testinal colonization.[3,4] It takes time for the neonate to build up and develop its immune system and expand its range of specificities in this way.

The neonate clearly needs help from the mother for immediate protection, for colonization with the mother's gut flora, and for the long-term build up of its own immune system. This immunological support arrives via the placenta and the milk.

TRANSPLACENTAL SUPPORT OF HOST DEFENSE OF THE OFFSPRING

The well-known active transport of maternal IgG to the fetus in man usually results in about 90% of the maternal serum level of IgG antibodies in the full-term newborn at delivery. This transport takes place via an Fcγ receptor, called the Brambell receptor after its discoverer.[5] Its binding to the Fc portion of the maternal IgG does not explain why IgG antibodies of certain specificities and of high avidity are more efficiently transferred than others.[6,7]

It is quite clear that such IgG antibodies are important for the protection of the infant during the first several months of life. These antibodies meet microbes both on mucosal membranes and in the circulation and tissues. In the latter situation, complement activation and influx of neutrophils will follow resulting in an inflammatory reaction with production of pro-inflammatory cytokines. Thus, clinical symptoms, tissue damage, and considerable energy cost will ensue from this form of protection. For infants with frequent infections, this high energy cost may impair growth. This is especially striking in poor countries.[8]

The presence of an idiotypic network has been extensively studied and discussed.[9,10] The significance of such a network in the relation between the immune systems of the mother and her offspring has been analyzed. Numerous studies give support to tolerogenic as well as imunogenic effects of maternal anti-idiotypic antibodies in the offspring.[11,12] Injection of minute amounts of monoclonal anti-idiotypic or idiotypic antibodies against the *Escherichia coli* K13 polysaccharide capsular antigen into newborn mice primed them for protective responses after vaccination with killed *E. coli* K13.[13] Neonates do not otherwise respond well to polysaccharide antigens. Several other studies have illustrated that neonatal exposure to monoclonal antibodies can induce selection and expansion as well as deletion of anti-idiotypic B cells and idiotypic antibody-specific regulatory or helper T cells given that the antibodies are specific for T cell dependent antigens.[14–18]

Secretory IgA (SIgA) and IgM antibodies against *E. coli* and poliovirus were found in Swedish newborns.[19] Such antibodies are not transported from the mother to the fetus. The suggestion that they had been produced by the fetus was supported by the finding that such antibodies also occurred in newborns of mothers lacking IgA and/or IgM because they had hypogammaglobulinemia or selective IgA deficiency.[20] The possibility was considered that the poliovirus antibodies could have been induced by anti-idiotypic antibodies to poliovirus from the immunoglobulin given prophylactically to the deficient mothers. We did find such anti-idiotypes both in the immunoglobulin preparations used and in the cord sera from the studied neonates.[20,21] It should be added that Sweden uses only inactivated poliovirus vaccine, which has successfully exterminated the wild virus. The vaccine virus does not exist in the country. Neither are any cross-reactive viruses known.

MOTHER'S MILK: PASSIVE AND ACTIVE NON-INFLAMMATORY SUPPORT OF HOST DEFENSE IN THE OFFSPRING

Passive Transfer of Defense

SIgA is the predominant isotype in milk, occurring at levels of about 0.5–1.5 g/L. There is only little IgG and IgM.[22] Because of the *enteromammaric link*, lymphocytes home from the gut lymphoid tissues, primarily the Peyer's patches, to the lactating mammary glands. Therefore, the milk SIgA provides protection against all the microbes the mother has or has had in her gut. Thus, the neonatal colonization with the mother's gut flora does not pose a threat even if pathogens are transferred, provided that the child is breastfed.[23–25]

The SIgA antibodies prevent microbes from attaching to mucosal surfaces, especially in the gut. In this way, tissue engagement, inflammation, and an energy-costly defense reaction are avoided. Such protection by milk SIgA antibodies has been proven against *Shigella, Vibrio cholerae, Campylobacter, ETEC,* and *Giardia liamblia.*[26–30]

Lactoferrin (LF) is the major protein in mature milk at about 1–4 g/L. Like SIgA, it is relatively resistant to enzymatic degradation. It is striking that the SIgA and LF, both important for host defense, make up about 30% of the milk proteins although milk is to cover the needs of nutrients for the baby. This is totally different from the bovine, for instance, where these defense factors correspond to just 5% of the milk protein.[31] LF carries a number of effects being microbicidal, immunostimulatory, and efficiently anti-inflammatory, by turning off the production of numerous pro-inflammatory cytokines such as IL-1β, IL-6, TNF-α, and IL-8.[32–36] Especially during neonatal colonization and the subsequent expansion of the intestinal microflora, it might be important to have considerable amounts in the gut of a protein such as LF, which is both bactericidal and prevents the induction of cytokines that cause clinical symptoms, energy consumption, and inflammation. It may also be important to prevent the appearance of such cytokines at each encounter with new colonizers, many of which carry lipopolysaccharides, which efficiently activate the innate immune reactivity characterized by production of pro-inflammatory cytokines. The cytokines induce increased production of leptin, which adds to a reduction in appetite.[37] This is one reason why repeated infections in infants often cause undernutrition.

The oligosaccharide fraction in milk contains analogues to various receptors for microbes on mucosal epithelium. Such milk oligosaccharides can prevent mucosal attachment of several pathogens, that is, pneumococci and *Haemophilus influenzae.*[38] In addition, milk contains numerous other components such as the anti-secretory lectins,[39] and lysozyme presumably supports the host defense of the offspring.

Active Stimulation of the Immune System of the Offspring via Milk

It is well evidenced that during lactation human milk protects against numerous infections such as otitis media, upper and lower respiratory tract infections, diarrhea, urinary tract infections, neonatal septicemia, and also necrotizing enterocolitis (for a review, see Ref. 40).

More recently, it has been recognized that breastfed children are also significantly better protected against a number of infections for several years compared to non-breastfed children. This is true for otitis media, respiratory tract infections, diarrhea, infection-induced wheezing bronchitis, and invasive *H. influenzae* type b infections.[40]

Furthermore, it has been noted that vaccination during or after breastfeeding often results in better antibody and T cell responses than in non-breastfed children.[41–43] A few vaccine studies have not seen these effects. Some of these have used oral vaccines with live viruses, which may be neutralized by the milk antibodies in breastfed infants (for a review, see Ref. 40).

Several mechanisms have been proposed to explain the active stimulation of the infant's immune response by breastfeeding. One is based on animal experiments showing that transfer of anti-idiotypic antibodies against a bacterial and a viral antigen via milk resulted in priming of the offspring.[13,44] In a preliminary study in germ-free newborn piglets, either anti-polio or anti-polio anti-idiotypic monoclonals, or both, were given orally at birth. A poliovirus vaccination at two weeks of age showed that those piglets who had been given the mix of idiotypic and anti-idiotypic antibodies had the highest number of IgM anti-poliovirus secreting cells, followed by those who had obtained the idiotypic antibodies only.[45] Such oral transfer of idiotypic and anti-idiotypic antibodies may take place in humans because human milk is known to contain anti-idiotypic antibodies.[21] In experimental systems such anti-idiotypic antibodies delivered orally could induce active protection.[46] The possible effect of monoclonal anti-idiotypic antibodies from mouse against the *E. coli* K13 polysaccharide over two subsequent generations when given orally was studied in neonatal rats.[47] Two-day-old pups were tube-fed with either 1 or 10 μg of IgG1 anti-idiotypic anti-K13. Six weeks later they were colonized in the intestine with *E. coli* O6K13 bacteria genetically manipulated to produce ovalbumin (OVA). At 4 months of age, the female rats were mated with naive males. The offspring, being the second generation, were colonized with the OVA-producing *E. coli* O6K13 at birth by their mothers. Whereas the rats in the first generation had not shown any major differences in their immune response to the bacteria, quite profound differences were noted in the second generation. Thus, there was a greatly increased antibody response not only to the idiotypically connected antigen K13, but also to the O6 LPS and the OVA of the bacteria. At this time, the proliferative response of spleen cells to OVA and the wild-type bacteria was lowered. The higher dose of the anti-idiotypic antibodies gave greater effects. These observations suggest that exposure of the neonate perorally to idiotypically connected antibodies against one bacterial component can affect and direct the immune response to other antigens on the same bacterium. In the neonatal rat, the Brambell Fcγ receptor protects the IgG from catabolism and may be crucial for the above results.[5]

The reason for the enhanced effect in the second generation is not clear, but might relate to that the animals in the first generation were exposed to the anti-idiotypic antibodies on only one occasion, whereas the second generation was presumably exposed continuously by their mothers. Another factor of likely importance is that the first generation was colonized at six weeks of age, whereas the second generation was colonized already from birth. We have noticed that neonatal colonization with bacteria leads to tolerance, while colonization at a later time induces a more intense immune response.[48] Thus, it might have been easier to note modulations of the immune response by the bacteria in the second generation. Nevertheless, it is striking that a relatively small amount of anti-idiotypic antibodies given in the neonatal period can influence the immune system in a manner so profound that the effects could last for two generations.

The IgG2 antibody response to *H. influenzae* type b infection in children was enhanced by previous long-term (compared to short-term) breastfeeding.[49] This might well be an effect of exposure to idiotypic and anti-idiotypic antibodies in the milk together with effects attributable to the IFN-γ in the milk on the switching to IgG2 during lactation.

Quite remarkably, it has been found that lymphocytes from the milk are taken up and found in the intestinal mucosa and local lymph glands in a number of experimental models (for a review, see Ref. 50). The offspring becomes tolerized to the maternal major histocompatibility antigens. It has been noted that a kidney transplant to a person from the mother does significantly better if that person had been breastfed by the mother as a child.[51,52] Breastfed infants have fewer cytotoxic T cells directed against the maternal HLA than non-breastfed infants.[53] Such milk lymphocytes from a sheep vaccinated against tetanus transferred to her lamb's intestine resulted in priming the lamb against the vaccine.[54] Using B cell deficient mice, it could be shown the functional Ig-secreting B cells were transferred from the mother via the milk.[55] Partial reconstitution occurred with B cells populating spleen and bone marrow.

Our ongoing studies in rat dams have shown that tolerance against ovalbumin can be transferred to the pups via the milk.[56] However, this effect was abrogated by a high maternal intake of polyunsaturated fatty acids. The ratio of n-6/n-3 fatty acids in the diet of the dams seemed to be of importance.

Exclusively breastfed infants have a thymus twice the size of that in non-breastfed infants, as measured with ultrasound.[57] This observation has yet to be explained.

Does Breastfeeding Protect against Immunological Diseases?

A recent critical review of the literature on whether breastfeeding protects against allergic diseases gave a surprisingly positive answer.[58] A few studies have indicated that breastfeeding also protects against celiac disease, at least in childhood,[59] and also ulcerous colitis, Crohn's disease, diabetes mellitus type I, rheumatoid arthritis, and multiple sclerosis, but this needs more study.[60]

ACKNOWLEDGMENTS

This work was supported by the Swedish Medical Research Council and the Hesselman, Vårdal, and Bergvall Foundations.

REFERENCES

1. SCHELONKA, R.L. & A.J. INFANTE. 1998. Neonatal immunology. Semin. Perinatol. **22:** 2–14.
2. ADKINS, B. 1999. T-cell function in newborn mice and humans. Immunol. Today **20:** 330–335.
3. CRABBE, P.A. *et al.* 1970. Immunohistochemical observations on lymphoid tissues from conventional and germ-free mice. Lab. Invest. **22:** 448–457.

4. HORSFALL, D.J., J.M. COOPER & D. ROWLEY. 1978. Changes in the immunoglobulin levels of the mouse gut and serum during conventionalisation and following administration of *Salmonella typhimurium*. Aust. J. Exp. Biol. Med. Sci. **56:** 727–735.
5. JUNGHANS, R.P. 1997. Finally! The Brambell receptor (FcRB). Mediator of transmission of immunity and protection from catabolism for IgG. Immunol. Res. **16:** 29–57.
6. SOTO, H. *et al.* 1979. Transfer of measles, mumps and rubella antibodies from mother to infant. Am. J. Dis. Child. **133:** 1240–1243.
7. AVANZINI, M.A. *et al.* 1998. Placental transfer favours high avidity IgG antibodies. Acta Paediatr. **87:** 180–185.
8. HANSON, L.Å. *et al.* 1998. Undernutrition, immunodeficiency, and mucosal infections. *In* Mucosal Immunology. P. Ogra *et al.*, Eds. Academic Press. New York.
9. VARELA, F.J. & A. COUTINHO. 1991. Second generation immune networks. Immunol. Today **12:** 159–166.
10. LUNDKVIST, I. *et al.* 1989. Evidence for a functional idiotypic network among natural antibodies in normal mice. Proc. Natl. Acad. Sci. USA **86:** 5074–5078.
11. WIKLER, C. *et al.* 1980. Immunoregulatory role of maternal idiotopes: ontogeny of immune networks. J. Exp. Med. **152:** 1024–1035.
12. SEEGER, M. *et al.* 1998. Antigen-independent suppression of the IgE immune response to bee venom phospholipase A2 by maternally derived monoclonal IgG antibodies. Eur. J. Immunol. **28:** 2124–2130.
13. STEIN, K.E. & T. SÖDERSTRÖM. 1984. Neonatal administration of idiotype or antiidiotype primes for protection against *Escherichia coli* K13 infection in mice. J. Exp. Med. **160:** 1001–1011.
14. MARTINEZ, C. *et al.* 1985. Establishment of idiotypic helper T-cell repertoires early in life. Nature **317:** 721–723.
15. MARTINEZ, C. *et al.* 1987. A functional idiotypic network of T helper cells and antibodies, limited to the compartment of "naturally" activated lymphocytes in normal mice. Eur. J. Immunol. **17:** 821–825.
16. CERNY, J., R. CRONKHITE & C. HEUSSER. 1983. Antibody response of mice following neonatal treatment with a monoclonal anti-receptor antibody: evidence for B cell tolerance and T suppressor cells specific for different idiotopic determinants. Eur. J. Immunol. **13:** 244–248.
17. AUGUSTIN, A. & H. COSENZA. 1976. Expression of new idiotypes following neonatal idiotypic suppression of a dominat clone. Eur. J. Immunol. **6:** 497–501.
18. RETH, M., G. KELSOE & K. RAJEWSKY. 1981. Idiotypic regulation by isologous monoclonal anti-idiotope antibodies. Nature **290:** 257–259.
19. MELLANDER, L., B. CARLSSON & L.Å. HANSON. 1986. Secretory IgA and IgM antibodies to *E. coli* O and poliovirus type I antigens occur in amniotic fluid, meconium and saliva from newborns. A neonatal immune response without antigenic exposure: a result of anti-idiotypic induction? Clin. Exp. Immunol. **63:** 555–561.
20. HAHN-ZORIC, M. *et al.* 1992. Presence of non-maternal antibodies in newborns of mothers with antibody deficiencies. Pediatr. Res. **32:** 150–154.
21. HAHN-ZORIC, M. *et al.* 1993. Anti-idiotypic antibodies to poliovirus antibodies in commercial immunoglobulin preparations, human serum, and milk. Pediatr. Res. **33:** 475–480.
22. GOLDBLUM, R., L.Å. HANSON & P. BRANDTZAEG. 1996. The mucosal defense system. *In* Immunological Disorders in Infants and Children. 3rd edit. E. Stiehm, Ed.: 159–199. Saunders. Philadelphia, PA.
23. MATA, L.J. *et al.* 1969. *Shigella* infection in breast-fed Guatemalan Indian neonates. Am. J. Dis. Child. **117:** 142–146.
24. ADLERBERTH, I. *et al.* 1991. Intestinal colonization with Enterobacteriaceae in Pakistani and Swedish hospital-delivered infants. Acta Paediatr. Scand. **80:** 602–610.
25. ADLERBERTH, I., L.Å. HANSON & A.E. WOLD. 1999. Ontogeny of the intestinal flora. *In* Development of the Gastrointestinal Tract. I.R. Sanderson & W.A. Walker, Eds.: 279–292. B.C. Decker. Hamilton, Ontario.
26. GLASS, R.I. *et al.* 1983. Protection against cholera in breast-fed children by antibodies in breast milk. N. Engl. J. Med. **308:** 1389–1392.

27. CRUZ, J.R. *et al.* 1988. Breast milk anti-*Escherichia coli* heat-labile toxin IgA antibodies protect against toxin-induced infantile diarrhea. Acta Paediatr. Scand. **77:** 658–662.
28. RUIZ-PALACIOS, G.M. *et al.* 1990. Protection of breast-fed infants against *Campylobacter* diarrhea by antibodies in human milk. J. Pediatr. **116:** 707–713.
29. HAYANI, K.C. *et al.* 1992. Concentration of milk secretory immunoglobulin A against *Shigella* virulence plasmid-associated antigens as a predictor of symptom status in *Shigella*-infected breast-fed infants. J. Pediatr. **121:** 852–856.
30. WALTERSPIEL, J.N. *et al.* 1994. Secretory anti-*Giardia lamblia* antibodies in human milk: protective effect against diarrhea. Pediatrics **93:** 28–31.
31. LARSON, B.L. 1985. Determination of specific milk proteins. *In* Human Lactation. Vol. 1. R.G. Jensen & M.C. Neville, Eds.: 33–38. Plenum Press. London.
32. MATTSBY-BALTZER, I. *et al.* 1996. Lactoferrin or a fragment thereof inhibits the endotoxin-induced interleukin-6 response in human monocytic cells. Pediatr. Res. **40:** 257–262.
33. ELASS, E. *et al.* 2002. Lactoferrin inhibits the lipopolysaccharide-induced expression and proteoglycan-binding ability of interleukin-8 in human endothelial cells. Infect. Immun. **70:** 1860–1866.
34. HÅVERSEN, L.A. *et al.* 2000. Human lactoferrin and peptides derived from a surface-exposed helical region reduce experimental *Escherichia coli* urinary tract infection in mice. Infect. Immun. **68:** 5816–5823.
35. HÅVERSEN, L.A. *et al.* Anti-inflammatory activities of human lactoferrin and synthetic peptides thereof in experimental dextran-sulphate induced colitis in mice. Submitted for publication.
36. HÅVERSEN, L.A. *et al.* Lactoferrin inhibits the LPS-induced production of cytokines in THP-1 cells. In preparation.
37. BARBIER, M. *et al.* 1998. Elevated plasma leptin concentrations in early stages of experimental intestinal inflammation in rats. Gut **43:** 783–790.
38. ANDERSSON, B. *et al.* 1986. Inhibition of attachment of *Streptococcus pneumoniae* and *Haemophilus influenzae* by human milk and receptor oligosaccharides. J. Infect. Dis. **153:** 232–237.
39. HANSON, L.Å. *et al.* 2000. Nutrition resistance to viral propagation. Nutr. Rev. **58:** S31–S37.
40. HANSON, L.Å. 1998. Breastfeeding provides passive and likely long-lasting active immunity. Ann. Allergy Asthma Immunol. **81:** 523–533.
41. PABST, H.F. *et al.* 1989. Effect of breast-feeding on immune response to BCG vaccination. Lancet **1:** 295–297.
42. PABST, H.F. *et al.* 1997. Differential modulation of the immune response by breast- or formula-feeding of infants. Acta Paediatr. **86:** 1291–1297.
43. HAHN-ZORIC, M. *et al.* 1990. Antibody responses to parenteral and oral vaccines are impaired by conventional and low protein formulas as compared to breast-feeding. Acta Paediatr. Scand. **79:** 1137–1142.
44. OKAMOTO, Y. *et al.* 1989. Effect of breast feeding on the development of anti-idiotype antibody response to F glycoprotein of respiratory syncytial virus in infant mice after post-partum maternal immunization. J. Immunol. **142:** 2507–2512.
45. LUNDIN, B.S. 1998. Regulation of the immune response to intestinal antigens. Ph.D. thesis, Göteborg University, Göteborg.
46. LUCAS, G.P., C.L. CAMBIASO & J.P. VAERMAN. 1991. Protection of rat intestine against cholera toxin challenge by monoclonal anti-idiotypic antibody immunization via enteral and parenteral route. Infect. Immun. **59:** 3651–3658.
47. LUNDIN, B.S. *et al.* 1999. Antibodies given orally in the neonatal period can affect the immune response for two generations: evidence for active maternal influence on the newborn's immune system. Scand. J. Immunol. **50:** 651–656.
48. KARLSSON, M.R. *et al.* 1999. Neonatal colonization of rats induces immunological tolerance to bacterial antigens. Eur. J. Immunol. **29:** 109–118.
49. SILFVERDAL, S.E. *et al.* 2002. Long-term enhancement of the IgG2 antibody response to *Haemophilus influenzae* type b by breastfeeding. Pediatr. Inf. Dis. J. **21:** 816–821.

50. HANSON, L.Å. *et al.* 2002. Immune system modulation by human milk. *In* Research Agenda for the Millenium: Integrating Population Outcomes, Biological Mechanisms and Research Methods in the Study of Human Milk and Lactation. C. Isaacs *et al.*, Eds.: 99–106. Kluwer Academic/Plenum.
51. CAMPBELL, D.A., JR. *et al.* 1984. Breast feeding and maternal-donor renal allografts: possibly the original donor-specific transfusion. Transplantation **37:** 340–344.
52. KOIS, W.E. *et al.* 1984. Influence of breast feeding on subsequent reactivity to a related renal allograft. J. Surg. Res. **37:** 89–93.
53. ZHANG, L. *et al.* 1991. Influence of breast feeding on the cytotoxic T cell allorepertoire in man. Transplantation **52:** 914–916.
54. TUBOLY, S. *et al.* 1995. Intestinal absorption of colostral lymphocytes in newborn lambs and their role in the development of immune status. Acta Vet. Hung. **43:** 105–115.
55. ARVOLA, M. *et al.* 2000. Immunoglobulin-secreting cells of maternal origin can be detected in B cell-deficient mice. Biol. Reprod. **63:** 1817–1824.
56. KOROTKOVA, M. *et al.* Modulation of neonatal oral tolerance to ovalbumin by maternal essential fatty acids intake during lactation in rats. Submitted.
57. HASSELBALCH, H. *et al.* 1996. Decreased thymus size in formula-fed infants compared with breastfed infants. Acta Paediatr. **85:** 1029–1032.
58. VAN ODIJK, J. *et al.* Breast feeding and allergy: a multidisciplinary review of the literature (1966–2001) on feeding mode in infancy and its implications on later allergic manifestations. Submitted.
59. IVARSSON, A. *et al.* 2002. Breast-feeding protects against celiac disease. Am. J. Clin. Nutr. **75:** 914–921.
60. DAVIS, M.K. 2001. Breastfeeding and chronic disease in childhood and adolescence. Pediatr. Clin. North. Am. **48:** 125–141.

Novel Method to Control Pathogenic Bacteria on Human Mucous Membranes

VINCENT A. FISCHETTI

The Rockefeller University, 1230 York Avenue, New York, New York 10021, USA

ABSTRACT: Nearly all infections begin at a mucous membrane site. Also, the human mucous membranes are a reservoir for many pathogenic bacteria found in the environment (that is, pneumococci, staphylococci, streptococci), some of which are resistant to antibiotics. Clearly, if this human reservoir can be reduced or eliminated, the incidence of disease will be markedly reduced. However, compounds designed to eliminate this reservoir are not available. Towards this goal, we have exploited the highly lethal effects of bacteriophage lytic enzymes (lysins) to specifically destroy disease bacteria on mucous membranes. Such lysins are used by the phage to release their progeny at the end of their replicative cycle. We have identified and purified these enzymes and found that when applied externally to gram-positive bacteria, they are killed seconds after contact. For example, 10^7 *S. pyogenes* are reduced to undetectable levels 10 s after enzyme addition. A feature of these enzymes is their high specificity; that is, streptococcal lysins kill streptococci and pneumococcal lysins kill pneumococci without effects on the normal flora organisms. *In vivo*, an oral colonization model for *S. pyogenes* and a nasal colonization model for *S. pneumoniae* were developed to test the capacity of the lysins to kill organisms on these surfaces. In both cases, when the animals were pre-colonized with their respective bacteria then treated with a small amount of lysin, specific for the colonizing organism, all the animals were found to be free of colonizing bacteria shortly after lysin treatment. Thus, lysins may be added to our armamentarium to control antibiotic-resistant bacteria.

KEYWORDS: bacteriophage; lytic enzyme; lysin; mucous membranes; carrier state; streptococci; pneumococci; bacillus

INTRODUCTION

The human mucous membranes are a reservoir for many pathogenic bacteria found in the environment (that is, pneumococci, staphylococci, streptococci), some of which are resistant to antibiotics. Generally, it is this reservoir that is the focus of infection in the population.[1–3] To date, except for polysporin and mupirocin ointments, there are no anti-infectives that are designed to control pathogenic bacteria on mucous. Because of the fear of developing resistance, antibiotics are not indicated to control the carrier state of disease bacteria. It is clear, however, that reducing

Address for correspondence: Vincent A. Fischetti, Ph.D., The Rockefeller University, 1230 York Avenue, New York, NY 10021. Voice: 212-327-8166; fax: 212-327-7584.
vaf@mail.rockefeller.edu

Ann. N.Y. Acad. Sci. 987: 207–214 (2003).

or eliminating this human reservoir will markedly reduce the incidence of disease in the community. Our laboratory has developed enzymes that are designed to prevent infection by safely and specifically destroying disease bacteria on mucous membranes. For example, enzymes specific for *S. pneumoniae* and *S. aureus* may be used nasally to control these organisms in daycare centers, hospitals, and nursing homes to prevent or markedly reduce both transmission and serious infections caused by these bacteria.

We accomplish this by capitalizing on the efficient system developed by bacteriophage to kill bacteria. When bacteriophage infect their host bacteria to produce progeny virus particles, they are faced with a problem: to release the progeny phage particles trapped in the bacterium at the end of the replicative cycle. They solve this problem by producing an efficient enzyme termed lysin that rapidly degrades the cell wall of the infected bacteria to release the phage progeny.[4]

We have identified and purified these enzymes and found that when applied externally to Gram-positive bacteria, they are killed seconds after contact.[5,6] For example, 10^7 group A streptococci could be reduced to undetectable levels 10 s after enzyme addition. To date, except for chemical agents, there is no biological compound known that can kill bacteria this quickly. Such phage lytic enzymes are the culmination of thousands of years of development by the bacteriophage during their association with bacteria.

ENZYME STRUCTURE

Most phage lytic enzymes that have been characterized so far possess a two-domain structure.[7,8] Generally, the N-terminal domain contains the catalytic activity of the enzyme that will cleave one of the four major bonds in the peptidoglycan of the bacterial cell wall. This activity may be an endo-β-*N*-acetylglucosaminidase or an *N*-acetylmuramidase (lysozymes), either of which acts on the sugar moiety; an endopeptidase, which acts on the peptide cross-bridge; or an *N*-acetylmuramyl-L-alanine amidase (or amidase), which hydrolyzes the amide bond connecting the sugar and peptide moieties.[4] Of the phage lytic enzymes that have been reported thus far, the great majority are amidases. The C-terminal domain of phage lytic enzymes has a specificity for a cell wall substrate.[9–11] Thus, unless the binding domain binds to its wall substrate, the catalytic domain will not cleave, offering specificity to the enzyme. The reason for this specificity was not apparent at first, since it seemed counterintuitive that the phage would specifically design an enzyme that was lethal for its host organism. However, as we learned more about how these enzymes function, the possible reason for this specificity became apparent (see below, Resistance).

MODE OF ACTION

By thin-section electron microscopy of phage enzyme-treated bacteria, it appears that the enzymes exert their lethal effects by digesting the peptidoglycan and forming holes in the cell wall, resulting in extrusion of the cytoplasmic membrane and, ultimately, hypotonic lysis (FIG. 1). Isolated cell walls treated with lysin are cut into pieces, verifying the results seen with intact bacteria (data not shown).

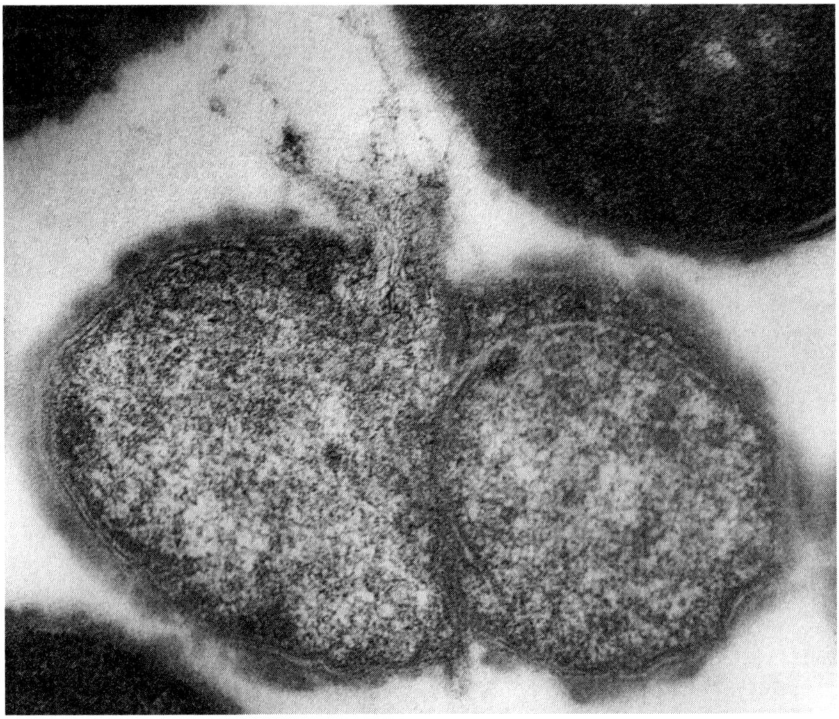

FIGURE 1. Effects of lysin on whole bacteria and cell walls. Thin-section electron micrographs of whole *S. pneumoniae* treated with Pal phage lytic enzyme. A bacterial cell is seen in which a hole has been digested in the cell wall, allowing the cytoplasmic membrane to become externalized, resulting in osmotic lysis and death.

TARGETED KILLING

An interesting feature of these enzymes is that they kill the species of bacteria from which they were produced. For instance, enzymes produced from streptococcal bacteriophage kill streptococci, and enzymes produced by pneumococcal bacteriophage kill pneumococci.[5,6] Specifically, the group A streptococcal lysin will kill group A streptococci efficiently while having but a small effect on groups C and G streptococci, and essentially no effect on normal oral streptococci (FIG. 2).

Similar results are seen with a pneumococcal-specific lysin. In this case, however, the enzyme was also tested against strains of pneumococci that were resistant to penicillin, and the killing efficiency was the same (FIG. 3). Unlike antibiotics, which are usually broad spectrum and kill many different bacteria found in the human body, some of which are beneficial, the phage enzymes kill only the disease bacteria with little to no effect on the normal human bacterial flora. Thus, the phage enzymes result in a targeted killing strategy to destroy bacteria. When the group A streptococcal enzyme was tested for safety in two animal model systems, one mucosal and the oth-

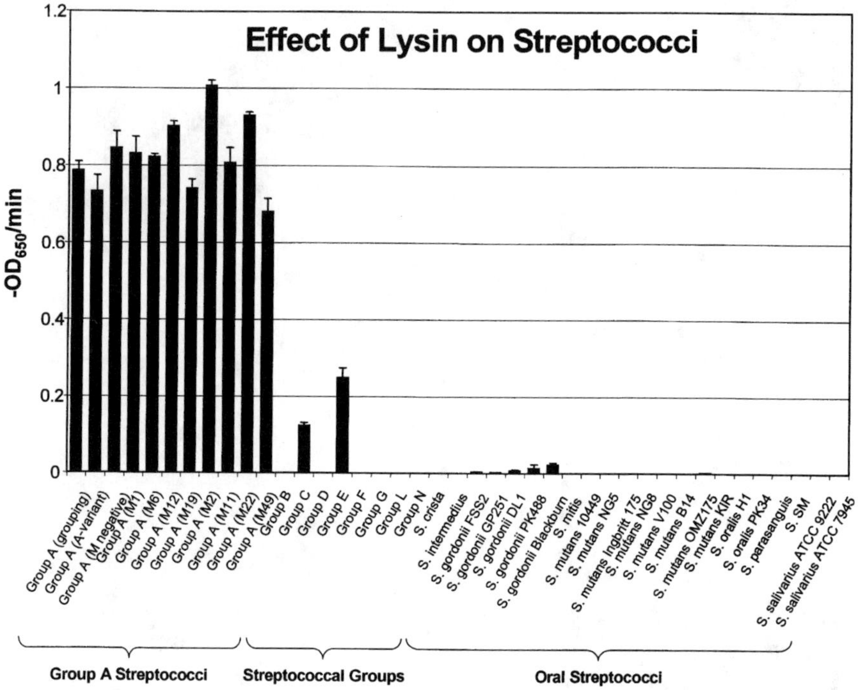

FIGURE 2. Lysin activity on various streptococci. Representative streptococcal strains were exposed to 250 U of purified lysin, and the OD_{650} was monitored. The activity of lysin for each strain was reported as the initial velocity of lysis, in $-OD_{650}$/min, based on the time it took to decrease the starting OD by half (that is, from an OD_{650} of 1.0 to 0.5). All assays were performed in triplicate, and the data are expressed as means ± standard deviations.

er skin, in which enzyme was added to these surfaces daily for 7 days, and the tissues examined both visually and histologically, nothing unusual was observed. This was not surprising since the bonds cleaved by the phage enzymes are found only in bacteria and not human tissues. Thus, it is anticipated that these enzymes will be well tolerated by the human mucous membranes.

IN VIVO EXPERIMENTS

Two *in vivo* animal models of mucosal colonization were developed to test the capacity for the lysins to kill organisms on these surfaces. An oral colonization model was developed for group A streptococci, and a nasal model was developed for pneumococci.[5,6] In both cases, when the animals were colonized with their respective bacteria and treated with a small amount of lysin specific for the colonizing organism, the animals were found to be free of colonizing bacteria 2–5 h after lysin treatment (FIG. 4). In the group A streptococcal experiment, animals were also swabbed 24 and 48 h after lysin treatment. During that time, most animals remained negative

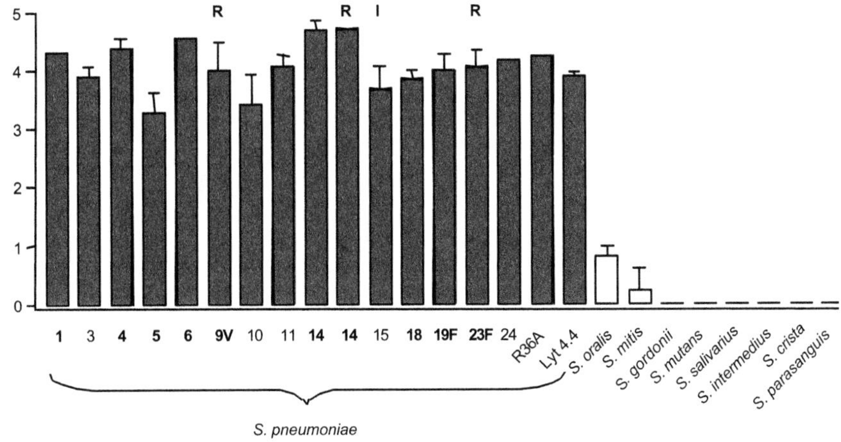

FIGURE 3. Rapid killing of pneumococci. *In vitro* killing of 14 clinical *S. pneumoniae* capsular types, 2 pneumococcal mutants, and 5 oral streptococcal strains with 100 U/mLPal for 30 s, expressed as the decrease of bacterial titers in powers of 10. Numbers above "*S. pneumoniae*" indicate serotypes. Error bars show standard deviation of triplicates. I: intermediate susceptibility to penicillin (MIC 0.1–1.0); R: highly penicillin resistant (MIC 2.0).

for streptococci, but one animal had died and two others showed positive colonies. We interpret these results to mean that the positive animals became infected during the first four days of colonization where some organisms became intracellular. Thus, while the lysin is able to clear organisms found on the surface, it was unable to kill organisms that had initiated an infection. To rule out the possibility that the organisms that appeared in 24 and 48 h did so because they became resistant to the lysin, we confirmed their sensitivity to the lysin.

KILLING BIOWARFARE BACTERIA

Because phage enzymes are so efficient in killing pathogenic bacteria, they may be a valuable tool in controlling biowarfare bacteria. To determine the feasibility of the approach, we identified a lytic enzyme from the gamma phage that is specific for *Bacillus anthracis*.[12] The gamma lysin, referred to as PlyG, was purified to homogeneity by a two-step chromatographical procedure and tested for its lethal action on gamma phage-sensitive bacilli. Three seconds after contact, as little as 100 units of PlyG mediated a 5,000-fold decrease in viable counts of a ~10^7 *Bacillus* culture. This lethal activity was observed in growth media, phosphate buffer, and even in human blood. When the enzyme was then tested against five mutant *B. anthracis* strains, it was found to be lethal for them all. When 10^6 bacilli were administered intraperitoneally (ip) to 10 mice, all except one died (10% survival) of rapidly fatal septicemia within 4 h. When a second set of 18 mice were also challenged ip with bacilli but given only 100 units of PlyG 15 min later by the same route, only five animals died

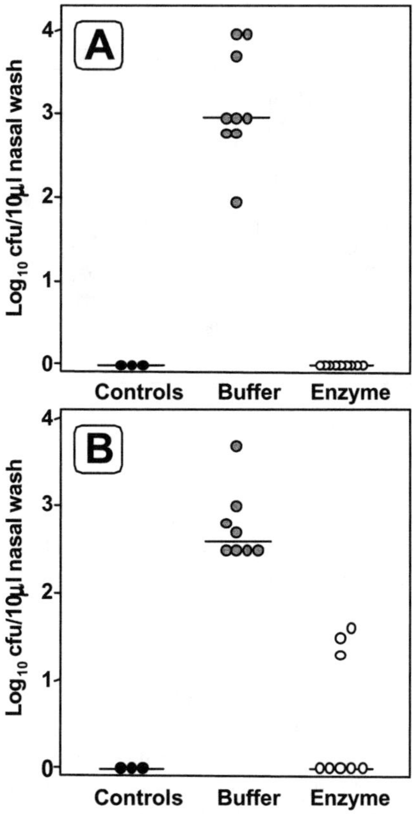

FIGURE 4. Killing pneumococci *in vivo*. Elimination of *S. pneumoniae* serotype 14 in the mouse model of nasopharyngeal carriage. (**A**) After nasal and pharyngeal treatment with a total of 1,400 U Pal, no pneumococci were retrieved in the nasal wash, compared to buffer-treated colonized mice ($P < 0.001$). No pneumococci were isolated from non-colonized control mice. (**B**) After treatment with a total of 700 U Pal, pneumococci were completely eliminated in five of eight colonized mice ($P < 0.001$) and significantly reduced overall.

(72% survival), and in three of these animals death was delayed more than 24 h. We anticipate that higher doses of enzyme will result in greater protection.

RESISTANCE TO ENZYMES

Repeated exposure to low concentrations of lysin to bacteria grown on agar plates did not lead to the recovery of resistant strains. Nor were we able to identify resistant bacteria after several cycles of exposure to low concentrations of enzyme in liquid.[5] This may be explained by the fact that the cell wall receptor for the pneumococcal lysin is choline,[13] a molecule that is necessary for pneumococcal viability. For group

A streptococci, we find that polyrhamnose, a cell wall component of the bacteria, is necessary for lysin binding, and polyrhamnose has also been shown to be important for streptococcal growth. While not yet proven, it is possible that during a phage's association with bacteria over the millennia, to avoid becoming trapped inside the host, the binding domain of their lytic enzymes has evolved to target a unique and essential molecule in the cell wall, making resistance to these enzymes a rare event.

CONCLUDING REMARKS

Phage lytic enzymes represent a new reagent for the control of bacterial pathogens. Since this capability has not been previously available, its acceptance may take a while. For the first time, we are able to specifically kill pathogens without affecting the surrounding normal bacteria. Whenever there is a need to kill bacteria, and contact can be made with the organism, phage enzymes may be freely utilized. Not only may they be used to control pathogenic bacteria on human mucous membranes, they may also find utility in the food industry to control disease bacteria without the extensive use of antibiotics in feed or harsh reagents to decontaminate. Because of the serious problems of resistant bacteria in hospitals, daycare centers, and nursing homes, particularly staphylococci and pneumococci, such enzymes will be of immediate benefit in these environments. The enzymes isolated thus far are remarkably heat stable (up to 60°C) and are very easy to produce in a purified state, which suggests that manufacturing costs will be pennies per dose. Thus, we may add phage enzymes to our armamentarium against pathogenic bacteria. They are molecules that have been in development for millions of years by bacteriophage in their battle to survive within bacteria. All we have done is to exploit them.

ACKNOWLEDGMENTS

I wish to acknowledge the members of my laboratory who are responsible for much of the work I described. My thanks to Daniel Nelson, Jutta Loeffler, and Raymond Schuch. Also, I wish to recognize the excellent technical assistance of Shiwei Zhu, Mary Windels, and Ryann Russell. I also wish to thank Abraham Turetsky at the Aberdeen Proving Grounds and Leonard Mayer of the Centers for Disease Control for testing the gamma lysin against authentic *B. anthracis* strains. This work was supported by a grant from the Defense Advance Research Project Agency (DARPA) and New Horizons Diagnostics.

REFERENCES

1. EIFF, C.V., K. BECKER, K. MACHKA, *et al.* 2001. Nasal carriage as a source of *Staphlococcus aureus* bacteremia. N. Engl. J. Med. **344:** 11–16.
2. COELLO, R. *et al.* 1994. Prospective study of infection, colonization and carriage of methicillin-resistant *Staphylococcus aureus* in an outbreak affecting 990 patients. Eur. J. Clin. Microbiol. Infect. Dis. **13:** 74–81.
3. DE LENCASTRE, H. *et al.* 1999. Carriage of respiratory tract pathogens and molecular epidemiology of *Streptococcus pneumoniae* colonization in healthy children attending day care centers in Lisbon, Portugal. Microb. Drug Resist. **5:** 19–29.

4. YOUNG, R. 1992. Bacteriophage lysis: mechanism and regulation. Microbiol. Rev. **56:** 430–481.
5. LOEFFLER, J.M., D. NELSON & V.A. FISCHETTI. 2001. Rapid killing of *Streptococcus pneumoniae* with a bacteriophage cell wall hydrolase. Science **294:** 2170–2172.
6. NELSON, D., L. LOOMIS & V.A. FISCHETTI. 2001. Prevention and elimination of upper respiratory colonization of mice by group A streptococci by using a bacteriophage lytic enzyme. Proc. Natl. Acad. Sci. USA **98:** 4107–4112.
7. DIAZ, E., R. LOPEZ & J.L. GARCIA. 1990 Chimeric phage-bacterial enzymes: a clue to the modular evolution of genes. Proc. Natl. Acad. Sci. USA **87:** 8125–8129.
8. GARCIA, P., J.L. GARCIA, J.M. SANCHEZ-PUELLES & R. LOPEZ. 1990. Modular organization of the lytic enzymes of *Streptococcus pneumoniae* and its bacteriophages. Gene **86:** 81–88.
9. LOPEZ, R., E. GARCIA, P. GARCIA & J.L. GARCIA. 1997. The pneumococcal cell wall degrading enzymes: a modular design to create new lysins? Microb. Drug Resist. **3:** 199–211.
10. LOPEZ, R., J.L. GARCIA, E. GARCIA, *et al.* 1992. Structural analysis and biological significance of the cell wall lytic enzymes of *Streptococcus pneumoniae* and its bacteriophage. FEMS Microbiol. Lett. **79:** 439–447.
11. GARCIA, E., J.L. GARCIA, A. ARRARAS, *et al.* 1988. Molecular evolution of lytic enzymes of *Streptococcus pneumoniae* and its bacteriophages. Proc. Natl. Acad. Sci. USA **85:** 914–918.
12. WATANABE, T., A. MORIMOTO & T. SHIOMI. 1975. The fine structure and the protein composition of gamma phage of *Bacillus anthracis*. Can. J. Microbiol. **21:** 1889–1892.
13. GARCIA, P., E. GARCIA, C. RONDA, *et al.* 1983. Inhibition of lysis by antibody against phage-associated lysin and requirement of choline residues in the cell wall for progeny phage release in *Stretococcus pneumoniae*. Curr. Microbiol. **8:** 137–140.

Peptide Mimics of a Major Lupus Epitope of SmB/B′

KENNETH M. KAUFMAN,[a,b,c] MONICA Y. KIRBY,[a,b]
JOHN B. HARLEY,[a,b,c] AND JUDITH A. JAMES[a,b]

[a]*Arthritis and Immunology Program, Oklahoma Medical Research Foundation, Oklahoma City, Oklahoma 73104, USA*

[b]*Departments of Medicine and Pathology, University of Oklahoma Health Sciences Center, Oklahoma City, Oklahoma 73106, USA*

[c]*Department of Veterans Affairs Medical Center, Oklahoma City, Oklahoma, USA*

ABSTRACT: One of the most distinguishing features of systemic lupus erythematosus is the presence of high concentrations of autoantibodies that recognize a limited number of self-antigens. Even though many lupus autoantigens have been identified, the inciting triggers of these abnormal immune responses are not fully understood. One mechanism that could generate these autoantibodies is a normal immune response toward a foreign epitope that mimics a common antigenic target of an autoantigen. Antibody generated toward the foreign epitope could also bind the autoantigen. This "cross-reactivity" would result in the presentation of the autoantigen to the immune system. Under autoimmune-prone conditions, tolerance toward the native protein is broken and an autoimmune response is initiated. Previously, it was suggested that Epstein-Barr virus might use such a mechanism to initiate an autoimmune response. Cross-reactive epitopes may have a similar amino acid sequence or a similar tertiary structure that is independent of amino acid sequence. A major, and likely initial, target of the lupus anti-SmB′ response is a repeated, proline-rich sequence, PPPGMRPP. To identify potential cross-reactive targets, we used affinity-purified autoantibodies specific for PPPGMRPP to screen a random heptapeptide phage display library. Eighty-five clones were isolated and sequenced with eleven distinct sequence motifs being identified. Two of these motifs were homologous to the SmB′ epitope, while the other nine were not. Interestingly, one of the peptide motifs that mimicked the SmB′ epitope is identical to a peptide sequence found in the Epstein-Barr virus major DNA binding protein.

KEYWORDS: systemic lupus erythematosus (SLE); lupus; phage; SmB/B′; autoantigen

Address for correspondence: Judith A. James, M.D., Ph.D., Arthritis and Immunology Program, Oklahoma Medical Research Foundation, 825 N.E. 13th Street, Oklahoma City, OK 73104. Voice: 405-271-4987; fax: 405-271-4110.
jamesj@omrf.ouhsc.edu

Ann. N.Y. Acad. Sci. 987: 215–229 (2003). © 2003 New York Academy of Sciences.

INTRODUCTION

Systemic lupus erythematosus (SLE) is a clinically heterogeneous autoimmune disorder of unknown etiology. However, existing data are consistent with genetic, hormonal, and environmental factors, each contributing to pathogenesis. Two patients may be diagnosed with SLE but share none of the laboratory or clinical symptoms, according to the diagnostic criteria.[1,2] Estimates of the number of Americans with SLE range from 150,000 to 400,000.[3] Over 98% of all SLE patients produce antinuclear autoantibodies. Many of these autoantibodies bind RNA-protein particles such as Sm, nRNP (nuclear ribonucleoproteins), Ro, and La at such high concentrations that precipitins form. Nearly half of the lupus patients produce precipitating levels of anti-nRNP antibodies; however, anti-nRNP autoantibodies are also found in other autoimmune disorders. Antibodies directed toward Sm are often found associated with anti-nRNP autoantibodies in SLE patients. Anti-Sm is so specific for SLE that the presence of these autoantibodies is considered a diagnostic criterion.[1,2] Anti-Sm autoantibodies are more common in African-American patients,[4] and several clinical manifestations have been associated with these autoantibodies. These clinical findings include late onset renal disease and poor prognosis,[5] malar rash, hematological involvement, and hypocomplementemia.[6]

Previous work has shown that a primary target of anti-Sm autoantibodies is the proline-rich, repeated sequence PPPGMRP(G)P.[7–14] In fact, all Sm-positive lupus patients that we have tested bind this sequence.[7] This region has also been shown to be the major epitope in murine models of lupus.[15] Recently, human monoclonal antibodies to this epitope have been cloned and characterized.[9] The initial anti-Sm response appears to be toward the PPPGMRPP epitope and spreads to other early epitopes by the sequential generation of cross-reactive antibodies[8] and then to new specificities. Animals immunized with the PPPGMRPP peptide have been shown to develop an autoimmune response that is characterized by both inter- and extramolecular epitope spreading and the development of lupus-like symptoms.[10,15,16] Interestingly, a number of lupus autoantigens have similar proline-rich sequences that are major autoantibody epitopes.[7] In fact, several studies have reported cross-reactivity between anti-Sm autoantibodies and these other autoantigens. This cross-reactivity has been observed with nRNP-A[17,18] and nRNP-C.[19]

Recently, we identified a high frequency of previous exposure to Epstein-Barr virus in pediatric lupus patients that was not observed in case-matched controls and suggests that a prerequisite for the development of lupus is infection with Epstein-Barr virus.[20] Interestingly, there are regions of the Epstein-Barr virus nuclear antigen 1 (EBNA-1) that are highly homologous to SmB′ epitopes. Antibodies toward the EBNA-1 peptide sequence (PPPGRRP) can also bind SmB′. Thus, these antibodies have the ability to cross-react with a known lupus autoantigen.[21] Rabbits immunized with the EBNA-1 proline-rich peptide have also been shown to generate an immune response that cross-reacts with SmB′.[21]

Random peptide phage display libraries have been used to study a number of protein–protein interactions. The use of these libraries is particularly useful in identifying peptides that mimic non-protein epitopes recognized by autoantibodies. Several studies have identified peptides that bind murine anti-dsDNA monoclonal antibodies[22–24] and polyreactive anti-DNA lupus antibodies.[23] Not only do these peptides bind autoantibodies but in some cases have also been shown to develop anti-dsDNA and lupus autoimmunity.[23,25]

Identification of the peptide sequences that interact with autoantibodies can provide information on possible etiological agents of lupus and identify structures that lead to epitope spreading, diagnostic reagents, and possible therapeutic interventions. To increase our understanding of the anti-Sm specificity, we have used affinity-purified human lupus autoantibodies to screen a random heptapeptide phage display library. Peptides that interact with anti-Sm autoantibodies have been identified and characterized.

MATERIALS AND METHODS

Patient Selection and Autoantibody Profile

The SLE patient used to screen the phage display library was chosen based on the presence of high concentrations of anti-PPPGMRPP autoantibodies. Significant binding to other regions of SmB/B′ was also present. However, following affinity purification, reactivity was limited to the PPPGMRPP epitope. In addition to a long-standing and well-monitored course of lupus, the patient used in this phage display screening process produced an autoantibody profile that included precipitating levels of Sm, nRNP, antinuclear antibodies by immunofluorescence (titer and pattern determined), and dsDNA autoantibodies.

Affinity-Purified PPPGMRPP

Affinity-purified human autoantibodies specific for the PPPGMRPP peptide were required to examine the epitopes recognized by anti-Sm sera. Multiple antigenic peptide (MAP)-PPPGMRPP synthesized by the University of Oklahoma Health Sciences Molecular Biology Resource Facility was coupled to CNBr preactivated Sepharose-4B according to manufacturer's directions (Sigma Chemical, St. Louis, MO).[26,27] Each MAP molecule contains eight copies of the PPPGMRPP peptide on a poly-lysine backbone. Sera (1 mL) from an Sm-precipitin-positive African-American female lupus patient was passed over the column and extensively washed. Bound antibodies were eluted with 3 M guanidine and then dialyzed against 25 mM tris-HCl, pH 8.0. The column affinity purification was repeated using the first round bound material. Purified antibodies were concentrated and quantitated by UV absorption.

ELISA of Purified Anti-PPPGMRPP Antibodies

The ELISA testing the specificity of column-purified anti-PPPGMRPP antibodies was performed as previously described.[28] Briefly, 1 μg per well of PPPGMRPP-MAP was coated and incubated with purified anti-PPPGMRPP antibodies (1:100 dilution). After washing, the alkaline-phosphatase anti-human secondary antibody at a dilution of 1:10,000 (Jackson Immunoresearch Laboratories) was incubated and developed with PNPP substrate. Plates were read at 405 on a Dynatech MicroELISA plate reader and standardized based upon a common positive control.

Development and Description of the Phage Library

To identify the peptide epitopes recognized by human anti-PPPGMRPP antibodies, we screened a random heptapeptide phage display library from New England Biolabs (Bar Harbor, MA). A heptapeptide library was chosen to enable all 1.28×10^9 seven amino-acid possibilities to be represented. In this library, each random heptapeptide is expressed at the N-terminus of the pIII minor phage coat protein followed by a Gly-Gly-Gly-Ser spacer. There are, on average, five copies of the pIII protein per phage particle. Theoretically, all possible combinations of seven amino-acid sequences should be expressed. However, there are some constraints on this library. The first amino acid of the random peptide cannot be a proline. This amino acid will inhibit pIII processing and prevent formation of the phage particle. Also, arginines are underrepresented in the library. The basic charge on this amino acid has an inhibitory effect on phage secretion.

Phage Library Screening Process

To enrich for phage particles that recognize the PPPGMRPP epitope, 300 ng of purified antibody were incubated with 2×10^{11} phage particles or plaque-forming units (pfu). By screening this number of phage clones, there is a 99.99% chance of any one peptide being represented. Antibody–phage complexes were isolated by incubation with protein-A agarose. The samples were spun down in a microcentrifuge tube and washed 10 times. Bound phage particles were eluted from the antibody by incubation with 100 mM glycine, pH 2.2, followed by neutralization with 1 M tris-HCl, pH 9.2. The titer of the enriched phage was determined by plating out serial dilutions. The remaining phage stock was amplified by infecting *E. coli* cells (ER2537), growing for 5 h, and recovering the supernatant. The amplified phage stock was titered, and a second round of enrichment was performed using 2×10^{11} pfu and protein-G agarose instead of protein-A agarose to capture antibody–phage complexes. This procedure was repeated two additional times for a total of four enrichments. We alternated with protein-A and protein-G agarose to avoid enrichment for peptides that bound to these specific proteins.

Sequencing

Sequencing of all isolated phage clones was performed on a Li-Cor 4200 automated DNA sequencing system following the manufacturer's recommendations (Lincoln, NE). Sequencing reactions were completed using long-range thermocycling kits with end-labeled M13 forward and M13 reverse universal primers (Epicentre Technologies, Madison, WI).

Computational and Statistical Analysis

We searched the current releases of Genbank and the Swiss protein database with the peptide sequences obtained from the phage clones. Densitometry calculations of the Western and dot blotting were obtained using an image-analysis software package from the National Institutes of Health.

Western Blotting of Denatured Phage Particles

For Western blot analysis, 1×10^{10} pfu of phage was separated on a 10% SDS-PAGE and transferred onto nitrocellulose. A phage type from each of the consensus peptide sequence motifs was run in individual wells. A 1:100 dilution of the anti-PPPGMRPP antibodies was reacted with each of the phage types as well as a 1:1000 dilution of a monoclonal anti-pIII antibody (Mo Bi Tec) to verify the concentration of the phage particles added. The procedure for the Western blotting is as previously described.[18]

Dot Blotting of Native Phage Particles

To test the binding of the anti-PPPGMRPP antibodies to a non-denatured form of the phage, a dot blot system was utilized. Then, 1×10^{10} pfu of each phage type was vacuum transferred onto nitrocellulose. The blot was then treated as discussed earlier for the denatured form of the phage particles in Western blotting.

Inhibition Experiments

To show phage specificity, aqueous phase inhibition experiments were performed by preincubating 100 ng of anti-PPPGMRPP antibodies with 0, 1, 10, 100, or 1000 ng of PPPGMRPP-MAP. The antibodies were then incubated with 20,000 pfu of the GPPPGMRPP control phage or various test phages. Phage–antibody complexes were isolated using protein-A agarose. The amount of phage bound by the antibodies was determined by titering the bound and unbound material.

RESULTS

Affinity Purification and Screening of Phage Library

Serum was selected from a systemic lupus erythematosus patient with precipitating levels of anti-Sm autoantibodies. The patient was within 2 years of diagnosis and was not on systemic immunosuppression at the time. Serum was passed over a PPPGMRPP-MAP affinity column for a total of two passes. To verify the specificity of the affinity-purified anti-PPPGMRPP antibodies, a standard ELISA was performed following two rounds of purification (FIG. 1A). Affinity-purified anti-PPPGMRPP antibodies cross-reacted with PPPGMRGP of SmB′, but did not react with any of the other overlapping octapeptides of SmB′ (data not presented).

The screening of a random heptapeptide phage display library allowed for the identification of peptide epitopes recognized by human anti-PPPGMRPP antibodies. To verify the enrichment of phage recognized by these anti-PPPGMRPP antibodies, a titer of the phage particles was determined following each screening step (FIG. 1B). The number of phages is held constant for each screening. This protocol results in the same amount of non-specific binding clones being isolated after each screen. When the number of clones isolated increases in each subsequent screening, it suggests that phages specifically interacting with the autoantibodies are being isolated. The increase in phage particles isolated after each enrichment and amplification suggested that we were enriching phage clones that specifically bound the anti-PPPGMRPP antibodies.

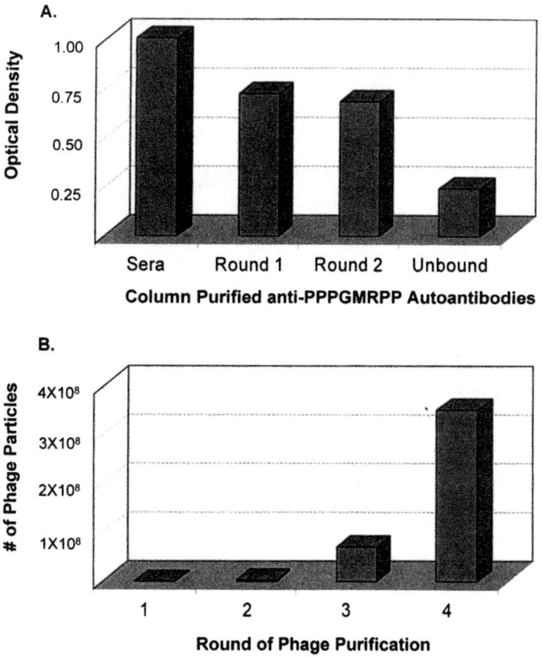

FIGURE 1. ELISA of affinity-purified anti-PPPGMRPP autoantibodies and phage enrichment. (**A**) Sera (1 mL) from an Sm-precipitin-positive black female lupus patient were passed over the PPPGMRPP column and extensively washed. Purified antibodies were concentrated and quantitated by UV absorption. ELISA was performed at a 1:100 dilution with wells coated with 1 μg PPPGMRPP-MAP. Results shown indicate sera before purification, first round bound material, second round bound material, and unbound material. (**B**) To verify the enrichment of phage recognized by the anti-PPPGMRPP antibodies, a titer of the phage particles was determined following each enrichment step. The increase in phage particles isolated after each enrichment and amplification suggests that we were enriching phage clones that specifically bound the anti-PPPGMRPP antibodies rather than nonspecific binding phage.

Phage Sequencing and Peptide Sequence Analysis

Following the fourth round of amplification, 100 clones were isolated. DNA sequence was obtained for 85 clones. Phages were sequenced in groups of 16 clones. After the last group was sequenced, no new sequences were obtained (TABLE 1). Peptide sequences were initially aligned using Wisconsin Package Version 10.0, Genetics Computer Group Pileup program and then visually separated into 11 distinct motifs. Both type I and type II motifs share obvious homology with the PPPGMRPP peptide. Interestingly, these motifs correspond to either the N-terminal PPPG sequence or the C-terminal RPP sequence of the parent PPPGMRPP peptide of purification. No motifs were identified that represented the middle GMR sequence. This would suggest that the PPPGMRPP epitope may contain two distinct epitopes that are recognized by an autoantibody or that the critical region required for antibody recognition does not include the middle residues.

TABLE 1. Sequence of phage peptides recognized by anti-PPPGMRPP autoantibodies

Type I	Type II	Type III	Type IV	TypeV	Type VI	Type VII	Type VIII	Type IX	Type X	Type XI	Misc.
	ARILYPP										CXLSVLK
	ATIYYPN										MPYMMYQ
	ALIQRPP	VPLTVLL	QLPLSLV								AGRLQRT
QLPPPGY	ALIQRPP	VPLTVLL	SPLSTLL	QHFKHPP							XXIQRPR
QLPPPGY	ALIQRPP	VPLTVLL	SPLSTLL	QHFKHPP							RQPCYAP
ILPPSGY	AVIHRPP	VPLTVLL	SPLSTLI	MKLKHPP	KIGFPIL						QPTYPTP
ILPPSGY	AVIHRPP	VPLTVLL	SPLTTLL	MKLKHPP	KIGFPIL						ATTQXTW
ILPPPGY	AVINRPP	VPLTVLL	SPLSTLR	MKLKHPP	KIGFPHI						ILPLRG
ILPPPGY	ASILRPP	VPLTVLL	SPISTLA	MQKVKHP	KIGFPHI	YLTPLQI					XXLAPPX
VLPPPGY	ASILRPP	VPLTVLL	SPLSSLT	ALKDKLP	KIGFPHI	KFLAPLQ	IPRPLDY	NHSLPLP	SPPEWLK	NHSLPLP	AKPFKTK
VLAPPGY	ATIFRPS	VPLSVLL	SPHTTLW	ANLDKLP	KIGFPHI	AFLPTLQ	IPRPLDY	NHSLPLP	SPPEWLK	NHSLPLP	MPNPVSG
TLPPPGR	AQILRPL	SPLNVLM	SPYTILT	AAGIKLP	KIGFPHI	SLFPWQR	VPRPLDI	NHSLPLP	SPPSWLK	NHSLPLP	HPHHLPP
Q	T QY		T T	QHF H	IL	L					
ILPPPGY	ALILRPP	VPLTVLL	SPLSTLL	AMKLKLPP	KIGFPHI	XFLXPLQ	IPRPLDY	NHSLPLP	SPPEWLK	NHSLPLP	
V	V			D							
	S										

NOTE: The consensus sequence for each peptide motif is shown at the bottom of each column.

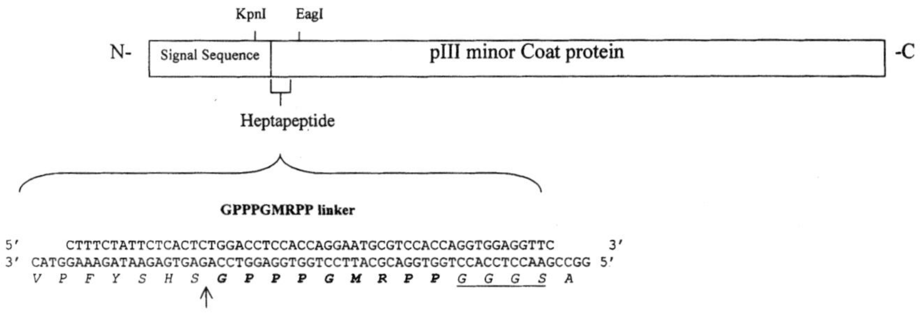

FIGURE 2. Construction of PPPGMRPP-phage. A linker encoding the PPPGMRPPP peptide was cloned into KpnI-EagI digested M13 phage DNA. Cleavage following the signal sequence (↑) results in the expression of the peptide sequence GPPPGMRPP (in *bold type*) at the amino terminus of the M13 pIII minor phage coat protein. The Gly-Ggly-Gly-Ser linker is also indicated by an *underscore*. This construct was used as a positive control to analyze binding of affinity-purified anti-PPPGMRPP autoantibodies.

Binding Characteristics and Affinity of Phage Clones

A positive control set of primers was used to examine the binding characteristics of these phage clones. Two complementary oligonucleotide primers were synthesized that encoded the amino acids GPPPGMRPP (FIG. 2). The oligonucleotides were then used to replace the peptide coding sequence from one of the M13 phage clones. The resulting clone was sequenced to verify that it contained the correct sequence and maintained the proper open reading frame. There are two differences between this peptide and the peptides expressed by the phage clones. This control phage peptide is nine amino acids long, whereas the phage peptides are seven amino acids long. We also included a glycine at the beginning of this sequence, because the amino acid proline inhibits cleavage of the phage pIII protein signal sequence. However, these differences should not limit the usefulness of this phage as a positive control.

Utilizing this generated positive control phage (GPPPGMRPP), anti-PPPGMRPP antibody binding to the different types of peptides displayed on the phage was characterized. Phage particles were obtained by polyethylene glycol precipitation. The purified phage particles were titered by plating out serial dilutions on lawns of *E. coli*. Initially, Western blots were used to characterize the binding of anti-PPPG-MRPP antibodies to the phage peptides (FIG. 3). Equivalent amounts of phage particles were separated on a 10% SDS gel and transferred to a nitrocellulose membrane. The results using anti-M13 pIII monoclonal antibody showed that approximately equivalent amounts of protein were present in each lane.

However, when purified anti-PPPGMRPP antibodies were utilized, differences in the intensity of the bands were found. These data suggest that these anti-PPPGMRPP antibodies have different affinities for the various peptides. Another possibility is that when the proteins are denatured, structural epitopes are partially lost, leading to different affinities for various peptides. To investigate this possibility, we utilized a dot-blot system in which phages were vacuum transferred to nitrocellulose under

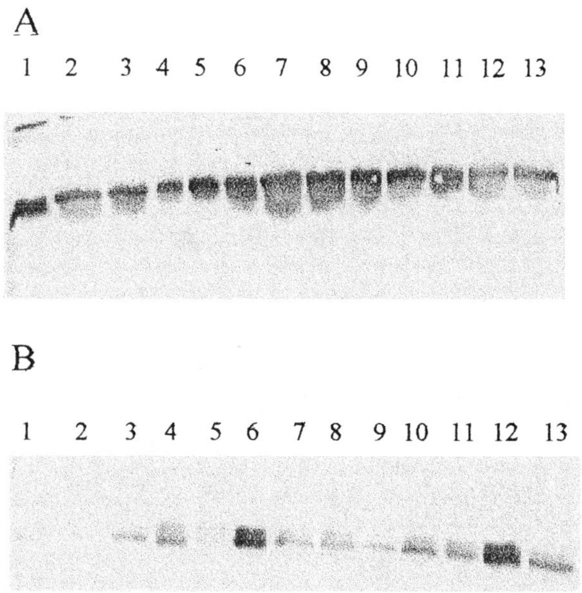

FIGURE 3. In the Western blot analysis shown, 1×10^{10} pfu of a phage from each type represented was separated on a 10% SDS gel and transferred to a nitrocellulose membrane. The lane designations are as follows: Lane 1: wt M13; 2: SPLSTLL; 3: KIGFPHI; 4: GPPPGMRPP; 5: IPRPLDY; 6: MKLKHPP; 7: ILPPPGY; 8: AVIHRPP; 9: ALIQRPP; 10: BPLTVLL; 11: KFLAPLQ; 12: SPPEWLK; 13: IPRPLDY. (**A**) Western blotting using anti-pIII monoclonal antibody. (**B**) Western blotting using affinity-purified anti-PPPGMRPP antibodies.

non-denaturing conditions (FIG. 4). Intensities of the dots were measured using the image analysis software package from the National Institutes of Health. The wild-type M13 clone had the lowest binding intensity. Six of the clones had an apparent affinity lower than the GPPPGMRPP-positive phage clone (FIG. 4, dots, 3, 4, 5, 7, 9, and 10). Two of the clones had approximately equivalent affinity to the positive control (dots 6 and 11), while the remaining two clones showed significantly higher signals (dots 8 and 12). These results were verified by dilution experiments (data not shown) and are summarized in TABLE 2.

Specificity of Phage Particles

Aqueous phase inhibition experiments were performed to indicate phage affinity and specificity. A 100-ng amount of PPPGMRPP-MAP blocked approximately 40% of the antibody–phage binding (FIG. 5). Similar experiments were then performed using six different phage clones (FIG. 6). The PPPGMRPP-MAP inhibited binding to five of the six clones. These results suggest that the antibody–phage binding is due to the sequence of the expressed peptides, at least in five of six examples. The results obtained in these experiments were consistent with the results obtained in the dot

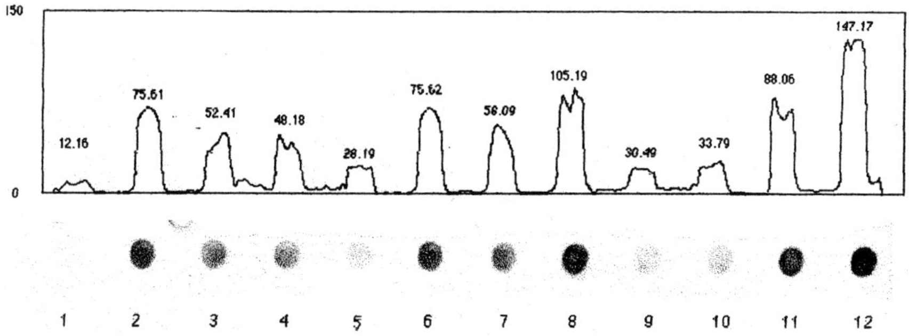

FIGURE 4. In this dot-blot analysis, 1×10^{10} pfu of each phage was vacuum transferred to nitrocellulose. Membranes were then probed with affinity-purified anti-PPPGMRPP antibodies. The graph shows the mean density as calculated using an image analysis software package provided by the National Institutes of Health. Peptides represented are as follows: 1: wt M13; 2: GPPPGMRPP; 3: SPLSTLL; 4: KIGFPHI; 5: IPRPLDY; 6: MKLKHPP; 7: ILPPPGY; 8: AVIHRPP; 9: ALIQRPP; 10: VPLTVLL; 11: SPPELK; 12: KFLAPLQ.

TABLE 2. Apparent peptide affinity of anti-PPPGMRPP antibodies under denatured or native conditions

	Affinity	
Peptide Sequence	Denatured	Non-Denatured
PPPGMRPP	+++	+++
VPLTVLL	+++	++
SPLSTLL	+	+++
KIGFPHI	++	++
IPRPLDY	+++	++
MKLKHPP	++++	+++
ILPPPGY	++	+++
AVIHRPP	+++	++++++
ALIQRPP	++	++
VPLTVLL	+++	++
KFLAPLQ	+++	++++++
SPPEWLK	++++	++++
IPRPLDY	+++	++

NOTE: Apparent affinity is based on denistometry measurements on Western blots (denatured conditions; FIG. 3) and dot blots (non-denatured conditions; FIG. 4). Densitometry measurements 0–25 (+), 26–50 (++), 51–75 (+++), 76–100 (++++), 101–125 (+++++), and 126–150 (++++++).

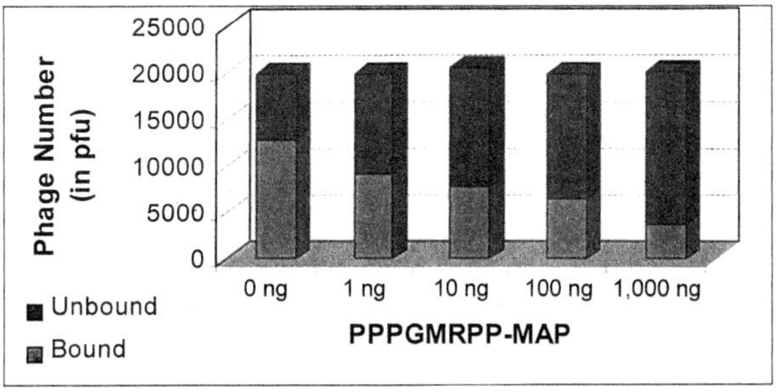

FIGURE 5. Inhibition of anti-PPPGMRPP autoantibodies. Affinity-purified anti-PPPGMRPP antibodies (100 ng) were incubated with 0, 1, 10, 100, or 1000 ng of PPPGMRPP-MAP. Antibody-phage complexes were recovered using protein-A agarose. The amount of bound and unbound material was determined by plating serial dilutions.

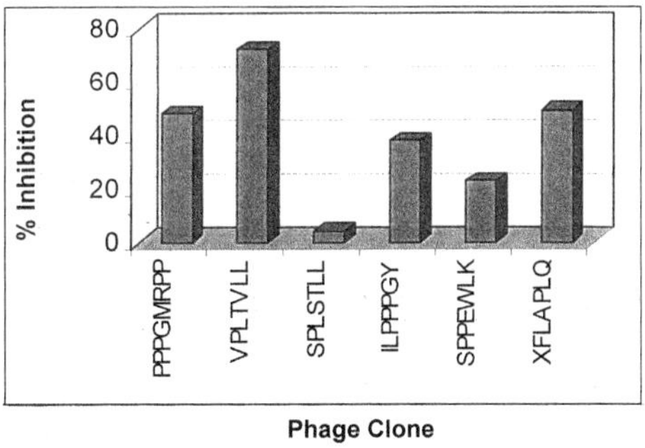

FIGURE 6. Inhibition of anti-PPPGMRPP-phage binding. Affinity-purified anti-PPPGMRPP antibodies (100 ng) were preincubated with 100 ng of PPPGMRPP-MAP and then mixed with 20,000 pfu of the indicated phage. Antibody-phage complexes were isolated using protein-A agarose and then plated in serial dilutions to determine the amount of bound and unbound material.

blots. Interestingly, in each of these experiments, the apparent affinity for the Epstein-Barr virus peptide SPPEWLK is higher than the GPPPMRPP-phage.

Sequence Homology with Database Entries

The current releases of GenBank and the Swiss protein database were then searched with the peptide sequences obtained from the phage clones. Three of the peptide sequences were identical to proteins contained in these databases. The *E. coli* ornithine aminotransferase contains a peptide sequence that is identical to the type I peptide ILPPPGY. Of the 85 clones sequenced, 9 of them contained sequences homologous to this sequence (2 were identical). This peptide contains the first four peptides found in the PPPGMRPP epitope. The mouse embryonic development control protein (NEDD1) contains an identical peptide to a type III peptide SPLNVLM. Ten of the 70 clones were homologous to this protein with only one being identical.

Perhaps the most interesting protein identified in the database search was the Epstein-Barr virus (EBV) major DNA binding protein. As mentioned earlier, our group has published data that show strong association of Epstein-Barr virus with pediatric and adult SLE.[20,28–32] This EBV protein contains a peptide sequence that is identical to the type X peptide SPPEWLK. The viral protein sequence is followed by a glycine (SPPEWLKG), which is identical to the phage peptide due to the Gly-Gly-Gly-Ser spacer separating the random peptide and the pIII protein. Two of the three phage clones were identical to this sequence, and the other contained a single amino acid substitution. The EBV major DNA binding protein is a 135-kDa protein that is encoded by the BALF2 open reading frame. This protein is required for viral DNA synthesis and is expressed immediately prior to the virus entering the lytic state. Homologues of this protein are found in other herpes viruses such as the cytomegalovirus and herpes simplex virus. However, the peptide sequence bound by anti-PPPGMRPP antibodies is not conserved in any of the other herpes viruses.

DISCUSSION

Nearly 30% of all SLE patient sera contain anti-Sm autoantibodies with SmB′ serving as the primary target. Characterization of the anti-SmB′ fine specificity indicates a proline-rich peptide sequence of PPPGMRPP to be a key antigenic target[7] and the initial recognition sequence.[8,10] There have been no previous data supporting the idea that each anti-Sm response is generated by unique autoantibodies to PPPGMRPP via different mechanisms. Therefore, the mechanism responsible for the generation of these autoantibodies remains an issue of speculation.

The use of a random peptide phage display library acts as an exceptional tool to identify epitopes that cross-react with human SLE patient autoantibodies, such as those directed against SmB′ and specifically PPPGMRPP. The ability to identify and provide information of cross-reactive antigens for antibody molecules is unusual. In the case of rheumatoid arthritis, a new peptide epitope RASFp1 has recently been identified from utilizing an IgG1 lambda monoclonal antibody to rheumatoid factor. In this case, it was concluded that the RASFp1-binding antibody's subtle conformational changes may very well convert a normal IgG molecule to recognize autoantigen that it normally would ignore.[33,34] Also noted was the possibility for differing specificities of antibodies in recognizing the identical epitope. This was shown to be true with the peptides derived from our screening of a phage library. When native dot blots and inhibition experiments of our phage were tested, it was evident that there were unique specificities of the various peptides expressing phage for the PPPGMRPP sequence. In some cases, data show that the affinity of the non-homologous peptides found by the phage display approach had higher affinity for the autoantibody of interest than the original peptide.

The utilization of a phage display library can also strongly support the molecular mimicry question by its ability to identify potential initial antibody targets when the inciting trigger may remain unknown. In the case of dsDNA and lupus, peptide mimitopes of reactive portions to dsDNA from a monoclonal antibody elicited an anti-dsDNA antibody response in BALB/c mice.[23,25] Upon immunization with autoantigen-derived peptides, these mice induced anti-dsDNA antibody production with renal IgG deposits, as well as a lupus-like syndromes.[23,25]

In this study, we identified 11 distinct anti-PPPGMRPP epitopes that have little to no homology with the original screening sequence. Analysis of antibody-bound peptide indicates that different peptide motifs can react at a single binding site, yet not have an obvious homologous sequence. The ability of an antibody to cross-react with multiple proteins could be due to epitopes that are similar in amino acid sequence or tertiary structure totally independent of amino acid sequence. A foreign pathogen could possibly induce an autoimmune response if the antibody produced is capable of cross-reacting with an epitope of a known autoantigen. This has been proven to be so in immunized rabbits with a PPPGMRPP homologous sequence from EBNA-1, an Epstein-Barr virus nuclear antigen. These PPPGRRP-immunized rabbits begin producing antibodies to the peptide of immunization, but also generate an anti-spliceosomal autoimmune response. Additional results from this study suggest that regions of another EBV protein may cross-react with human SLE anti-SmB′ autoantibodies. This peptide phage display technique has allowed us to identify additional potential binding sequences of a common, initial target of lupus anti-Sm antibodies.

ACKNOWLEDGMENTS

Supported by the Oklahoma Center for the Advancement of Sciences and Technology (Grant HN%-041 to K.M.K.), the National Institutes of Health (Grants AR01981, AR45084, and AR45451 to J.A.J.; Grants GM08237 and A131584 to J.B.H.), and the U.S. Department of Veterans Affairs (a grant to J.B.H).

REFERENCES

1. TAN, E.M., A.S. COHEN, J.F. FRIES, *et al.* 1982. The 1982 revised criteria for the classification of systemic lupus erythematosus. Arthritis Rheum. **25:** 1271–1277.
2. HOCHBERG, M.C. 1997. Updating the American College of Rheumatology revised criteria for the classification of systemic lupus erythematosus. Arthritis Rheum. **40:** 1725.
3. STEINBERG, A.D. 1994. Systemic lupus erythematosus: theories of pathogenesis and approaches to therapy. Clin. Immun. Immunopath. **72:** 171–176.
4. ARNETT, F.C., R.G. HAMILTON, M.G. ROEBBER, *et al.* 1988. Increased frequencies of Sm and nRNP autoantibodies in American blacks compared to whites with systemic lupus erythematosus. J. Rheumatol. **15:** 1773–1776.
5. HOMMA, M., T. MIMORI, Y. TAKEDA, *et al.* 1987. Autoantibodies to the Sm antigen: immunological approach to clinical aspects of systemic lupus erythematosus. J. Rheum. Suppl. **13:** 188–193.
6. THOMPSON, D., A. JUBY & P. DAVIS. 1993. The clinical significance of autoantibody profiles in patients with systemic lupus erythematosus. Lupus **2:** 15–19.
7. JAMES, J.A. & J.B. HARLEY. 1992. Linear epitope mapping of an Sm B/B′ polypeptide. J. Immunol. **148:** 2074–2079.
8. ARBUCKLE, M.R., M. REICHLIN, J.B. HARLEY & J.A. JAMES. 1999. Shared early autoantibody recognition events in the development of anti-Sm B/B′ in human lupus. Scand. J. Immunol. **50:** 447–455.
9. JAMES, J.A., M.T. MCCLAIN, G. KOELSCH, *et al.* 1999. Side-chain specificities and molecular modeling of peptide determinants for two anti-Sm B/B′ autoantibodies. J. Autoimmun. **12:** 43–49.

10. JAMES, J.A., T. GROSS, R.H. SCOFIELD & J.B. HARLEY. 1995. Immunoglobulin epitope spreading and autoimmune disease after peptide immunization: Sm B/B′-derived PPPGMRPP and PPPGIRGP induce spliceosome autoimmunity. J. Exp. Med. **181:** 453–461.
11. DEL RINCON, I., M. ZEIDEL, E. REY, *et al.* 2000. Delineation of the human systemic lupus erythematosus anti-Smith antibody response using phage-display combinatorial libraries. J. Immunol. **165:** 7011–7016.
12. PETROVAS, C.J., P.G. VLACHOYIANNOPOULOS, A.G. TZIOUFAS, *et al.* 1998. A major Sm epitope anchored to sequential oligopeptide carriers is a suitable antigenic substrate to detect anti-Sm antibodies. J. Immunol. Methods **220:** 59–68.
13. TSIKARIS, V., P.G. VLACHOYIANNOPOULOS, E. PANOU-POMONIS, *et al.* 1996. Immunoreactivity and conformation of the P-P-G-M-R-P-P repetitive epitope of the Sm autoantigen. Int. J. Pept. Protein Res. **48:** 319–327.
14. ROKEACH, L.A., M. JANNATIPOUR & S.O. HOCH. 1990. Heterologous expression and epitope mapping of a human small nuclear ribonucleoprotein-associated Sm-B′/B autoantigen. J. Immunol. **144:** 1015–1022.
15. JAMES, J.A. & J.B. HARLEY. 1998. A model of peptide-induced lupus autoimmune B cell epitope spreading is strain specific and is not H-2 restricted in mice. J. Immunol. **160:** 502–508.
16. JAMES, J.A. & J.B. HARLEY. 1998. B-cell epitope spreading in autoimmunity. Immunol. Rev. **164:** 185–200.
17. ARBUCKLE, M.R., A.R. SCHILLING, J.B. HARLEY & J.A. JAMES. 1998. A limited lupus anti-spliceosomal response targets a cross-reactive proline-rich motif. J. Autoimmun. **11:** 1–8.
18. JAMES, J.A. & J.B. HARLEY. 1996. Human lupus anti-spliceosome A protein autoantibodies bind contiguous surface structures and segregate into two sequential epitope binding patterns. J. Immunol. **156:** 4018.
19. JAMES, J.A. & J.B. HARLEY. 1995. Peptide autoantigenicity of the small nuclear ribonucleoprotein C. Clin. Exp. Rheumatol. **13:** 299–305.
20. JAMES, J.A., K.M. KAUFMAN, D.A. FARRIS, *et al.* 1997. The increased prevalence of Epstein-Barr virus infection in young patients suggests an etiology for systemic lupus erythematosus J. Clin. Invest. **100:** 3019–3026.
21. JAMES, J.A., R.H. SCOFIELD & J.B. HARLEY. 1995. Lupus autoimmunity after short peptide immunization. Ann. N.Y. Acad. Sci. **815:** 124–127.
22. SUN, Y., K.Y. FONG, M.C. CHUNG & Z.J. YAO. 2001. Peptide mimicking antigenic and immunogenic epitope of double-stranded DNA in systemic lupus erythematosus. Int. Immunol. **13:** 23–32.
23. SIBILLE, P., T. TERNYNCK, F. NATO, *et al.* 1997. Mimotopes of polyreactive anti-DNA antibodies identified using phage-display peptide libraries. Eur. J. Immunol. **27:** 1221–1228.
24. GAYNOR, B., C. PUTTERMAN, P. VALADON, *et al.* 1997. Peptide inhibition of glomerular deposition of an anti-DNA antibody. Proc. Natl. Acad. Sci. USA **94:** 1955–1960.
25. PUTTERMAN, C. & B. DIAMOND. 1998. Immunization with a peptide surrogate for double-stranded DNA (dsDNA) induces autoantibody production and renal immunoglobulin deposition. J. Exp. Med. **188:** 29–38.
26. AXEN, R., J. PORATH & S. ERNBACK. 1967. Chemical coupling of peptides and proteins to polysaccharides by means of cyanogen halides. Nature (London) **214:** 1302–1305.
27. KOHN, J. & M. WILCHECK. 1984. The use of cyanogen bromide and other novel cyanylating agents for the activation of polysaccharide resins. Appl. Biochem. Biotechnol. **9:** 285–304.
28. JAMES, J.A. & R.H. SCOFIELD. 2001. Role of viruses in systemic lupus erythematosus and Sjogren's syndrome. Curr. Opin. Rheum. **13:** 370–376.
29. MCCLAIN, M.T., J.B. HARLEY & J.A. JAMES. 2001. The role of Epstein-Barr virus in systemic lupus erythematosus. Frontiers Biosci. **6:** 137–147.
30. JAMES, J.A., B.R. NEAS, K.L. MOSER, *et al.* 2001. Systemic lupus erythematosus in adults is associated with previous Epstein-Barr virus exposure. Arthritis Rheum. **44:** 1122–1126.

31. HARLEY, J.B. & J.A. JAMES. 1999. EBV infection may be an environmental factor for SLE in children and teenagers. Arthritis Rheum. **42:** 1782–1783.
32. HARLEY, J.B. & J.A. JAMES. 1998. Is there a role for Epstein-Barr virus in lupus? Immunologist **6:** 79–83.
33. VON LANDENBERG, P., C. VON LANDENBERG, M. GRUNDL, *et al.* 1999. A new antigenic epitope localized within human kappa light chains specific for rheumatoid arthritis and systemic lupus erythematosus. J. Autoimmun. **13:** 83–87.
34. VON LANDENBERG, P., R. RZEPKA & I. MELCHERS. 1999. Monoclonal autoantibodies from patients with autoimmune diseases: synovial fluid B lymphocytes of a patient with rheumatoid arthritis produced an IgG lambda antibody recognizing J-sequences of Ig kappa chains in a conformation-dependent way. Immunobiology **200:** 205–214.

CD137-Mediated T Cell Co-Stimulation Terminates Existing Autoimmune Disease in SLE-Prone NZB/NZW F1 Mice

JUERGEN FOELL,[a] MEGAN McCAUSLAND,[a] JENNIFER BURCH,[a] NICHOLAS CORRIAZZI,[b] XIAO-JIE YAN,[b] CAROLYN SUWYN,[c] SHAWN P. O'NEIL,[c] MICHAEL K. HOFFMANN,[d] AND ROBERT S. MITTLER[a,e]

[a]*Department of Surgery, Emory University School of Medicine, Atlanta, Georgia 30329, USA*

[b]*North Shore-LIJ Research Institute, Manhasset, New York 11021, USA*

[c]*Department of Microbiology and Immunology, Yerkes Regional Primate Research Center, Emory University, Atlanta, Georgia 30329, USA*

[d]*Departments of Microbiology and Immunology, and Pathology, New York Medical College, Valhalla, New York 10595, USA*

[e]*Emory Vaccine Center, Emory University School of Medicine, Atlanta, Georgia 30329, USA*

ABSTRACT: T cell receptor recognition of antigen and major histocompatibility complex (signal 1) and T cell co-stimulation (signal 2) are essential for full T cell activation, differentiation, and survival of naïve and activated T cells. The prototypical T cell co-stimulatory receptor, CD28, is a constitutively expressed type I integral transmembrane protein and member of the Ig superfamily. Since its discovery, additional T cell co-stimulatory receptors have been identified, a number of which belong to the tumor necrosis factor receptor superfamily. Included within this group is CD137 (4-1BB), an activation-inducible, type I transmembrane protein. Co-stimulation of T cells through CD137 effectively up-regulates CD8 T cell activation and survival. Although $CD4^+$ T cells are efficiently activated through the T cell receptor and CD137 receptor, it provokes $CD4^+$ T cell anergy and blockade of T-dependent humoral immune responses. Therefore, we tested whether agonistic anti-CD137 monoclonal antibodies (mAbs) would be effective in blocking the induction or progression of B cell dependent autoimmune disease. Herein, we demonstrate the protective effect of agonistic anti-CD137 mAbs in blocking systemic lupus erythematosus (SLE) disease progression in NZB/W F1 mice. Protection from SLE following anti-CD137 mAb treatment is not confined to rescuing mice from disease progression; rather, it fully protects young mice from developing any symptoms of disease. We further found that treatment of proteinuric mice with anti-CD137 blocks ongoing anti-dsDNA autoantibody production.

KEYWORDS: CD137; 4-1BB; systemic lupus erythematosus; glomerulonephritis

Address for correspondence: R.S. Mittler, Department of Surgery and Emory Vaccine Center, Emory University School of Medicine, Atlanta, GA 30329. Voice: 404-727-9425; fax: 404-727-8199.

rmittler@rmy.emory.edu

Ann. N.Y. Acad. Sci. 987: 230–235 (2003). © 2003 New York Academy of Sciences.

INTRODUCTION

Systemic lupus erythematosus (SLE) is a chronic multi-organ disease that primarily affects women of childbearing age. The hallmark of SLE is dysregulated B cell function and the uncontrolled production of anti-nuclear autoantibodies including anti-dsDNA.[1] Clinical manifestations of disease include intermittent fatigue, joint pain, swelling, vasculitis, lymphopenia, and skin rashes and can include more serious symptoms such as central nervous system vasculitis, nephritis, pulmonary hypertension, interstitial lung disease, and stroke.[2] Disease development is a consequence of autoantibody-mediated tissue injury as well as vascular inflammation.[3,4] While the causes of disease initiation are not well understood, genetic predisposition to disease is clearly a major factor in lupus.[5,6] We have chosen the NZB/W F1 model for testing the effectiveness of anti-CD137 mAbs in treating SLE-like disease because these mice best reflect a number of pathogenic events that transpire during the development of SLE in humans.[3,6]

Injection of agonistic anti-CD137 mAbs into normal and tumor-bearing mice in a variety of model systems has demonstrated remarkable efficacy in activating $CD8^+$ T cells and protecting them from activation-induced cell death.[7,8] In this context, tumor-bearing mice reject their tumors,[9] and virus-infected mice rapidly clear their infection when they are immunized against the appropriate viral peptide in the presence of anti-CD137 mAbs.[10] In contrast, despite the fact that anti-CD137 mAbs efficiently co-stimulate $CD4^+$ T cells, they effectively block the development of T-dependent humoral immunity when administered early in the immunization by inducing a state of $CD4^+$ T cell anergy.[11] Here we demonstrate that anti-CD137 mAbs effectively block the onset of SLE disease in young NZB/W F1 mice and block its progression in animals that have advanced disease.

RESULTS AND DISCUSSION

Periodic injection of anti-CD137 mAbs into NZB/W F1 mice had a profound effect on preventing the development (FIG. 1a) or progression of established SLE disease (FIG. 1b) in these mice. The latter effect, unlike humoral immunity to non-self-antigens, seems to be related to the continued requirement for $CD4^+$ T cell help (FIG. 1c). In mice that received multiple injections of anti-CD137 mAbs (8–25 weeks of age), or those treated between weeks 26–35 of age after disease onset, virtually all of the mice remained free of proteinuria 78 weeks later (FIG. 1d). To determine whether the effect of anti-CD137 mAb treatment was directed towards $CD4^+$ T cell help, we depleted $CD4^+$ or $CD8^+$ T cells in these mice, or treated them with isotype-matched control mAbs. Our results demonstrated that a similar reduction in autoantibody production seen following anti-CD137 mAb treatment was also observed following $CD4^+$ T cell depletion, although the effect lasted for a much shorter period of time (FIG. 1c).

To determine the health of the kidneys from mice that had been treated or not with anti-CD137 mAbs, we prepared frozen sections of kidneys from mice that had commenced treatment at 14 weeks of age and that continued every third week until the mice were 25 weeks old. The sections were stained with an FITC-conjugated goat anti-mouse IgG and analyzed and photographed by fluorescence microscopy. In con-

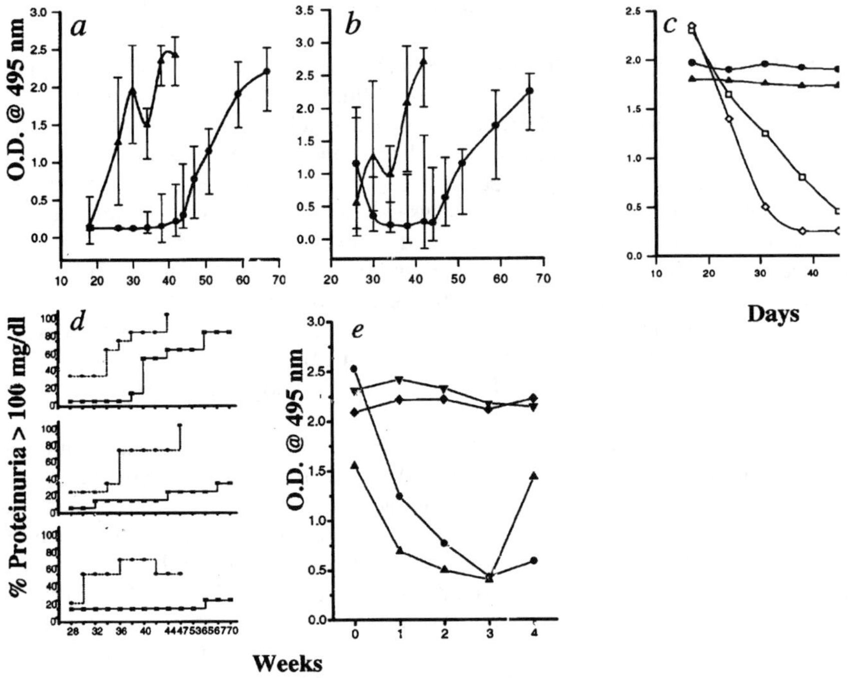

FIGURE 1. Inhibition of anti-dsDNA autoantibody production by anti-CD137 mAbs. NZB/W F1 mice received intraperitoneal (i.p.) injections with anti-CD137 or control mAb, and serum samples were collected and assayed in triplicates as indicated by ELISA for anti-dsDNA-reactive antibodies. IgG anti-dsDNA autoantibody titers are expressed as the mean O.D. at 490 nm of triplicates for individual mice using a 1/100 diluted serum sample. (**a**) Kinetics of the appearance of IgG anti-dsDNA antibodies in NZB/W F1 mice that received i.p. injections of anti-CD137 (*circles*) or control mAb (*triangles*) from 8 to 25 weeks of age every third week. (**b**) Kinetics of IgG anti-dsDNA autoantibody production in mice in which mAb treatment commenced at week 26 and continued every third week through week 35. Ten animals per group received i.p. injections of 200 μg of anti-CD137 mAb (*circles*) or control mAb (*triangles*). Each value is a mean ± standard deviation (see error bars). (**c**) Inhibition of anti-dsDNA autoantibody (*open symbols*) but not sheep red blood cells (SRBC) antibody production (*closed symbols*) in mice that received anti-CD137 mAb treatment after they had developed antibodies against SRBC or autoantibodies. (**d**) Protection from proteinuria in NZB/W F1 mice treated with anti-CD137 mAb (*squares*) or control mAb (*circles*) injections—a single injection (*top*), multiple early injections (*middle*), or multiple injections after disease onset (*bottom*) were evaluated. (**e**) $CD4^+$ T cell depletion temporarily decreases serum anti-dsDNA autoantibodies. NZB/W F mice 36–40 weeks old with high titers of anti-dsDNA antibodies were depleted for $CD4^+$ T cells or $CD8^+$ T cells with four i.p. injections of anti-CD4 mAb (*triangles*) or anti-CD8 mAb (*diamonds*) every 3 days, or treated with anti-CD137 mAb (*circles*) or control mAb (*inverted triangles*). After the last injection of antibody, serum samples were collected as indicated and assayed in triplicate for anti-dsDNA-reactive antibodies. The kinetics of the IgG anti-dsDNA autoantibody titers are expressed as the mean O.D. measured at 495 nm of each individual mouse using a 1/100 diluted serum sample. Data represent one of two independent experiments with similar results.

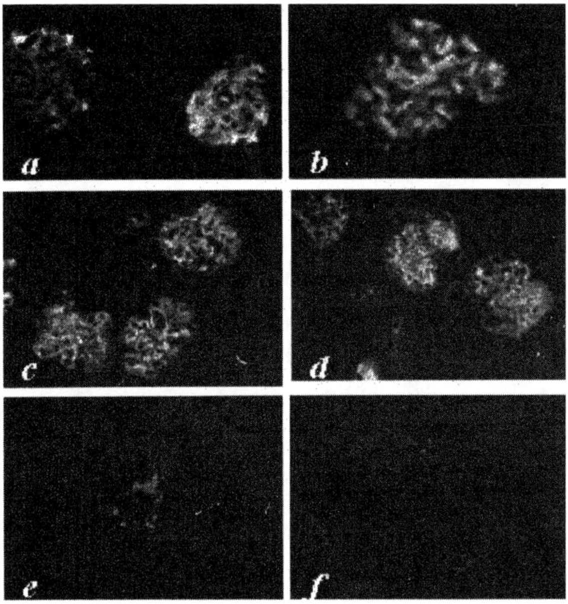

FIGURE 2. Anti-CD137 mAb treatment blocks immune complex deposition in kidneys of NZB/W F1 mice. NZB/W F1 mice were left untreated (**a**, **b**) or given i.p. injections of isotype-matched control mAb (**c**, **d**) or anti-CD137 mAb (**e**, **f**) every 3 weeks from week 14 through week 26. At week 35 (**a**, **c**, **e**) and week 45 (**b**, **d**, **f**), kidneys were removed and frozen sections were stained with an anti-mouse IgG-FITC mAb. The photomicrographs present representative immunohistological analyses of the kidneys from the three different treatment groups (magnification: 40×).

trast to control mice, those that had received anti-CD137 mAbs showed little evidence of immune complex deposition when examined 10 and 20 weeks after cessation of treatment (FIG. 2).

To examine kidney architecture in the treated and untreated mice, histological sections were prepared, hematoxylin and eosin stained, and then examined and photographed by light microscopy. The result of this study demonstrated that the microscopic anatomy of the glomeruli in kidneys from anti-CD137 mAb treated mice looked remarkably like those of normal BALB/c mice (FIG. 3D versus 3A). In contrast, the glomeruli in kidneys of mice injected with an isotype-matched control mAb displayed severe anatomical destruction that included diffuse proliferative glomerulonephritis, focal extra-capillary proliferation, and mesangial thickening with eosinophilic deposits with scattered germinal centers and focal plasma cell differentiation, as observed in the untreated mice (FIG. 3C versus 3B).

The results of this study demonstrate the remarkable capacity of an agonistic anti-T cell co-stimulatory receptor mAb (anti-CD137) to not only block SLE disease in young mice, but also to completely suppress its progression in animals that have established disease. The ability of this reagent to affect disease can be traced to its capacity to anergize $CD4^+$ T cell help and the fact that autoimmune humoral immune

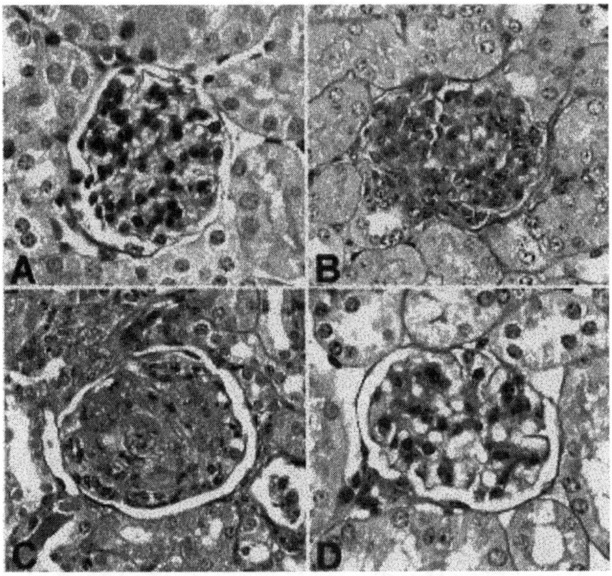

FIGURE 3. Anti-CD137 mAb treatment blocks the development of kidney disease. Mice were left untreated (**B**) or given i.p. injections of control mAb (**C**) or anti-CD137 mAb (**D**). Sections were microscopically examined for basic histopathology. Photomicrographs of hematoxylin-eosin–stained sections are shown at a magnification of 100×. An example of healthy renal histology of a BALB/c mouse kidney is shown (**A**).

(unlike humoral) responses to non-self-antigens in the NZB/W F1 mouse require continuous, life-long T helper functions. The biological differences between the regulation of non-autoimmune and autoimmune humoral responses in these mice is currently under investigation.

ACKNOWLEDGMENTS

This work was supported by funding from the National Institutes of Health (R21 AI48471-01 and 1RO1 1AI/RR49155-01A1), the Carlos and Marguerite Mason Trust (grant awarded to R.S.M.), and the Deutsche Forschungsgemeinschaft (FO318/1-1 and 1RO 1AI/RR49155-01A1, awarded to J.F.).

REFERENCES

1. Stott, D.I., J. McLearie & L. Neilson. 1986. Analysis of the clonal origins of autoantibodies against thyroglobulin and DNA in autoimmune thyroiditis and systemic lupus erythematosus. Clin. Exp. Immunol. **65:** 520–533.
2. Mills, J.A. 1994. Systemic lupus erythematosus. N. Engl. J. Med. **330:** 1871–1879.
3. Channing, A.A., T. Kasuga, R.E. Horowitz, *et al.* 1965. An ultrastructural study of spontaneous lupus nephritis in the NZB-BL-NZW mouse. Am. J. Pathol. **47:** 677–694.

4. CLYNES, R., C. DUMITRU & J.V. RAVETCH. 1998. Uncoupling of immune complex formation and kidney damage in autoimmune glomerulonephritis. Science **279:** 1052–1054.
5. WAKELAND, E.K., A.E. WANDSTRAT, K. LIU & L. MOREL. 1999. Genetic dissection of systemic lupus erythematosus. Curr. Opin. Immunol. 11: 701–707.
6. WAKELAND, E.K., K. LIU, R.R. GRAHAM & T.W. BEHRENS. 2001. Delineating the genetic basis of systemic lupus erythematosus. Immunity **15:** 397–408.
7. SHUFORD, W.W., K. KLUSSMAN, D.D. TRITCHLER, *et al.* 1997. 4-1BB costimulatory signals preferentially induce CD8+ T cell proliferation and lead to the amplification in vivo of cytotoxic T cell responses. J. Exp. Med. **186:** 47–55.
8. TAKAHASHI, C., R.S. MITTLER & A.T. VELLA. 1999. Cutting edge: 4-1BB is a bona fide CD8 T cell survival signal. J. Immunol. **162:** 5037–5040.
9. MELERO, I., W.W. SHUFORD, S.A. NEWBY, *et al.* 1997. Monoclonal antibodies against the 4-1BB T-cell activation molecule eradicate established tumors. Nat. Med. **3:** 682–685.
10. TAN, J.T., J.K. WHITMIRE, K. MURALI-KRISHNA, *et al.* 2000. 4-1BB Costimulation is required for protective anti-viral immunity after peptide vaccination. J. Immunol. **164:** 2320–2325.
11. MITTLER, R.S., T.S. BAILEY, K. KLUSSMAN, *et al.* 1999. Anti-4-1BB monoclonal antibodies abrogate T cell-dependent humoral immune responses in vivo through the induction of helper T cell anergy. J. Exp. Med. **190:** 1535–1540.

Using Single-Gene Deletions to Identify Checkpoints in the Progression of Systemic Autoimmunity

K. MICHAEL POLLARD,[a] PER HULTMAN,[c] AND DWIGHT H. KONO[b]

[a]*Department of Molecular and Experimental Medicine and* [b]*Department of Immunology, Scripps Research Institute, 10550 North Torrey Pines Road, La Jolla, California 92037, USA*

[c]*Division of Molecular Immunology and Pathology, Department of Molecular and Clinical Medicine, Linköping University, University Hospital, SE-581 85, Linköping, Sweden*

ABSTRACT: Systemic lupus erythematosus is a multigenic disorder of unknown etiology. To investigate the roles that specific genes play in lupus, we have examined the disease profiles in mice with single-gene deletions. In total, some 17 genes have been studied. Absence of certain genes, such as CD40L, CD28, or Igh6, abrogated induction of autoimmunity. Other genes, such as Igh5, IL-4, or ICAM-1, had little effect on the development of disease. Intermediate effects were observed in IL-6-deficient mice, while absence of β2-microglobulin resulted in loss of hypergammaglobulinemia and IgG1 autoantibodies, but produced little change in anti-chromatin antibodies or glomerular deposits. The most interesting observations were obtained with genes related to the expression or function of interferon-γ (IFN-γ). Reductions in IFN-γ levels in murine lupus are associated with reductions in both autoantibody levels and immune-complex-mediated pathology. Genes involved in up-regulation of IFN-γ expression, such as IL-12, STAT-4, or ICE, did not significantly influence autoimmunity, whereas absence of IFN-γ or IFN-γ receptor led to greatly reduced autoantibody response and immunopathology. Absence of IRF-1, a gene expressed in response to IFN-γ, resulted in selective retention of anti-chromatin antibodies but little glomerular pathology. These studies suggest that the presence of a baseline level of IFN-γ, rather than increased expression, is important for autoimmunity. Furthermore, as the IRF-1 knockout demonstrates, specific defects in signaling pathways and gene expression subsequent to IFN-γ/IFN-γ receptor interaction may influence only certain disease parameters. It has not escaped our attention that IFN-γ influences the expression and function of other immunologically relevant genes, such as IL-4, IL-6, and β2-microglobulin. Thus, these genes may be part of the downstream events following IFN-γ/ IFN-γ receptor interaction that promote the development of autoimmunity.

KEYWORDS: autoimmunity; IFN-γ; systemic lupus erythematosus; interleukin

Address for correspondence: Department of Molecular and Experimental Medicine and Department of Immunology, Scripps Research Institute, 10550 North Torrey Pines Road, La Jolla, CA 92037. Voice: 858-784-9214; fax: 858-784-2131.
mpollard@scripps.edu

Ann. N.Y. Acad. Sci. 987: 236–239 (2003). © 2003 New York Academy of Sciences.

INTRODUCTION

Systemic lupus erythematosus is a multigenic disorder of unknown etiology. To investigate the role that specific genes play in lupus, we have examined the disease profiles in mice with single-gene deletions. In total, some 17 genes have been studied in a model of induced systemic autoimmunity in which exposure to the heavy metal mercury elicits a lupus-like disease characterized by lymphadenopathy, hypergammaglobulinemia, autoantibodies, and immune-complex deposits.[1] Absence of certain genes, such as the co-stimulatory molecules CD40L and CD28, abrogated induction of autoimmunity. Igh6-deficient mice, which lack B cells, were also resistant to induction of autoimmunity. Other gene deletions, such as Igh5 (IgD), IL-4 or ICAM-1, had little effect on the development of disease. Some gene deletions produced intermediate effects. For example, IL-6-deficient mice lacked hypergammaglobulinemia and anti-chromatin autoantibodies but retained anti-nucleolar antibodies (ANoA) and immune-complex deposits. Absence of β2-microglobulin resulted in loss of hypergammaglobulinemia, IgG1 autoantibodies, and immune-complex deposits in the spleen, and reduced ANoA titers. However, such mice had little change in IgG anti-chromatin autoantibody titers or glomerular deposits. The anti-chromatin autoantibodies in β2-microglobulin-deficient mice were primarily of the IgG2 subclass, suggesting that IFN-γ was contributing to the autoimmunity in these mice. This was confirmed by *in vitro* studies showing that $HgCl_2$ stimulated increased IFN-γ expression in splenocyte cultures compared to wild-type controls.

IFN-γ AND IFN-γ RECEPTOR IN SYSTEMIC AUTOIMMUNITY

IFN-γ plays a significant role in systemic autoimmunity. We have previously shown that systemic autoimmunity induced by mercury is dependent upon the presence of IFN-γ.[2] In addition, reductions in IFN-γ levels in spontaneous murine lupus are associated with reductions in both autoantibody levels and immune-complex-mediated pathology.[3] Absence of IFN-γ receptor was also associated with a lack of mercury-induced autoimmunity (HgIA). The presence of serum IgG2a, anti-chromatin autoantibodies, ANoA, and immune-complex deposits was significantly reduced in IFN-γ-receptor-deficient mice even though such mice had intact serum IgG and IgG1 levels.

GENES THAT REGULATE THE EXPRESSION OF IFN-γ

To determine whether up-regulation of IFN-γ expression contributes to systemic autoimmune disease we examined the effect of deficiencies in IL-12, interleukin-1 converting enzyme (ICE), and signal transducer and activator of transcription 4 (STAT4). Absence of either STAT4 (which activates IFN-γ expression via IL-12) or ICE (which cleaves proIL-18 to activate IL-18, leading to IFN-γ induction) had little effect on induction of autoimmunity. IL-12, a heterodimer of p35 and p40 subunits, is required for the differentiation of Th1 cells, which produce IFN-γ. Deficiencies of either p35 or p40 did not affect induction of HgIA, although $p35^{-/-}$ mice did show a reduction in anti-chromatin antibodies, and fewer $p40^{-/-}$ mice were ANoA-positive

TABLE 1. Effect of individual gene deficiences on systemic autoimmunity

Hypergamma-globulinemia	Autoantibody		Immune Complex Deposits
	ANoA	Anti-chromatin	
IL-6↓		IL-6↓↓	
β2M↓	β2M↓		β2M↓
IRF-1↓	IRF-1↓↓	IRF-1↓	IRF-1↓
		IL-12p35↓	

NOTE: Suppression of all parameters was found in the absence of IFN-γ, IFN-γ R, CD28, CD40L, or Igh6. No effects were observed in the absence of STAT4, ICE, IL-12p40, ICAM-1, IL-4, or Igh5. (Double arrows indicate absence of antibody response.)

than wild-type controls. Thus the absence of genes that increase IFN-γ expression does not significantly affect induction of systemic autoimmunity. It therefore seems likely that the requirement for IFN-γ lies not with increased expression, but rather with events that follow IFN-γ/IFN-γ receptor interaction.

GENES REGULATED BY IFN-γ EXPRESSION

Among the first genes to be expressed in response to IFN-γ is interferon regulatory factor 1 (IRF-1). IRF-1 induces transcription of genes with interferon-stimulated response element (ISRE) in their promoters. IRF-1 is also induced by IL-12[4] and IL-6.[5] Mice lacking IRF-1 developed few of the features of systemic autoimmunity. These mice lacked hypergammaglobulinemia, ANoA, and glomerular deposits. However IRF-$1^{-/-}$ mice were ANA positive and did produce anti-chromatin antibodies. In additional, several mice were positive for immune-complex deposits in the spleen. Thus, gene expression initiated by IRF-1 is responsible for most but not all the features of induced systemic autoimmunity.

IFN-γ affects the expression of over 200 different genes.[6] It is, therefore, technically challenging to examine the individual roles that such a large number of genes might play in induced systemic autoimmunity. This problem is being approached by custom DNA microarrays containing probes for genes that are regulated, positively or negatively, by IFN-γ.

CONCLUSIONS

Using gene knockout mice to examine the role of specific genes in the development of systemic autoimmunity induced by mercury, we have begun to define the critical molecular events in this disease. Our findings reveal that genetic deficiencies can affect development of systemic autoimmunity in various ways (TABLE 1). Some genes may completely abrogate induction of all disease parameters, while deficiencies in others may affect different features of disease. Significantly, our studies confirm the important role of a single cytokine, IFN-γ, in systemic autoimmunity, and suggest that it is not increased expression of this pleiotropic cytokine that is important, but rather a baseline level of expression. Absence of IRF-1 demonstrates that defects in signaling pathways and gene expression subsequent to IFN-γ/IFN-γ receptor interaction may influence expression of individual disease parameters. Deficiencies in other genes, such as IL-6, and β2-microglobulin, also influenced expression of some but not all parameters of autoimmunity. Some of these genes may play a role in the biological responses that follow IFN-γ/IFN-γ receptor interaction.

ACKNOWLEDGMENTS

This work was supported by grants ES07511, ES09802, ES08080, and ES08666 from the National Institutes of Health (awarded to K.M.P. and D.H.K.) and a grant from the Swedish Research Council (project 09453, awarded to P.H.). This is publication number 14926-MEM from the Department of Molecular and Experimental Medicine of the Scripps Research Institute.

REFERENCES

1. POLLARD, K.M. & P. HULTMAN. 1997. Effects of mercury on the immune system. Met. Ions Biol. Syst. **34:** 421–440.
2. KONO, D.H., D. BALOMENOS, D.L. PEARSON, *et al.* 1998. The prototypic Th2 autoimmunity induced by mercury is dependent on IFN-gamma and not Th1/Th2 imbalance. J. Immunol. **161:** 234–240.
3. LAWSON, B.R., G.J. PRUD'HOMME, Y. CHANG, *et al.* 2000. Treatment of murine lupus with cDNA encoding IFN-gammaR/Fc. J. Clin. Invest. **106:** 207–215.
4. GALON, J., C. SUDARSHAN, S. ITO, *et al.* 1999. IL-12 induces IFN regulating factor-1 (IRF-1) gene expression in human NK and T cells. J. Immunol. **162:** 7256–7262.
5. TANIGUCHI, T., K. OGASAWARA, A. TAKAOKA & N. TANAKA. 2001. IRF family of transcription factors as regulators of host defense. Annu. Rev. Immunol. **19:** 623–655.
6. BOEHM, U., T. KLAMP, M. GROOT & J.C. HOWARD. 1997. Cellular responses to interferon-gamma. Annu. Rev. Immunol. **15:** 749–795.

Activation of the Ets Transcription Factor Elf-1 Requires Phosphorylation and Glycosylation

Defective Expression of Activated Elf-1 Is Involved in the Decreased TCR ζ Chain Gene Expression in Patients with Systemic Lupus Erythematosus

GEORGE C. TSOKOS,[a,b] MADHUSOODANA P. NAMBIAR,[a,b] AND YUANG-TAUNG JUANG[a,b]

[a]*Department of Medicine, Uniformed Services University of the Health Sciences, Bethesda, Maryland 20814, USA*

[b]*Department of Cellular Injury, Walter Reed Army Institute of Research, Silver Spring, Maryland 20910, USA*

ABSTRACT: Elf-1, a member of the Ets transcription factor family with an estimated molecular mass of 68 kDa, is involved in the transcriptional regulation of several hematopoietic cell genes. It is shown that following O-GlcNAc glycosylation and phosphorylation by PKC ϑ, the cytoplasm-located, 80-kDa Elf-1 translocates to the nucleus as a 98-kDa protein. In the nucleus, Elf-1 binds to the promoter of the *TCR* ζ gene and promotes its transcription in Jurkat and fresh human T cells. It is also shown that in the majority of patients with systemic lupus erythematosus (SLE), who are known to express decreased levels of T cell receptor (TCR) ζ chain and mRNA, the 80-kDa Elf-1 protein does not undergo proper post-transcriptional modification, which results in low levels of the 98-kDa protein, lack of Elf-binding to the TCR ζ promoter, and decreased gene transcription. Therefore, a novel activation pathway for a member of the Ets family of transcription factors, which is defective in patients with systemic autoimmunity, has been revealed.

KEYWORDS: Ets transcription factor; systemic lupus erythematosus; human T cells; T cell receptor; Elf-1; TCR ζ chain

INTRODUCTION

Immune cells respond to external antigens following their engagement to the antigen receptor through a series of biochemical processes. These processes involve protein tyrosine phosphorylation, calcium mobilization, and activation of transcription factors, which together are known as cell signaling. The antigen receptor, T cell receptor (TCR)/CD3, is a multisubunit complex consisting of α and β chains; the

Address for correspondence: George C. Tsokos, M.D., Department of Medicine, Uniformed Services University of the Health Sciences, Bethesda, MD 20814. Voice: 301-319-9911); fax: 301-319-9133.
gtsokos@usuhs.mil

Ann. N.Y. Acad. Sci. 987: 240–245 (2003).

CD3 γ, δ, and ε chains; and a ζ–ζ homodimer. Each subunit of the ζ chain homodimer possesses three immune receptor tyrosine-activated motifs (ITAMs), whereas other CD3 chains contain one ITAM each. Upon TCR activation, the tyrosine residues within the ITAMs become phosphorylated by Lck and Fyn, leading to the association and activation of ZAP-70.

ABERRANT ANTIGEN-INITIATED CELL SIGNALING AND DEFECTIVE TCRζ CHAIN EXPRESSION IN SLE T CELLS

Following engagement of the antigen receptor, T and B lymphocytes from systemic lupus erythematosus (SLE) patients respond rapidly by hyperphosphorylating a number of cytosolic signaling protein intermediates and increasing their concentration of free calcium. This abnormal response occurs in both principal T cell subsets and in cell cultures propagated *in vitro*. In SLE T cells, the more sustained supranormal calcium response has faster kinetics and is observed under deficient TCR ζ chain expression.[1] A vast majority of the SLE patients also display decreased expression of TCRζ chain mRNA. The TCR ζ chain transcript is generated as the spliced product of eight exons that are separated by distances of 0.7 kb to more than 8 kb. *TCRζ* gene is located in chromosome 1q23. Genes encoded in chromosome 1q have been suggested to contribute to genetic predisposition and susceptibility to SLE by genome-wide scan of multiplex SLE families.[2]

Although the precise molecular mechanisms underlying TCRζ chain deficiency are still being examined, current evidence supports the possibility of a transcriptional defect. In SLE T cells that expressed low levels of TCR ζ chain transcripts, cloning, and sequencing revealed more frequent heterogeneous polymorphisms/mutations and alternative splicing of the TCR ζ chain.[3,4] Most of these mutations are localized to the three ITAMs or the GTP-binding domain, and could functionally affect the ζ chain, providing a molecular basis to the known T cell signaling abnormalities in SLE T cells. Absence of the mutations/polymorphisms in the genomic DNA suggests that these are the consequence of irregular RNA editing. SLE patients also showed a significant increase in the splice variants of the ζ chain. RT-PCR analysis of the TCR ζ chain 3′-untranslated region showed an alternatively spliced 344-bp product with both splicing donor and acceptor sites, resulting from deletion of nucleotides from 672 to 1233 of the TCR ζ chain mRNA. Unlike the normal TCR ζ chain, the expression of the TCR ζ chain with the alternatively spliced 344-bp, 3′-untranslated region was higher in SLE T cells compared to non-SLE controls.[5]

How does the TCR engagement induce hyperphosphorylation of cytosolic proteins and supranormal $[Ca^{2+}]_i$ response in the milieu of deficient TCR ζ chain? In addition to the possible gain-of-function mutations of the TCR ζ chain, our investigation has proposed two mechanisms. One involves increased expression of the FcεRIγ chain, and the other involves increased membrane lipid–raft association of the residual TCR ζ chain, and either mechanism could explain the supranormal TCR/CD-mediated $[Ca^{2+}]_i$ response in SLE T cells. The FcεRIγ chain is expressed at increased levels of the TCR, and following engagement of the TCR, it becomes phosphorylated and associates with the *Syk* kinase.[6] These data have shown that the TCR in SLE T cells has been rewired, and this results in abnormal signaling events.

The decreased levels of TCR ζ chain mRNA in SLE T cells and the absence of mutations in the proximal promoter of the TCR ζ chain provided the impetus to study the expression of the transcription factors that have been suggested to regulate the expression of the ζ chain gene.

Ets PROTEIN BIOCHEMISTRY

Elf-1 is a transcription factor that belongs to the Ets (E twenty-six specific) family of proteins. In hematopoeitic cells, Elf-1 mediates the induction of various genes, including the CD4, the GM-CSF, and the IgH enhancer genes and the TCR α and ζ chains. Point mutations, introduced into the two Elf-1 binding sites (–147/–119 and –66/–33) of the TCR ζ chain promoter, abolish its transcriptional activity in Jurkat cells.[7]

Elf-1 cDNA encodes a protein with an expected molecular weight of 68 kDa. However, most of the published work indicates that Elf-1 exists as a protein with an apparent molecular weight of 98 kDa. Proteins that contain a high ratio of either positively or negatively charged amino acids are known to display discrepancies between the estimated molecular weight and that observed by SDS gel electrophoresis. However, Elf-1 does not contain a high ratio of charged amino acids, suggesting that charge alone cannot be responsible for the dichotomy between the predicted and observed molecular weights. We have therefore hypothesized that this discrepancy is likely to be caused by posttranslational modifications, which are pivotal in determining the subcellular localization, protein metabolism, and transcriptional activity of Elf-1 (FIG. 1).[8]

POSTTRANSLATIONAL MODIFICATION OF ELF-1 IN NORMAL HUMAN T CELLS

Using cytoplasmic and nuclear extracts from primary and Jurkat T cells, we found that the 80-kDa form of Elf-1 was prevalent in the cytoplasmic extracts, whereas the 98-kDa form of Elf-1 was prevalent in the nuclear extracts. We transfected Jurkat and fresh peripheral T cells with the pcDNA/Elf-1 plasmid, which encodes the full length Elf-1 cDNA and established that it encodes a protein that resides in the cytoplasm as an 80-kDa protein, and in the nucleus of T cells as a 98-kDa protein.

Because PKC θ has been recently shown to be involved in the phosphorylation of several transcription factors involved in the regulation of the expression of T lymphocyte genes,[9,10] we considered that similar PKC isoforms may not only be involved in the phosphorylation of Elf-1, but may also be responsible for the increase in the apparent molecular weight of Elf-1. In addition, sequence analysis of Elf-1 revealed the existence of multiple PKC phosphorylation sites. Jurkat cells cultured in the presence of PMA and calphostin C, which inhibits all the activities of PKC isoforms, displayed diminished expression of 98-kDa Elf-1. The PKC θ inhibitor, rottlerin, also decreased the expression of 98-kDa Elf-1, although not to the same extent as calphostin C, suggesting that multiple isoforms of PKC are involved in the conversion of the 80- to the 98-kDa form. Interestingly, treatment of Jurkat cells with

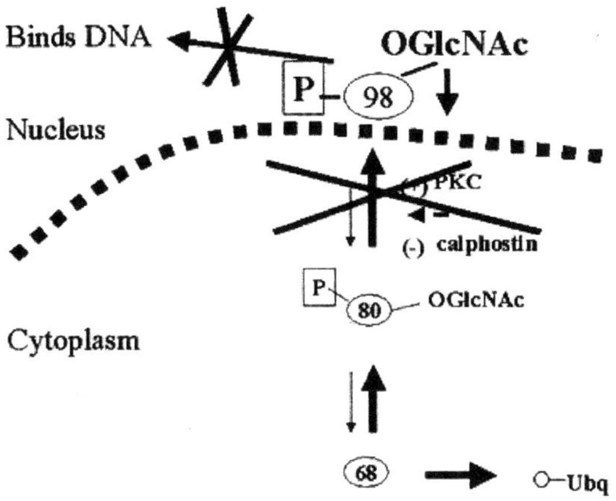

FIGURE 1. Elf-1 is produced as a 68-kDa protein that undergoes partial phosphorylation and O-glycosylation to convert to a 80-kDa protein. This form stays in the cytoplasm, where it can be further phosphorylated and O-glycosylated to convert to a 98-kDa form, which travels to the nucleus, where it can bind to the promoter of the TCR ζ promoter. The 68- and 90-kDa forms, but not the 80-kDa form, may undergo proteasome degradation. In T cells from patients with SLE, there is a defect in the phosphorylation and O-glycosyaltion of the 80-kDa form, which results in decreased levels or absence of the 98-kDa form. Decreased binding of Elf-1 to the TCR ζ chain promoter is associated with decreased levels of the ζ chain.

calphostin C enhanced the expression of 80-kDa Elf-1, indicating that the 98-kDa form can be derived directly from the 80-kDa form.

To test directly which of the various PKC isoforms are involved in the phosphorylation of Elf-1, we transiently transfected Jurkat cells with a series of plasmids expressing different active PKC isoforms. Transient expression of either PKC θ or PKC ε increased the expression of the 98-kDa form, whereas overexpression of PKC α did not affect the relative levels of the 80- and 98-kDa forms of Elf-1. These data show that the so-called novel PKC isoforms θ and ε, but not the conventional PKC isoform α, are involved in the posttranslational modification of Elf-1.

These conclusions are supported by additional data in which cells treated with the phosphatase inhibitor, okadaic acid, displayed enhanced expression of the 98-kDa band, whereas it minimally enhanced the intensity of the upper 80-kDa Elf-1 band. In addition, treatment of total cellular proteins with the bacteriophage λ phosphatase decreased the intensity of the 98-kDa form. Altogether, these experiments indicate that the 98-kDa Elf-1 is heavily phosphorylated and undergoes dephosphorylation in a dynamic manner.

We considered that glycosylation contributes to the appearance of the 98-kDa form of Elf-1, because phosphorylation alone cannot account for the rather large increase in molecular mass. Elf-1 has multiple potential sites for O-GlcNAc glycosylation; therefore, we investigated whether Elf-1 is also O-GlcNAcylated. Immunoprecipitation of nuclear proteins either with an antibody against Elf-1 or an antibody (RL-2) against the O-GlcNAc moiety[11] and blotting of the separated immunoprecipitates with the reverse antibody revealed that the 98-kDa form is O-GlcNAc glycosylated. Similar experiments using whole cellular extracts, which contain comparable amounts of both the 98- and 80-kDa forms, showed that the 80-kDa form is also recognized by the RL-2 antibody, and therefore it is also decorated with GlcNAc moieties.

To determine the functional consequence of both the phosphorylation and glycosylation on Elf-1, we performed shift assays using nuclear proteins from primary and Jurkat T cells and a labeled oligonucleotide spanning the Elf-1 binding site (–147/–119) of the TCR ζ chain promoter. Using anti-Elf-1 and anti-RL2 antibodies in supershift assays, we showed that the Elf-1 protein that binds to the TCR promoter contains O-GlcNAc moieties. Using extracts from cells treated with PKC or phosphatase inhibitors, we showed that the Elf-1 that binds to the ζ chain promoter is phosphorylated. Collectively, these experiments show that the Elf-1, which binds to the TCR ζ chain promoter, is both glycosylated and phosphorylated.

We considered that the phosphorylated 98-kDa Elf-1 undergoes degradation to prevent excessive accumulation. Because many transcription factors are degraded through the proteasome pathway, we treated Jurkat cells with different doses of the proteasome inhibitor, LLnL, and noted that the cytoplasmic levels of the 98-kDa Elf-1 increased significantly, indicating that the 98-kDa Elf-1 is constantly degraded through the proteasome pathway. In contrast, the 80-kDa Elf-1 did not change in the presence of LLnL, suggesting that this form is not degraded by proteasome pathway.

DEFECTIVE EXPRESSION OF THE FUNCTIONAL 98-kDa FORM OF ELF-1 IN PATIENTS WITH SLE IS ASSOCIATED WITH DECREASED BINDING TO THE TCR ζ CHAIN PROMOTER AND DECREASED TCR ζ GENE EXPRESSION

To determine whether structural or functional defects in the expression of Elf-1 are responsible for the reported decreased expression of the TCR ζ chain mRNA in SLE, we purified cytoplasmic and nuclear proteins from both normal and SLE T cells and conducted immunoblotting and shift assays. Cytoplasmic extracts from 83% of SLE patients expressed comparable levels of the 80-kDa form of Elf-1 to those from normal controls. In contrast, the nuclear extracts from one-third of SLE patients expressed the 98-kDa form of Elf-1 at a significantly decreased level compared to those from normal T cells. These data indicate that although there is no defect in the protein expression of Elf-1 in SLE patients, slightly less than one half of SLE patients display defective posttranslational modifications necessary for the conversion of Elf-1 from 80 to 98 kDa.

To determine whether the observed decrease in the presence of the 98-kDa form of Elf-1 had a functional effect on the ability of Elf-1 to interact with the TCR ζ chain promoter, we analyzed the ability of proteins from nuclear extracts of SLE T cells to bind to the –147/–119 TCR ζ chain promoter oligonucleotide in gel shift assays. Nuclear extracts from nearly all SLE patients who displayed decreased expression of the 98-kDa Elf-1 also had defective DNA binding compared to that of normal T cells and displayed decreased TCR ζ chain expression. In contrast, most of the SLE patients who displayed normal Elf-1–DNA binding complex had normal TCR ζ chain expression. Interestingly, we also noted decreased DNA binding despite normal levels of 98-kDa Elf-1 in some SLE patients. We observed that the expression of the TCR ζ chain (immunoblots) correlates with the levels of protein binding to the Elf-1 site (shift assays). In conclusion, patients with SLE have decreased amounts of the 98-kDa Elf-1 and decreased DNA binding that correlates strongly with the decreased expression of the TCR ζ chain.

Our studies provide evidence that decreased transcriptional activity due to defective activation of the involved transcription factor(s) can account for decreased expression of proteins involved in antigen-receptor-mediated cell signaling. The cause for the decreased formation of the active form of Elf-1 is apparently due to defective phosphorylation.

ACKNOWLEDGMENTS

This work was supported by grants RO1AI42269 and RO1 AI49954 from the U.S. Public Health Service.

The opinions and assertions contained herein are the private views of the authors and are not to be construed as official or as reflecting the views of the Department of the Army or the Department of Defense.

REFERENCES

1. TSOKOS, G.C. & S.N. LIOSSIS. 1999. Immune cell signaling defects in lupus: activation, anergy and death. Immunol. Today **20:** 123–128.
2. HARLEY, J.B., K.L. MOSER, P.M. GAFFNEY & T.W. BEHRENS. 1998. The genetics of human systemic lupus erythematosus. Curr. Opin. Immunol. **10:** 690–696.
3. NAMBIAR, M.P., E.J. ENYEDY, V.G. WARKE, *et al.* 2001. T cell signaling abnormalities in systemic lupus erythematosus are associated with increased mutations/polymorphisms and splice variants of T cell receptor zeta chain messenger RNA. Arthritis Rheum. **44:** 1336–1350.
4. NAMBIAR, M.P., E.J. ENYEDY, C.U. FISHER, *et al.* 2002. Abnormal expression of various molecular forms, fractions and splice variants of TCR ζ chain in T cells from patients with systemic lupus erythematosus. Arthritis Rheum. **45:** 163–174.
5. NAMBIAR, M.P., E.J. ENYEDY, V.G. WARKE, *et al.* 2001. Polymorphisms/mutations of TCR zeta-chain promoter and 3′ untranslated region and selective expression of TCR zeta-chain with an alternatively spliced 3′ untranslated region in patients with systemic lupus erythematosus. J Autoimmun. **16:** 133–142.
6. ENYEDY, E.J., M.P. NAMBIAR, S.N. LIOSSIS, *et al.* 2001. Fc epsilon receptor type I gamma chain replaces the deficient T cell receptor zeta chain in T cells of patients with systemic lupus erythematosus. Arthritis Rheum. **44:** 1114–1121.
7. RELLAHAN, B.L., J.P. JENSEN & A.M. WEISSMAN. 1994. Transcriptional regulation of the T cell antigen receptor zcta subunit: identification of a tissue-restricted promoter J. Exp. Med. **180:** 1529–1534.
8. JUANG, Y.-T., E. SOLOMOU, B. RELLAHAN & G.C. TSOKOS. 2002. Phosphorylation and O-linked glycosylation of Elf-1 leads to its translocation to the nucleus and binding to the promoter of the T cell receptor ζ chain. J. Immunol. **168:** 2865–2871.
9. COUDRONNIERE, N., M. VILLALBA, N. ENGLUND & A. ALTMAN. 2000. NF-kappa B activation induced by T cell receptor/CD28 costimulation is mediated by protein kinase C-theta. Proc. Natl. Acad. Sci. USA **97:** 3394–3399.
10. SOLOMOU, E.E., Y.T. JUANG & G.C. TSOKOS. 2001. Protein kinase c-theta participates in the activation of cyclic amp-responsive element-binding protein and its subsequent binding to the -180 site of the il-2 promoter in normal human T lymphocytes. J. Immunol. **166:** 5665–5674.
11. HART, G.W. 1997. Dynamic O-linked glycosylation of nuclear and cytoskeletal proteins. Annu. Rev. Biochem. **66:** 315.

Aberrant Activation of B Cells in Patients with Rheumatoid Arthritis

STEFFI LINDENAU,[a] SUSANN SCHOLZE,[b] MARCUS ODENDAHL,[a] THOMAS DÖRNER,[b] ANDREAS RADBRUCH,[a] GERD-RÜDIGER BURMESTER,[b] AND CLAUDIA BEREK[a]

[a]*Deutsches Rheumaforschungszentrum, Berlin, Gemany*

[b]*Rheumatologie, Charité, Berlin, Germany*

KEYWORDS: rheumatoid arthritis; peripheral blood B cells; systemic lupus erythematosus; CD27

INTRODUCTION

The analysis of peripheral blood B lymphocytes has demonstrated that CD27, a member of the TNF receptor family, is strictly regulated during B cell differentiation. Whereas CD27 is not detected on naïve cells, it is up-regulated on memory and plasma cells.[1,2] CD27 defines the memory compartment generated during the process of affinity maturation in the germinal center (GC).[3]

DIFFERENTIATION STATUS OF PERIPHERAL BLOOD B CELLS IN PATIENTS WITH RHEUMATOID ARTHRITIS

Fluorescence-activated cell sorting (FACS) analysis of peripheral blood mononuclear cells (PBMCs) was used to define the differentiation status of peripheral blood B cells. Staining with CD19, CD20, CD38, and CD27 allowed us to distinguish three different $CD19^+$ B cell subsets: $CD20^+$, $CD38^-$, $CD27^-$ naïve B cells; $CD20^+$, $CD38^-$, $CD27^+$ memory cells; and $CD20^-$, $CD38^+$, $CD27^{hi}$ plasma cells (FIG. 1). In the peripheral blood of healthy controls, approximately 60% of B cells have the phenotype of naïve cells, and 40% have that of memory cells. Typically, in such healthy donors, less than 2% of the peripheral B cells are plasma cells ($CD19^+$, $CD20^-$, $CD38^+$, and $CD27^{hi}$). The analysis of blood samples from patients with rheumatoid arthritis (RA) showed a clear change in the B cell populations with a shift toward cells expressing an activated and differentiated phenotype. A typical example is given in FIGURE 1. The majority of B cells are $CD27^+$. In addition to memory B cells, one sees a substantial population of $CD27^{hi}$ plasma cells.

Address for correspondence: Steffi Lindenau, AG Berek (B Cell Immunology), Deutsches Rheumaforschungszentrum, Schumannstrasse 21/22, 10117 Berlin, Germany. Voice: +49-03-28460-714.
lindenau@drfz.de

Ann. N.Y. Acad. Sci. 987: 246–248 (2003). © 2003 New York Academy of Sciences.

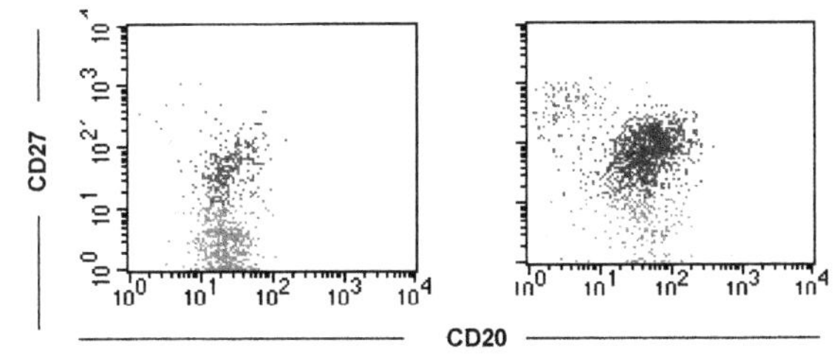

FIGURE 1. Differentiation status of peripheral blood B cells. The FACS analysis of blood samples from a healthy individual (**left panel**) and a patient with RA (**right panel**). Cells are gated on CD19.

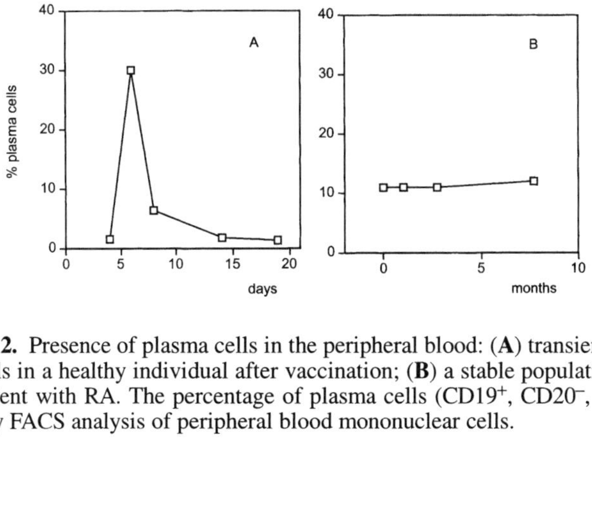

FIGURE 2. Presence of plasma cells in the peripheral blood: (**A**) transient appearance of plasma cells in a healthy individual after vaccination; (**B**) a stable population of plasma cells in a patient with RA. The percentage of plasma cells ($CD19^+$, $CD20^-$, $CD27^{hi}$) was determined by FACS analysis of peripheral blood mononuclear cells.

STABLE POPULATION OF PLASMA CELLS IN PATIENTS WITH RA

Vaccination experiments showed that in healthy individuals, plasma cells appear only transiently after vaccination. Within one week, a strong plasma cell population is seen in the peripheral blood, but by 14 days after immunization, the frequency of plasma cells returns to normal values (FIG. 2A). In contrast, in many patients with RA, we found stable populations of plasma cells over several months (FIG. 2B). It seems that the chronic inflammation in the joints of RA patients is paralleled by a continuous activation of B cells. As a result, increased frequencies of plasma cells are seen in the circulating B cell pool Lfor extended periods of time.

B CELL ACTIVATION IN AUTOIMMUNE DISEASES

Disturbances in peripheral blood B cell subpopulations have been described for a number of autoimmune diseases.[4–6] Increased numbers of plasma cells were found in children with systemic lupus erythematosus (SLE).[6] However, the appearance of plasma cells in the blood is only correlated with disease activity in adult SLE patients.[4] In patients with RA, we did not observe this strict correlation between the disease activity score and the frequency of plasma cells. Nevertheless, RA patients seem to have a generalized dysregulation of B cell homeostasis that is reflected in the fact that the majority of patients showed a distinct shift in the composition of their peripheral blood B cell population. In particular, the plasma cells represent a stably increased population. The increased frequencies of circulating plasma cells may be due to a continuous activation of B cells. In the initiation and maintenance of this mechanism, perhaps autoantigens participate. At present, the questions about their identity and their role in disease etiology remain open.

ACKNOWLEDGMENTS

This work has been supported by the Sonderforschungsbereich (Grant 421) to C.B. The Deutsches Rheumaforschungszentrum is supported by the Berlin Senate for Research and Education.

REFERENCES

1. MAURER, D., G.F. FISCHER, I. FAE, *et al.* 1992. IgM and IgG but not cytokine secretion is restricted to the $CD27^+$ B lymphocyte subset. J. Immunol. **148:** 3700–3705.
2. AGEMATSU, K., H. NAGUMO, F.C. YANG, *et al.* 1997. B cell subpopulations separated by CD27 and crucial collaboration of $CD27^+$ B cells and helper T cells in immunoglobulin production. Eur. J. Immunol. **27:** 2073–2079.
3. KLEIN, U., K. RAJEWSKY & R. KÜPPERS. 1998. Human immunoglobulin (Ig)M IgD^+ peripheral blood B cells expressing the CD27 cell surface antigen carry somatically mutated variable region genes: CD27 as a general marker for somatically mutated (memory) B cells. J. Exp. Med. **188:** 1679–1689.
4. ODENDAHL, M., A. JACOBI, A. HANSEN, *et al.* 2000. Disturbed peripheral B lymphocyte homeostasis in systemic lupus erythematosus. J. Immunol. **165:** 5970–5779
5. BOHNHORST, J.O., M.B. BJORGAN, J.E. THOEN, *et al.* 2001. Bm1-Bm5 classification of peripheral blood B cells reveals circulating germinal center founder cells in healthy individuals and disturbance in the B cell subpopulations in patients with primary Sjogren's syndrome. J. Immunol. **167:** 3610–3618.
6. ARCE, E., D.G. JACKSON, M.A. GILL, *et al.* 2001. Increased frequency of pre-germinal center B cells and plasma cell precursors in the blood of children with systemic lupus erythematosus. J. Immunol. **167:** 2361–2369.

The Role of *rel*B in Regulating the Adaptive Immune Response

MAURIZIO ZANETTI,[a] PAOLA CASTIGLIONI,[a] STEPHEN SCHOENBERGER,[b] AND MARA GERLONI[a]

[a]*Department of Medicine and the Cancer Center, University of California, San Diego, La Jolla, California 92093-0837, USA*

[b]*Division of Immune Regulation, The La Jolla Institute for Allergy and Immunology, La Jolla, California 92121, USA*

ABSTRACT: Dendritic cells (DCs), which represent a key type of antigen-presenting cell (APC), are important for the development of innate and adaptive immunity. DCs are involved in T cell activation in at least two main ways: priming via direct processing/presentation of soluble antigen taken up from the microenvironment (conventional priming), and processing/presentation of antigen released from other cells (cross-priming). *rel*B, a component of the NF-κB complex of transcription factors, is a critical regulator of the differentiation of DCs. In mice, lack of *rel*B impairs DCs derived from bone marrow both in number and function. Here *rel*B (–/–) bone marrow chimera mice is used to study the APC function of residual DCs in presentation of soluble antigen and cross-priming. It is found that the DCs in these mice are profoundly deficient in their ability to both prime and cross-prime T cell responses. It was concluded that the *rel*B gene is involved in regulating the APC function of DCs *in vivo*.

KEYWORDS: *rel*B; dendritic cells; T cell priming; cross-priming

INTRODUCTION

It has become increasingly evident that a large interface exists between innate and adaptive immunity. This interface rests on a complex array of interactions between cells, molecularly defined receptors, and signal transduction pathways and transcriptional events. This ensemble ensures that the initial innate response to stimuli from the outside world is converted into the defense mechanisms of the adaptive response. This coordinate functional organization between innate and adaptive immunity makes sure that the innate mechanisms fill the interval of time during which the more finely tuned adaptive immunity develops. Stimuli originating from the outside world are sensed by non-polymorphic structures or toll-like receptors (TLRs)[1,2] that recognize molecular patterns associated with pathogens (FIG. 1). These receptors are distributed at the surface of many cells, particularly antigen-presenting cells (APCs), among which dendritic cells (DCs) play a dominant if not an exclusive role. DCs re-

Address for correspondence: Maurizio Zanetti, M.D., University of California, San Diego, 9500 Gilman Drive, La Jolla, CA 92093-0837. Voice: 619-543-5733; fax: 619-543-5665. mzanetti@ucsd.edu

Ann. N.Y. Acad. Sci. 987: 249–257 (2003).

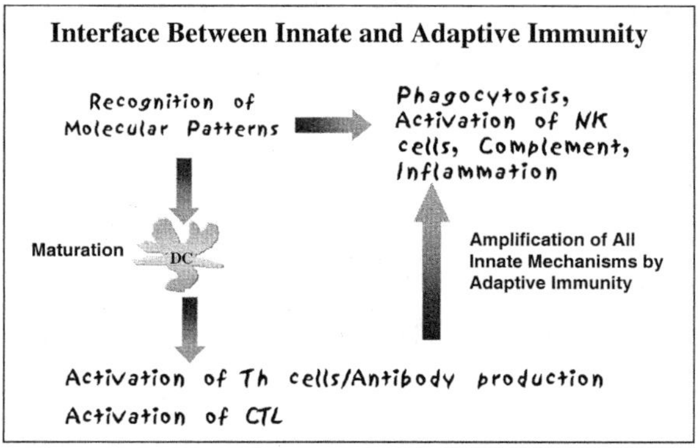

FIGURE 1. Interface between innate and adaptive immunity.

side in an immature state in lymphoid and non-lymphoid tissues where they phagocytose pathogens or other antigens. This process leads to their maturation, and they migrate from non-lymphoid tissues to draining lymph nodes (LNs).[3] The adaptive immune response initiates with up-regulation of costimulatory molecules and presentation of antigen to T cells.

The TLR family is composed of nine members, and these seem to interact with different stimuli. For instance, bacterial lipopolysaccharide (LPS), proteoglycans, and nucleic acid (DNA and RNA) can each target a different TLR.[1] However, the plasticity of DCs is such that these stimuli, albeit diverse, activate a common-core genetic program.[4] DCs, primary sentinels to the immune system, have evolved to utilize patterns of recognition and reactivity to enable antigen-specific and major-histocompatibility-complex-restricted (MHC) activation of T lymphocytes.

The NF-κB/*rel* complex of transcription factors is a collective name for inducible dimeric transcription factor composed of members of the rel family of DNA-binding proteins (*rel*A, c-*rel*, *rel*B, p100, and p105).[5,6] Upon activation, NF-κB/*rel* translocates to the nucleus, an event rich in effects for the immune system: activation of adaptive immunity, activation of cytokines, and induction of antiapoptotic effects.[7]

*rel*B has been implicated as a critical regulator of the differentiation of DC and is found in bone-marrow-derived DC and in follicular DC.[8] For this reason, mice that lack functional bone-marrow-derived DC as a result of a null (–/–) mutation in the *rel*B gene of the NFκB complex[9] consitute an ideal model system to probe DC function *in vivo*. *rel*B (–/–) mice have an atrophic thymic medulla, possess no LN, and lack bone-marrow-derived DC.[10] Bone-marrow chimeras (BMCs) constructed by transferring homozygous (–/–) *rel*B bone-marrow cells into lethally irradiated (1100 rads) hemizygous (+/–) *rel*B recipients carry the same DC defect as *rel*B (–/–) mice, but have a longer life span. In *rel*B (–/–) mice and *rel*B (–/–) BMC, the function of B lymphocytes is conserved relative to their ability to proliferate *in vivo* to LPS (our unpublished results), undergo isotype switch *in vitro*,[11] and form germinal centers and antibodies *in vivo*.[12]

In this paper we will present evidence that residual DC in *rel*B (–/–) BMCs are unable to process and present a model antigen (ovalbumin, OVA) *in vitro* and *in vivo*,

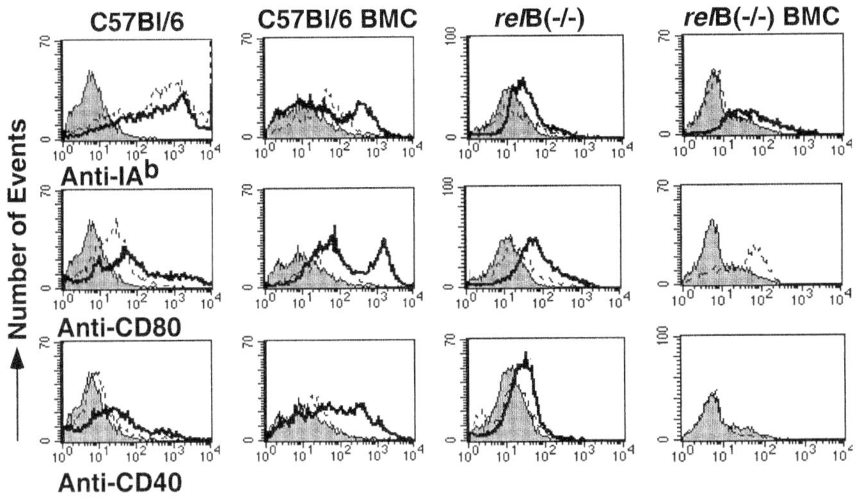

FIGURE 2. Surface phenotype analysis and quantitation of DCs sorted from spleens and LNs of C57Bl/6 mice, C57Bl/6 BMCs, *rel*B(–/–) and *rel*B(–/–) BMCs. Sorted CD11c$^+$ cells were analyzed for the expression of MHC Class II (IA^b), CD80, and CD40 surface molecules. The results are shown as *histograms* with fluorescence intensity on the *x*-axis and cell number on the *y*-axis. *Dotted lines* represent *ex vivo* staining of spleen cells, whereas *thick lines* represent staining after 4 days in culture with GM-CSF, IL-4, and LPS, which was added during the following 24 h. *Solid histograms* in each instance refer to the background (that is, autofluorescence). Cells were gated on the CD11c$^+$ population.

and that in *rel*B (–/–) BMCs, activation of CD8$^+$ cytotoxic T lymphocytes (CTLs) via cross-priming is not possible.

RESULTS

Spleen DCs of relB *(–/–) BMCs Fail to Activate T Cells*

We gathered initial evidence suggesting that CD11c$^+$ cells in the spleen of *rel*B(–/–) BMCs are very low in absolute number and have little if any constitutive expression of costimulatory and MHC Class II molecules (which are, moreover, not up-regulated after *in vitro* culture and stimulation). Analyses performed *ex vivo* (that is, immediately after tissue harvest) and *in vitro* following culture with IL-4 (4 ng/mL) and IFN-γ (1000 μg/mL) for 96 h, plus LPS (10 μg/mL) during the last 24 h, showed that DCs from *rel*B(–/–) BMCs have a much reduced expression of MHC Class II, CD40, and CD80 compared with C57Bl/6 mice or C57Bl/6 BM chimeras (FIG. 2). Importantly, up-regulation of these molecules after *in vitro* culture and stimulation with LPS was minimal, if present at all. In addition, LPS stimulation failed to expand residual CD11c$^+$ cells (TABLE 1).

OVA was used as a model antigen to probe the residual DCs in the spleen of *rel*B(–/–) BMCs *in vitro* and *in vivo*. In the *in vitro* experiments, purified spleen DCs

TABLE 1. Analysis of dendritic cells in the spleen of *rel*B (–/–} bone-marrow chimeras

	*rel*B(–/–) BMC	*rel*B(–/–)	C57B1/6 BMC[a]	C57B1/6
Total DC per spleen[b]	0.6 ± 0.4 (0.3%)	0.5 ± 0.1 (0.3%)	0.5 ± 0.2 (0.3%)	1.1 ± 0.2 (1%)
Total DC after culture[c]	0.06 (0.2%)	0.2 (1.25%)[d]	1.6 ± 1.4 (6.4%)	0.5 ± 0.1 (3%)

[a]Bone-marrow chimers (BMC).

[b]Each value represents the total number of DCs ($\times 10^6$) (and the corresponding percentage over the total number of cells) in the spleen *ex vivo*. Each result is expressed as a mean ± SD of three to five mice.

[c]Each value represents the total number of DCs ($\times 10^6$) (and the corresponding percentage) harvested after *in vitro* culture of 10^8 spleen cells with GM-CSF (1000 U/mL), IL-4 (4 ng/mL), and LPS (10 µg/mL).

[d]Value refers to a pool of three to five mice.

were pulsed overnight with OVA at 37°C and used to prime naïve C57Bl/6 CD4 and CD8 T lymphocytes. After a six-day culture, no CD4 T cell proliferation was observed in cultures seeded with *rel*B(–/–) BMC DCs (FIG. 3A). Lack of CD4 T cell priming was mirrored by the absence of IL-2 production in the corresponding cultures (data not shown). Similar results were obtained when OVA-pulsed DCs were used to prime naïve C57Bl/6 CD8 T cells (FIG. 3B). IL-2 and IFN-γ were detected only in cultures seeded with DCs derived from C57Bl/6 and C57Bl/6 BMCs but not *rel*B(–/–) BM chimeras. Similar deficits of DCs in *rel*B(–/–) BMC observed *in vitro* were also documented *in vivo*. To this end, C57Bl/6 mice, C57Bl/6 BMCs, and *rel*B (–/–) BMCs were injected intravenously with OVA (3 mg/mouse) to allow for antigen uptake by splenic DCs.[13] Twenty hours after injection, DCs were purified from the spleen and used as APCs to activate naïve OVA-specific CD4 (OT-II) and CD8 (OT-I) T lymphocytes.[14] No CD4 T cell proliferation was observed when DCs from *rel*B (–/–) BMCs served as APCs (FIG. 3C). Similarly, CD8 T cell proliferation was observed in cultures seeded with DCs from C57Bl/6 and C57Bl/6 BMCs, but not *rel*B (–/–) BMCs (FIG. 3D). No IFN-γ was produced in cultures where DCs from OVA-injected *rel*B (–/–) BMCs were used to prime OT-1 T cells (data not shown).

relB *Is Required for CTL Activation via Cross-Priming*

In the course of the adaptive immune response, cellular antigens can be transferred from one cell to host APCs to be processed and presented (cross-presentation), and to activate T cells (cross-priming).[15] This event has been shown to require T cell help[16] and the transporter associated with antigen presentation (TAP).[17] In addition, this event is at play in the response against viruses,[18] bacteria,[19] and tumor antigens.[20] Of two major categories of bone-marrow-derived DCs described in the mouse, $CD11c^+/CD8\alpha^-$ and $CD11c^+/CD8\alpha^+$ cross-priming appears to be mediated by $CD8\alpha^+$ DCs.[21,22]

To probe the role of *rel*B in cross-priming, *rel*B (–/–) BMCs were injected with 10^7 irradiated (900 rad) TAP-1(–/–) murine embryo cells expressing human adenovirus type 5 early region 1 (5E1).[23] In this model, CTL generation is wholly dependent on cross-priming host APCs, as murine embryo cells are TAP^- and cannot present

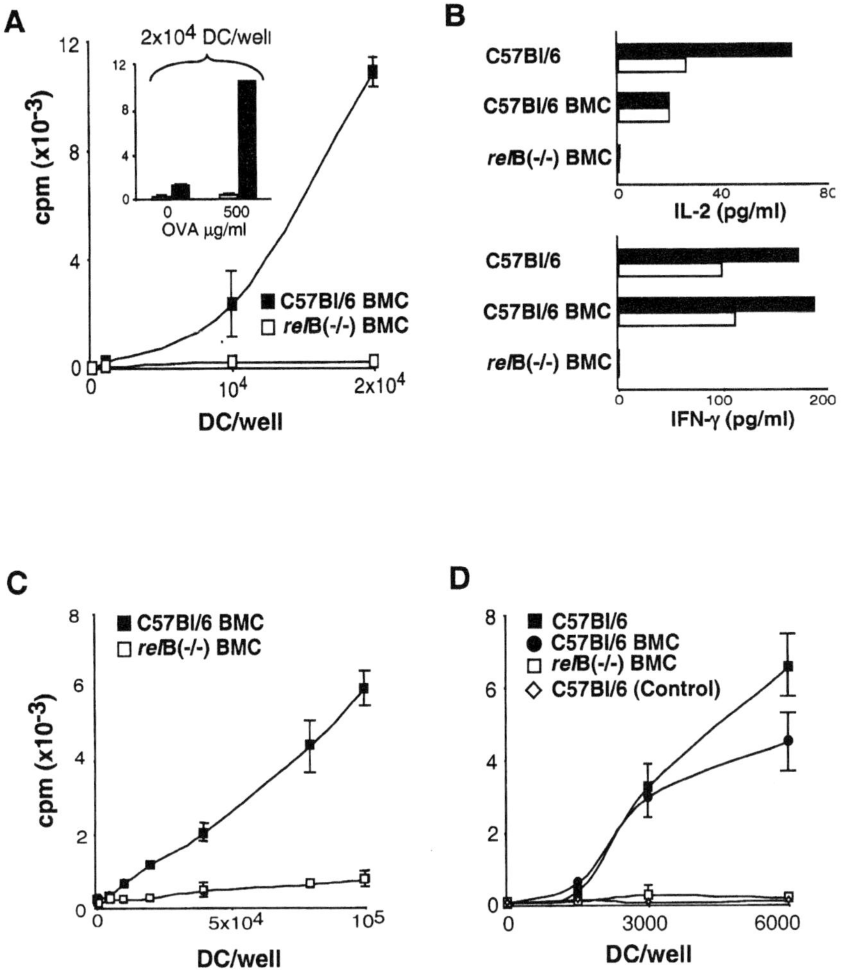

FIGURE 3. Spleen DCs of *rel*B (–/–) BMCs fail to prime CD4 and CD8 T cell responses against OVA. *Upper panels: In vitro* T cell responses. OVA-pulsed spleen DCs from C57Bl/6 mice, C57Bl/6 BMCs, and *rel*B (–/–) BMCs were used to prime naïve CD4 and CD8 T lymphocytes from the spleens of C57Bl/6 mice. (**A**) CD4 T cells (5×10^4) were cultured with different concentrations of DCs ($0–2 \times 10^4$). [^{3}H]Thymidine incorporation was measured on day 6. The *inset* shows that unpulsed 2×10^4 DCs do not elicit any proliferative response to exclude any effect due to a syngeneic mixed lymphocyte reaction. (**B**) CD8 T cells (5×10^4) were cultured with different concentrations of DCs (*solid bars*: 2×10^4; *open bars*: 10^4). For detection of IL-2 and IFN-γ in culture supernatants were harvested on day 5. The experiments shown were performed by pulsing DCs with OVA at 500 µg/mL. *Lower panels:* DC antigen presentation after OVA *in vivo* immunization. C57Bl/6 mice, C57Bl/6 BMCs, and *rel*B (–/–) BMCs were injected intravenously with OVA (3 mg). Twenty hours after injection, DCs were purified from a pool of two to four spleens and cultured at different concentrations with (**C**) 10^5 OVA TCR-transgenic (OT-II) CD4 T lymphocytes or with (**D**) $3 \times 4\ 10^4$ OVA transgenic (OT-I) CD8 T lymphocytes. [^{3}H]Thymidine incorporation was measured after 72 h. Proliferation assays were run in triplicate.

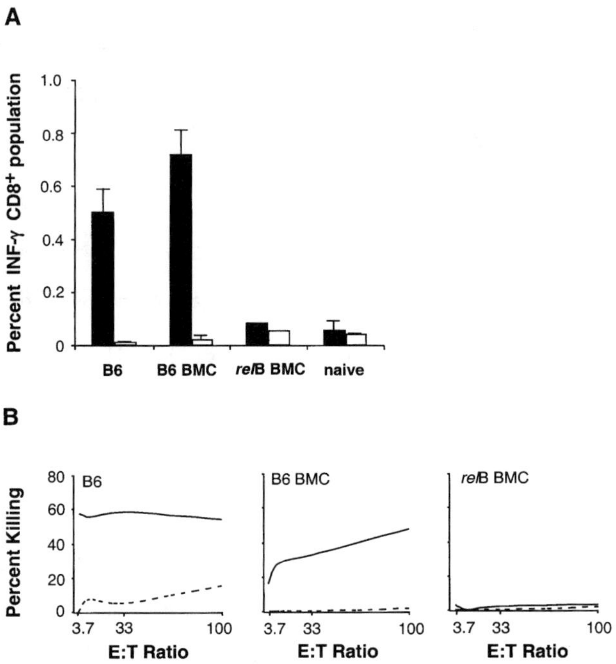

FIGURE 4. CTL responses in *rel*B and C57Bl/6 (B6) BMCs. (**A**) Detection of intracellular IFN-γ. Fourteen days after immunization, spleen and lymph nodes from individual mice were pooled and stimulated with specific ($E1B_{192–200}$) (*black columns*) or control ($OVA_{257–264}$) (*white columns*) peptide for 5 h prior to detection of intracellular IFN-γ. Tests were run in triplicate. Data are representative of two independent experiments that yielded similar results. (**B**) Representative example of cytotoxic cell function in restimulated bulk cultures. Six days after *in vitro* restimulation, viable lymphocytes were used as effectors in a JAM test using EL-4 target cells pulsed with $E1B_{192–200}$ (*solid line*) or $OVA_{257–264}$ (*broken line*) as a control.

the immunodominant $E1B_{192–200}$ epitope of adenovirus type 5 to CTL precursors. The immune response was evaluated 2 weeks after immunization. We found a strong $E1B_{192–200}$-specific IFN-γ production in control B6 mice and B6 BMCs (B6 bone marrow → B6 mice), but not in *rel*B (–/–) BMCs (FIG. 4A). Similarly, although the B6 mice and B6 BMCs mounted strong and specific CTLs, none of the *rel*B (–/–) BMCs generated detectable CTL effectors (FIG. 4B). Thus, two critical parameters of CTL function, IFN-γ secretion and cytotoxic activity, were not generated in *rel*B (–/–) BMCs. This could not be explained by a defect in the ability of T cells to respond because immunization by direct injection of DNA into the spleen generates a specific CTL response, suggesting that the T cells of *rel*B(–/–) BMCs are capable of responding to antigen (data not shown). We could also exclude a defect in antigen uptake or migration to lymphoid organs by Langherans cells of *rel*B(–/–) BMCs because their *in vivo* ability to take up FITC-conjugated OVA and migrate to the draining lymph nodes was normal.

TABLE 2. Functional DC defects in *rel*B (–/–) BMCs

Organ	Cell Type	Function Analyzed	Status
Spleen	Mainly CD8α$^+$	Maturation (*in vitro*)	Reduced (+++)
		Soluble antigen uptake (*in vivo*)	Reduced (+++)
		CD4/CD8 T cell priming to soluble antigen (*in vitro*)	Abolished
		CD4/CD8 T cell priming to soluble antigen (*in vivo*)	Abolished
		CD4 T cell priming to synthetic peptide (*in vitro*)	Reduced (+++)
		CD8 T cell cross-priming (*in vivo*)	Abolished
Lymph nodes	Mainly CD8α$^+$	Maturation (*in vitro*)	Reduced (+++)
		CD8 T cell cross-priming (*in vivo*)	Abolished
Skin	Langherans cells	Soluble antigen uptake (*in vivo*)	Normal
		Migration to LNs (*in vivo*)	Normal

DISCUSSION

The data presented indicate that the *rel*B gene is implicated in the APC function of DCs. Lack of *rel*B ultimately affects T cell priming and cross-priming, two key functions in the adaptive immune response. DCs in *rel*B (–/–) BMCs may be unable to mediate T cell priming because of the loss of several functions such as a diminished uptake of soluble antigen, reduced ability to undergo maturation, and greatly impaired presentation of processed antigen (TABLE 2). Deficiencies in T cell priming and cross-priming raise concerns about the actual role of CD8α$^+$ DCs in the *in vivo* generation of an adaptive T cell response. Whether CD8α$^+$ and CD8α$^-$ DCs represent two distinct lineages[24] or derive from a common progenitor cell[25] and represent different stages of differentiation, the data presented earlier suggest that to work as APCs, CD8α$^+$ DCs may require interaction with another cell, such as CD8α$^-$ DCs that are missing in *rel*B (–/–) BM chimeras.[26] Since both types of DCs possess phagocytic and pinocytic properties,[22] a defect would primarily underscore defective interaction. The demonstration that CTL responses *in vivo* may require a combination of CD8α$^+$ and CD8α$^-$ DCs, depending on the cytokine environment,[27,28] supports this view. The view is also consistent with our previous report that *in vivo* CD8α$^+$, DCs are refractory to GM-CSF, a typical DC maturation signal.[12] Thus, the *in vivo* experiments discussed here suggest that in a T cell response that involves DCs as the initial or main APCs, CD8α$^+$ DCs may require an organizational division of labor with CD8α$^-$ DCs. On the other hand, one cannot rule out that the absence of the *rel*B gene engenders abnormalities in stromal cells,[9,29] creating a situation in which CD8α$^+$ DCs no longer cooperate with non-DC cells present in the stroma. This would imply that *rel*B may be also necessary to create a permissive environment.

The foregoing discussion suggests that the activation of T cells via DCs, a process at the center of the adaptive immune response, is under the control of a gene that is itself part of a multicomponent transcriptional regulator, the NF-κB/*rel* complex.

The recent analysis of the human genome indicates that certain categories of genes have undergone preferential evolutionary expansion.[30] This is the case of genes coding for proteins involved in acquired immune functions and signaling pathways. The presence of a *rel*B equivalent in flies[31] suggests evolutionary conservation of this immunoregulatory pathway. Processes considered of key importance in the immune system are also primitive processes. For example, phagocytosis was discovered by studying floating larvae of starfish.[32] Therefore, since the function of DCs *in vivo* revolves around antigen capture, processing, and presentation, our finding on the requirement of *rel*B, a highly conserved member of a class of genes under evolutionary expansion, for DC-mediated activation of T cells is perhaps not surprising.

ACKNOWLEDGMENTS

This work was supported in part by the National Institutes of Health (grants NIH RO1 CA77427 and R21 AI49774).

REFERENCES

1. MEDZHITOV, R. 2001. Toll-like receptors and innate immunity. Nat. Rev. Immunol. **1:** 135–145.
2. BEUTLER, B. 2002. Toll-like receptors: how they work and what they do. Curr. Opin. Hematol. **9:** 2–10.
3. LARSEN, C.P., R.M. STEINMAN, M. WITMER-PACK, *et al.* 1990. Migration and maturation of Langerhans cells in skin transplants and explants. J. Exp. Med. **172:** 1483–1493.
4. HUANG, Q., D. LIU, P. MAJEWSKI, *et al.* 2001. The plasticity of dendritic cell responses to pathogens and their components. Science **294:** 870–875.
5. BALDWIN, A.S., JR. 1996. The NF-kappa B and I kappa B proteins: new discoveries and insights. Annu. Rev. Immunol. **14:** 649–683.
6. GHOSH, S., M.J. MAY & E.B. KOPP. 1998. NF-kappa B and rel proteins: evolutionarily conserved mediators of immune responses. Annu. Rev. Immunol. **16:** 225–260.
7. KARIN, M. & A. LIN. 2002. NF-kappaB at the crossroads of life and death. Nat. Immunol. **3:** 221–227.
8. CARRASCO, D., R.P. RYSECK & R. BRAVO. 1993. Expression of relB transcripts during lymphoid organ development: specific expression in dendritic antigen-presenting cells. Development **118:** 1221–1231.
9. BURKLY, L., C. HESSION, L. OGATA, *et al.* 1995. Expression of relB is required for the development of thymic medulla and dendritic cells. Nature **373:** 531–536.
10. LO, D., H. QUILL, L. BURKLY, *et al.* 1992. A recessive defect in lymphocyte or granulocyte function caused by an integrated transgene. Am. J. Pathol. **141:** 1237–1246.
11. SNAPPER, C.M., F.R. ROSAS, P. ZELAZOWSKI, *et al.* 1996. B cells lacking relB are defective in proliferative responses, but undergo normal B cell maturation to Ig secretion and Ig class switching. J. Exp. Med. **184:** 1537–1541.
12. GERLONI, M., D. LO & M. ZANETTI. 1998. DNA immunization in relB-deficient mice discloses a role for dendritic cells in IgM → IgG1 switch in vivo. Eur. J. Immunol. **28:** 516–524.
13. KAMATH, A.T., J. POOLEY, M.A. O'KEEFFE, *et al.* 2000. The development, maturation, and turnover rate of mouse spleen dendritic cell populations. J. Immunol. **165:** 6762–6770.
14. HOGQUIST, K.A., S.C. JAMESON, W.R. HEATH, *et al.* 1994. T cell receptor antagonist peptides induce positive selection. Cell **76:** 17–27.
15. BEVAN, M.J. 1976. Cross-priming for a secondary cytotoxic response to minor H antigens with H-2 congenic cells which do not cross-react in the cytotoxic assay. J. Exp. Med. **143:** 1283–1288.

16. BENNETT, S.R., F.R. CARBONE, F. KARAMALIS, *et al.* 1997. Induction of a CD8+ cytotoxic T lymphocyte response by cross-priming requires cognate CD4+ T cell help. J. Exp. Med. **186:** 65–70.
17. HUANG, A.Y., A.T. BRUCE, D.M. PARDOLL & H.I. LEVITSKY. 1996. In vivo cross-priming of MHC class I-restricted antigens requires the TAP transporter. Immunity **4:** 349–355.
18. SIGAL, L.J., S. CROTTY, R. ANDINO & K.L. ROCK. 1999. Cytotoxic T-cell immunity to virus-infected non-haematopoietic cells requires presentation of exogenous antigen. Nature **398:** 77–80.
19. LENZ, L.L., E.A. BUTZ & M.J. BEVAN. 2000. Requirements for bone marrow-derived antigen-presenting cells in priming cytotoxic T cell responses to intracellular pathogens. J. Exp. Med. **192:** 1135–1142.
20. WOLFERS, J., A. LOZIER, G. RAPOSO, *et al.* 2001. Tumor-derived exosomes are a source of shared tumor rejection antigens for CTL cross-priming. Nat. Med. **7:** 297–303.
21. DEN HAAN, J.M., S.M. LEHAR & M.J. BEVAN. 2000. CD8(+) but not CD8(–) dendritic cells cross-prime cytotoxic T cells in vivo. J. Exp. Med. **192:** 1685–1696.
22. POOLEY, J.L., W.R. HEATH & K. SHORTMAN. 2001. Cutting edge: intravenous soluble antigen is presented to CD4 T cells by $CD8^-$ dendritic cells, but cross-presented to CD8 T cells by $CD8^+$ dendritic cells. J. Immunol. **166:** 5327–5330.
23. SCHOENBERGER, S.P., E.I. VAN DER VOORT, G.M. KRIETEMEIJER, *et al.* 1998. Cross-priming of CTL responses in vivo does not require antigenic peptides in the endoplasmic reticulum of immunizing cells. J. Immunol. **161:** 3808–3812.
24. VREMEC, D. & K. SHORTMAN. 1997. Dendritic cell subtypes in mouse lymphoid organs: cross-correlation of surface markers, changes with incubation, and differences among thymus, spleen, and lymph nodes. J. Immunol. **159:** 565–573.
25. TRAVER, D., K. AKASHI, M. MANZ, *et al.* 2000. Development of CD8alpha-positive dendritic cells from a common myeloid progenitor. Science **290:** 2152–2154.
26. WU, L., A. D'AMICO, K.D. WINKEL, *et al.* 1998. relB is essential for the development of myeloid-related $CD8alpha^-$ dendritic cells but not of lymphoid-related $CD8alpha^+$ dendritic cells. Immunity **9:** 839–847.
27. KNIGHT, S.C., S. IQBALL, M.S. ROBERTS, *et al.* 1998. Transfer of antigen between dendritic cells in the stimulation of primary T cell proliferation. Eur. J. Immunol. **28:** 1636–1644.
28. SMITH, A.L. & B.F. DE ST GROTH. 1999. Antigen-pulsed CD8alpha dendritic cells generate an immune response after subcutaneous injection without homing to the draining lymph node. J. Exp. Med. **189:** 593–598.
29. WEIH, F., D. CARRASCO, S.K. DURHAM, *et al.* 1995. Multi-organ inflammation and hematopoietic abnormalities in mice with a targeted disruption of relB, a member of the NF-kappa B/rel family. Cell **80:** 331–340.
30. VENTER, J.C., M.D. ADAMS, E.W. MYERS, *et al.* 2001. The sequence of the human genome. Science **291:** 1304–1351; also Erratum in Science 2001. **292:** 1838.
31. DUSHAY, M.S., B. ASLING & D. HULTMARK. 1996. Origins of immunity: relish, a compound Rel-like gene in the antibacterial defense of Drosophila. Proc. Natl. Acad. Sci. USA **93:** 10343–10347.
32. METCHNIKOFF, E. 1884. Ueber eine sprosspilzkrankheit der Daphnien. Beitrag zur lehre uber den kampf der phagocyten gegan krakheitserreger. Arch. Pathol. Anat. Physiol. Clin. Med.

Tolerogenic Dendritic Cells Induced by Vitamin D Receptor Ligands Enhance Regulatory T Cells Inhibiting Autoimmune Diabetes

LUCIANO ADORINI

BioXell, Via Olgettina 58, I-20132 Milano, Italy

ABSTRACT: Dendritic cells (DCs) not only induce but also modulate T cell activation. 1,25-Dihydroxyvitamin D_3 [1,25-$(OH)_2D_3$] induces DCs with a tolerogenic phenotype, characterized by decreased expression of CD40, CD80, and CD86 co-stimulatory molecules, low IL-12, and enhanced IL-10 secretion. We have found that a short treatment with 1,25-$(OH)_2D_3$ induces tolerance to fully mismatched mouse islet allografts, and that this tolerance is stable to challenge with donor-type spleen cells and allows acceptance of donor-type vascularized heart grafts. This effect is enhanced by co-administration of mycophenolate mofetil (MMF), a selective inhibitor of T and B cell proliferation, that also has effects similar to 1,25-$(OH)_2D_3$ on DCs. Graft acceptance is associated with impaired development of type 1 $CD4^+$ and $CD8^+$ cells and an increased percentage of $CD4^+CD25^+$ regulatory cells expressing CD152 in the spleen and in the draining lymph node. Transfer of $CD4^+CD25^+$ cells from tolerant mice protects 100% of the syngeneic recipients from islet allograft rejection. $CD4^+CD25^+$ cells that are able to inhibit the T cell response to a pancreatic autoantigen and to significantly delay disease transfer by pathogenic $CD4^+CD25^-$ cells are also induced by treatment of adult nonobese diabetic (NOD) mice with a selected vitamin D receptor (VDR) ligand. This treatment arrests progression of insulitis and Th1 cell infiltration, and inhibits diabetes development at non-hypercalcemic doses. The enhancement of $CD4^+CD25^+$ regulatory T cells able to mediate transplantation tolerance and to arrest type 1 diabetes development by a short oral treatment with small organic compounds that induce tolerogenic DCs, like VDR ligands, suggests possible clinical applications of this approach.

KEYWORDS: dendritic cells; vitamin D receptor; type 1 diabetes

MAJOR TARGETS OF VDR LIGANDS IN THE IMMUNE SYSTEM: ANTIGEN-PRESENTING CELLS AND T CELLS

The active form of vitamin D, 1,25-dihydroxyvitamin D_3 [1,25-$(OH)_2D_3$], is a secosteroid hormone that binds to the vitamin D receptor (VDR), a member of the superfamily of nuclear receptors for steroid hormones, thyroid hormone, and

Address for correspondence: Luciano Adorini, BioXell, Via Olgettina 58, I-20132 Milano, Italy. Voice: +39-02-2884816; fax: +39-02-2153203.
Luciano.Adorini@bioxell.com

Ann. N.Y. Acad. Sci. 987: 258–261 (2003).

retinoic acid. VDR ligands regulate calcium and bone metabolism, control cell proliferation and differentiation, and exert immunoregulatory activities. An important property of VDR ligands is their capacity to modulate both antigen-presenting cells (APCs) and T cells. Thus, these agents can target T cells both directly and indirectly, via modulation of APC function. VDR ligands inhibit the differentiation and maturation of DCs,[1] a critical APC in the induction of T cell-mediated immune responses. These studies, performed either on monocyte-derived DCs from human peripheral blood or on bone-marrow-derived mouse DCs, have consistenly shown that *in vitro* treatment of DCs with 1,25-$(OH)_2D_3$ and its analogs leads to down-regulated expression of the co-stimulatory molecules CD40, CD80, and CD86, and to decreased IL-12 and enhanced IL-10 production, resulting in decreased T cell activation. The abrogation of IL-12 production and the strongly enhanced production of IL-10 highlight the important functional effects of 1,25-$(OH)_2D_3$ and its analogues on DCs and are, at least in part, responsible for the induction of DCs with tolerogenic properties.[2]

These effects are not limited to *in vitro* activity: VDR ligands can also induce DCs with tolerogenic properties *in vivo*, as demonstrated in models of allograft rejection by oral administration directly to the recipient[3] or by adoptive transfer of *in vitro*-treated DCs.[4] Tolerogenic DCs induced by 1,25-$(OH)_2D_3$ can, in turn, induce $CD4^+CD25^+$ regulatory T cells that are able to mediate transplantation tolerance.[3] This is a novel aspect of the multiple effects of VDR ligands on T cells. In addition, a combination of 1,25-$(OH)_2D_3$ and dexamethasone has been shown to induce human and mouse naive $CD4^+$ T cells to differentiate *in vitro* into regulatory T cells.[5] These agents induced the development of IL-10-producing T cells also in the absence of APCs, with IL-10 acting as a positive autocrine factor.[5]

IMMUNOMODULATORY EFFECTS OF 1,25-$(OH)_2D_3$ AND ITS ANALOGUES IN AUTOIMMUNE DIABETES

The nonobese diabetic (NOD) mouse spontaneously develops type 1 diabetes, a Th1-mediated chronic progressive autoimmune disease, by targeting pancreatic β cells with a pathogenesis similar to the human disease.[6] Agents such as VDR ligands, which are able to inhibit *in vivo* IL-12 production and Th1 development,[7] and to enhance $CD4^+CD25^+$ regulatory T cells,[3] may therefore be beneficial in the treatment of type 1 diabetes. 1,25-$(OH)_2D_3$ itself reduces the incidence of insulitis and prevents type 1 diabetes development, but only when administered to NOD mice starting from three weeks of age, before the onset of insulitis.[8]

In contrast, we have recently identified the VDR ligand 1,25-dihydroxy-16,23Z-diene-26,27-hexafluoro-19-nor vitamin D_3 (Ro 26-2198) that is able, as a monotherapy, to treat the ongoing type 1 diabetes in the adult NOD mouse, effectively blocking the disease course.[9] This property is likely due, at least in part, to the increased metabolic stability of this analogue against the inactivating C-24 and C-26 hydroxylations, and the C-3 epimerization, resulting in a 100-fold more potent immunosuppressive activity compared to 1,25-$(OH)_2D_3$. A short treatment with non-hypercalcemic doses of Ro 26-2198 inhibits IL-12 production and pancreatic infiltration of Th1 cells while increasing the frequency of $CD4^+CD25^+$ regulatory T cells in pancreatic lymph nodes, arresting the immunological progression and preventing the clinical onset of type 1 diabetes in the NOD mouse.[9] The frequency of

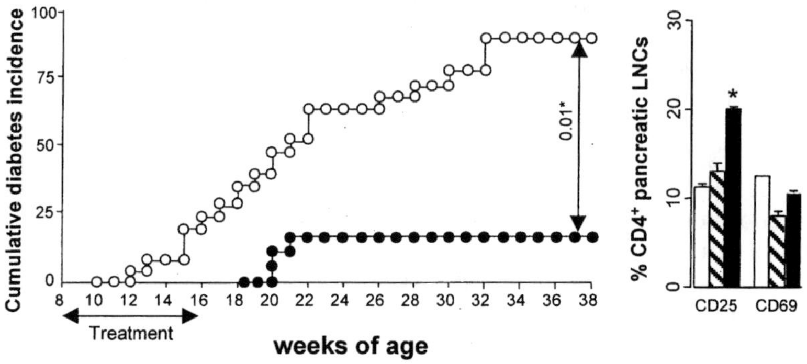

FIGURE 1. Ro 26-2198 administration to 8-week-old NOD mice inhibits the development of type 1 diabetes. NOD mice were treated five times each week with vehicle (*open circles*, $n = 16$) or with 0.03 μg/kg Ro 26-2198 p.o. (*filled circles*, $n = 12$) from 8 to 16 weeks of age. The development of diabetes was monitored twice weekly by measurement of blood glucose levels. Ro 26-2198 treatment enhances the frequency of $CD4^+CD25^+$ cells. Positively selected pancreatic lymph node $CD4^+$ T cells were stained with monoclonal antibodies specific for CD25 or CD69 molecules and analyzed by flow cytometry. Bars represent the percentage of lymph node $CD4^+$ T cells expressing the indicated surface molecules from untreated 8-week-old (*open bars*) or 20-week-old mice treated five times each week, from 8 to 16 weeks of age, with vehicle (*striped bars*) or with 0.03 μg/kg Ro 26-2198 (*filled bars*). Each value represents a mean ± SE of three separate experiments. The *P* values were calculated via the Mann-Whitney U test (* $P < 0.05$ vs. 8 week-old NOD mice).

$CD4^+CD25^+$ cells in the pancreatic lymph nodes of Ro 26-2198-treated NOD mice was twofold higher compared to untreated 8-week-old and to age-matched vehicle-treated controls. These cells were anergic, as demonstrated by their impaired capacity to proliferate and secrete IFN-γ in response to TCR ligation, inhibited the T cell response to the pancreatic autoantigen IA-2, and delayed disease transfer by pathogenic $CD4^+CD25^-$ cells.[9]

Immature DCs have been shown to induce $CD4^+$ cells with regulatory properties, and arrest of DCs at the immature stage induced by Ro 26-2198 treatment could account for the enhanced frequency of $CD4^+CD25^+$ cells. $CD4^+CD25^+$ regulatory T cells appear to play an important role in controlling the progression of type 1 diabetes in NOD mice, because a low level of $CD4^+CD25^+$ T cells correlates with exacerbation and acceleration of the disease. It is likely that this cell population is more relevant than Th2 cells in disease control, although both could contribute to protection. Indeed, 1,25-$(OH)_2D_3$ can induce regulatory cells with disease-suppressive activity in the NOD mouse,[8] and Ro 26-2198 could contribute to the deviation of pancreas-infiltrating cells to the Th2 phenotype.[9]

Polymorphysms of the vitamin D receptor gene have been associated with type 1 diabetes in different populations, and epidemiological studies have suggested a correlation with calcitriol levels. This is further supported by a large population-based case-control study showing that the intake of vitamin D contributes to a significantly decreased risk of type 1 diabetes development.[10] The observation that ongoing type 1 diabetes in the adult NOD mouse can be arrested by a relatively short course of treatment with a VDR ligand suggests that a similar treatment may also inhibit disease progression in prediabetic or newly diagnosed type 1 diabetes patients.

REFERENCES

1. PENNA, G. & L. ADORINI. 2000. 1,25-Dihydroxyvitamin D3 inhibits differentiation, maturation, activation and survival of dendritic cells leading to impaired alloreactive T cell activation. J. Immunol. **164:** 2405–2411.
2. MATHIEU, C. & L. ADORINI. 2002. The coming of age of 1,25-dihydroxyvitamin D(3) analogs as immunomodulatory agents. Trends Mol. Med. **8:** 174–179.
3. GREGORI, S. *et al.* 2001. Regulatory T cells induced by 1α,25-dihydroxyvitamin D_3 and mycophenolate mofetil treatment mediate transplantation tolerance. J. Immunol. **167:** 1945–1953.
4. GRIFFIN, M.D. *et al.* 2001. Dendritic cell modulation by 1α,25-dihydroxyvitamin D3 and its analogs: a vitamin D receptor-dependent pathway that promotes a persistent state of immaturity in vitro and in vivo. Proc. Natl. Acad. Sci. USA **98:** 6800–6805.
5. BARRAT, F.J. *et al.* 2002. In vitro generation of interleukin 10-producing regulatory CD4(+) T cells is induced by immunosuppressive drugs and inhibited by T helper type 1 (Th1)- and Th2-inducing cytokines. J. Exp. Med. **195:** 603–616.
6. ADORINI, L. *et al.* 2002. Understanding autoimmune diabetes: insights from mouse models. Trends Mol. Med. **8:** 31–38.
7. MATTNER, F. *et al.* 2000. Inhibition of Th1 development and treatment of chronic-relapsing experimental allergic encephalomyelitis by a non-hypercalcemic analogue of 1,25-dihydroxyvitamin D_3. Eur. J. Immunol. **30:** 498–508.
8. MATHIEU, C. *et al.* 1994. Prevention of autoimmune diabetes in NOD mice by 1,25-dihydroxyvitamin D3. Diabetologia **37:** 552–558.
9. GREGORI, G. *et al.* 2002. A 1α,25-dihydroxyvitamin D_3 analog enhances regulatory T cells and arrests autoimmune diabetes in NOD mice. Diabetes **51:** 1367–1374.
10. THE EURODIAB SUBSTUDY 2 STUDY GROUP. 1999. Vitamin D supplement in early childhood and risk for Type 1 (insulin-dependent) diabetes mellitus. Diabetologia **42:** 51–54.

Antibody Repertoire in a Mouse with a Simplified D_H Locus: The D-Limited Mouse

GREGORY C. IPPOLITO,[a] JUKKA PELKONEN,[b] LARS NITSCHKE,[c] KLAUS RAJEWSKY,[d] AND HARRY W. SCHROEDER, JR.[a]

[a]*Division of Developmental and Clinical Immunology, Departments of Medicine and Microbiology, University of Alabama at Birmingham, Birmingham, Alabama 35294-3300, USA*

[b]*Department of Clinical Microbiology, University of Kuopio, 70211 Kuopio, Finland*

[c]*Institute for Virology and Immunobiology, University of Würzburg, 97078 Würzburg, Germany*

[d]*The Center for Blood Research, Harvard Medical School, Boston, Massachusetts 02115, USA*

KEYWORDS: antibody; repertoire; CDR3

The humoral immune system is fundamentally reliant upon the broad expression of a large collection of B lymphocyte receptors so as to recognize the fullest universe of antigens that might be encountered by the organism. This collection, or repertoire, of antibody receptors is believed to be derived in a generally random fashion[1] from a discrete collection of discontinuous germline gene segments that are somatically recombined during B lymphocyte development.[2] Consequently, the lymphocyte repertoire of the species should be maximally diversified and should therefore meet or exceed the ensemble of permutations presented by the continually evolving antigenic universe.

Much has been described regarding the mechanisms that contribute to the diversity of the antibody repertoire. Here, conversely, we are interested in the mechanisms and significance of constraints that may limit this diversity. Because of the large and complex nature of the expressed antibody repertoire, it is imperative to evaluate antibody diversity in experimental models that minimize this complexity yet still allow a reasonably comprehensive survey of the repertoire expressed in the organism. One focus of regulation of the repertoire is the restriction in the sequence and structure of the third complementarity-determining region of the antibody heavy chain (HCDR3).[3]

Address for correspondence: Harry W. Schroeder, Jr., M.D., Ph.D., Division of Developmental and Clinical Immunology, WTI 378, 1530 3rd Avenue South, University of Alabama at Birmingham, Birmingham, AL 35294-3300. Voice: 205-934-1522; fax: 205-975-6352.
Harry.Schroeder@ccc.uab.edu

Ann. N.Y. Acad. Sci. 987: 262–265 (2003). © 2003 New York Academy of Sciences.

Initial diversity in the primary antibody repertoire is concentrated principally at the site of V_H, D_H, and J_H segment joining—that is, the HCDR3—which is positioned at the center of the antibody binding site and thus plays a critical role in defining the specificity of the antibody.[4,5] HCDR3 intervals of neutral polarity that are enriched for hydrophilic amino acids are overrepresented in the repertoires of vertebrate species, whereas those enriched for highly charged or hydrophobic amino acids are rare or absent.[6,7] This strikingly nonrandom restriction of the antibody repertoire may in large measure be predetermined by the germline evolution of the D_H elements that encode the core of HCDR3, and that, in one reading frame among six possible, will typically contain the hydrophilic residues tyrosine, glycine, and serine. We interpret these data to indicate not only that HCDR3 sequence is subject to constraints, but also that there must indeed be a functional reason why HCDR3 is constrained. Deviation from the normal restriction of HCDR3 can result in pathogenic anti-double-stranded DNA binding antibodies, for example, which typically contain positively charged amino acids such as arginine.

To test the significance of shared restrictions in HCDR3 diversity found in both mouse and human, we have created mice wherein the D_H locus has been simplified with precision using Cre-*loxP* gene targeting, forcing expression of antigen binding sites (HCDR3s) not normally seen in the healthy organism. In what we term a D-limited mouse, in one immunoglobulin IgH allele of a BALB/c ES cell line, we have eliminated all but one D_H gene segment, DFL16.1, from the immunoglobulin heavy-chain locus (DD_H); at the same time, the center of DFL16.1 was replaced with an inverted DSP2.2 gene segment, which encodes arginine, asparagine, and histidine

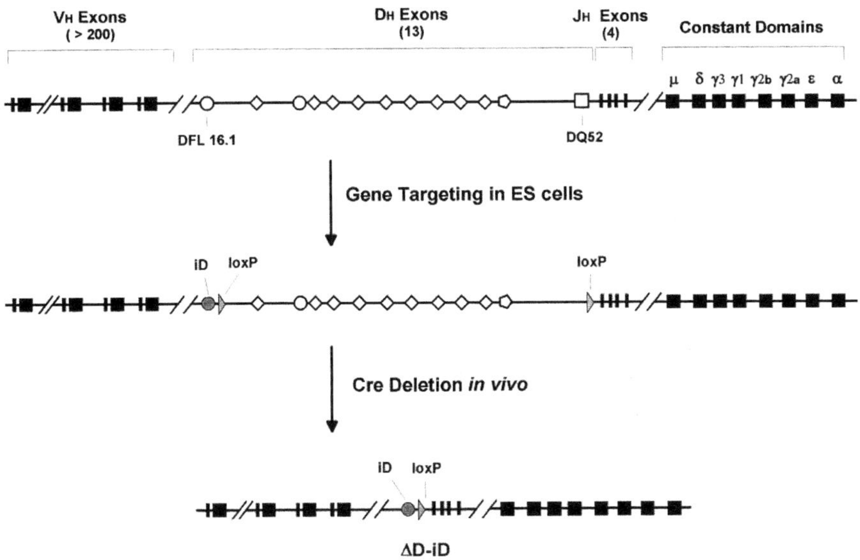

FIGURE 1. Generation of a single D_H-containing locus in the D-limited mouse. Illustration of the D_H locus: the wild-type (WT) locus, the locus after targeted replacement of DQ52 and DFL16.1, and the locus after Cre-mediated deletion leaving the single iD gene juxtaposed to the J_H genes (DD_H-iD).

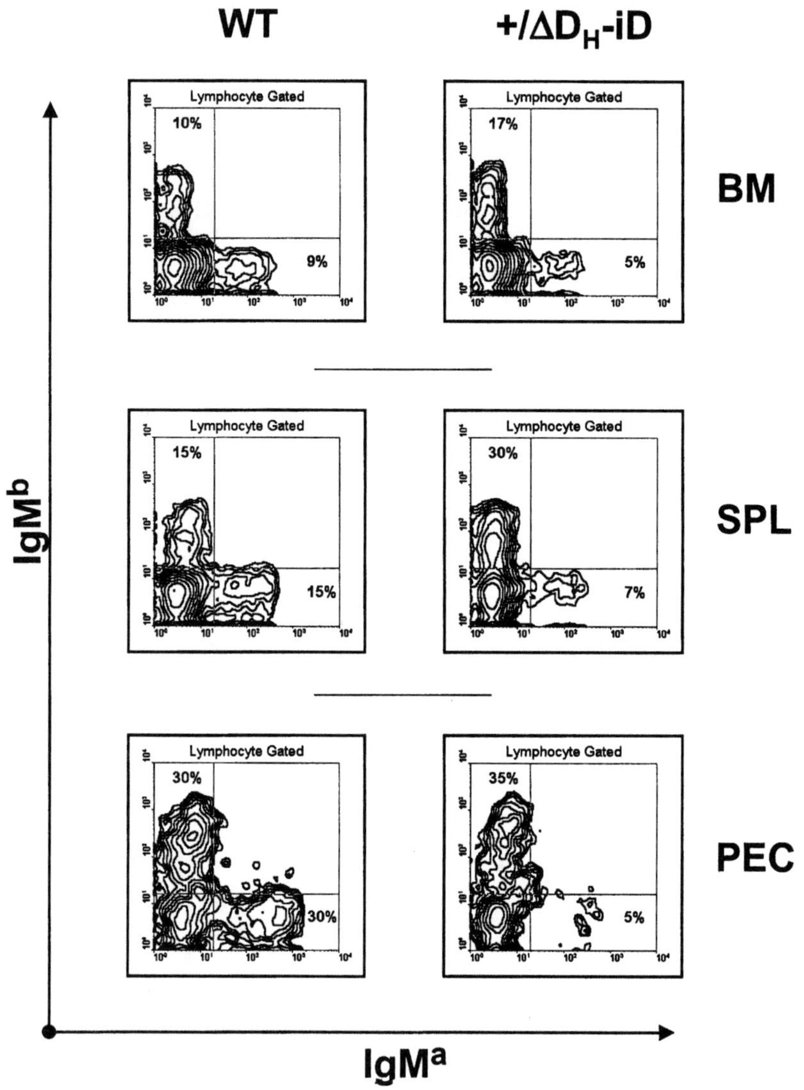

FIGURE 2. B cells bearing charged HCDR3s are disfavored compared to a wild-type repertoire. Flow cytometry indicates the relative proportion of B cells expressing a germline C57BL/6 allele (+, IgM^b), a germline BALB/c allele (+, IgM^a), or the altered D-limited allele (DD_H-iD, IgM^a). Results are shown for a heterozygous wild-type (WT) mouse and a heterozygous (+/DD-iD) D-limited mouse. Mice were 3-week-old littermates. The three tissues examined are bone marrow (BM), spleen (SPL), and peritoneal cavity (PEC).

residues (FIG. 1). We refer to this lone, quasi-native gene segment as iD, for inverted D. These alterations prejudice initial V(D)J rearrangement toward generation of HCDR3 intervals with charged amino acids (unpublished observations).

As shown in FIGURE 2, IgM allotype expression was evaluated in heterozygous F1 siblings, in a wild-type (+, IgM^b; +, IgM^a) mouse and a heterozygous "charged" D-limited mouse (+, IgM^b; DD_H-iD, IgM^a). B lymphocytes arising from the mutant DD_H-iD allele are increasingly reduced relative to wild-type as the cells progress from their site of generation (bone marrow) to a secondary lymphoid organ (spleen) and to a site where long-lived B cells accumulate (peritoneal cavity). In the wild-type littermate control, there is an equivalent representation of B cells bearing the IgM^b and IgM^a allotypes. In the bone marrow of the mutant, heterozygous +/DD_H-iD mouse, there is a 3:1 preference for use of the wild-type D_H locus. In the spleen, the ratio increases to 4:1. And in the peritoneal cavity, the wild-type allele is favored 6:1. Thus, use of the altered DD_H-iD allele is increasingly disfavored by more mature B cell populations.

These findings support the view that the antibody HCDR3 repertoire is evolutionarily depleted of charged amino acids because those B cells would be more likely to generate either ineffective or deleterious antibodies, and hence be subjected to negative selective mechanisms. Studies are currently ongoing to examine at the sequence level the population of immunogobulin produced at each stage of development and to evaluate responses to model antigens.

ACKNOWLEDGMENTS

This work was supported, in part, by the National Institutes of Health (AI 07051, AI42732, HD36292, and TW02130).

REFERENCES

1. MANSER, T., S.-Y. HUANG & M.L. GEFTER. 1984. Influence of clonal selection on the expression of immunoglobulin variable region genes. Science **226:** 1283–1288.
2. OKADA, A. & F.W. ALT. 1994. Mechanisms that control antigen receptor variable region gene assembly. Semin. Immunol. **6:** 185–196.
3. SCHROEDER, H.W., JR., G.C. IPPOLITO & S. SHIOKAWA. 1998. Regulation of the antibody repertoire through control of HCDR3 diversity. Vaccine **16:** 1383–1390.
4. KABAT, E.A. & T.T. WU. 1991. Identical V region amino acid sequences and segments of sequences in antibodies of different specificities: relative contributions of VH and VL genes, minigenes, and complementarity-determining regions to binding of antibody-combining sites. J. Immunol. **147:** 1709–1719.
5. XU, J.L. & M.M. DAVIS. 2000. Diversity in the CDR3 region of V(H) is sufficient for most antibody specificities. Immunity **13:** 37–45.
6. JOHNSON, G. & T.T. WU. 2000. Kabat database and its applications: 30 years after the first variability plot. Nucleic Acids Res. **28:** 214–218.
7. IVANOV, I. *et al.* 2002. Constraints on hydropathicity and sequence composition of HCDR3 are conserved across evolution. *In* The Antibodies. M. Zanetti & J.D. Capra, Eds.: 43–67. Taylor and Francis Group. London.

Long-Lived Plasma Cells in Immunity and Inflammation

A.E. HAUSER,[a,e] G. MUEHLINGHAUS,[a,e] R.A. MANZ,[a] G. CASSESE,[a] S. ARCE,[a] G.F. DEBES,[b] A. HAMANN,[b] C. BEREK,[a] S. LINDENAU,[a] T. DOERNER,[c] F. HIEPE,[c] M. ODENDAHL,[c] G. RIEMEKASTEN,[c] V. KRENN,[d] AND A. RADBRUCH[a]

[a]*Deutsches Rheumaforschungszentrum, Schumannstrasse 21/22, D-10117 Berlin, Germany*

[b]*Department of Experimental Rheumatology,* [c]*Department of Medicine, Rheumatology, and Clinical Immunology, and* [d]*Institute for Pathology, Charité University Hospital, Humboldt University, D-10117 Berlin, Germany*

KEYWORDS: plasma cells; survival; chemotaxis; autoimmunity

Plasma cells are terminally differentiated B lymphocytes that secrete antigen-specific antibodies and thus provide specific humoral immunity.[1] Upon provocation of an immune system with a "T lymphocyte dependent" antigen, B cells are activated to form either germinal centers or differentiate into plasma cells, the latter providing immediate protection by secreted antibodies of low affinity.[2] In the germinal centers, the variable region gene segments of the antibody genes undergo hypermutation, allowing the selection of cells expressing antibodies with increased affinities by memory B cells originating from those germinal centers. In memory responses, when reactivated by their cognate antigen, memory B cells start to proliferate quickly, increasing the numbers of memory B cells, but they also begin differentiating into plasmablasts (that is, the proliferating precursors of non-dividing plasma cells) and finally into plasma cells. These cells secrete antibodies at high rates. Long-lasting protective antibody titers against antigens, such as polio and tetanus, are found in sera of humans for decades.[3] While for a long time it had been anticipated that the plasma cells providing the long-lasting humoral protection were short-lived and constantly replaced by plasma cells newly generated from memory B cells, recent evidence suggests that long-lasting humoral immunity is provided by long-lived plasma cells. Antigen-specific plasma cells can be detected for at least one year after immunization in the bone marrow of immunized mice.[4] These plasma cells are long-lived and survive for more than three months without DNA synthesis,[5] with an estimated half-life of more than six months.[4–6] These long-lived plasma cells are not responsive to antigen. Co-transfer of antigen does not influence the generation of persistent

Address for correspondence: Andreas Radbruch, Deutsches Rheumaforschungszentrum, Berlin, Schumannstrasse 21/22, D-10117 Berlin, Germany. Voice: +49-30-28460-601; fax: +49-30-28460-603.

radbruch@drfz.de

[e]A.E.H. and G.M. have contributed equally to this work.

Ann. N.Y. Acad. Sci. 987: 266–269 (2003). © 2003 New York Academy of Sciences.

antibody titers provided by transferred murine bone marrow plasma cells.[7] The new concept that long-lived plasma cells provide humoral immunity and thus protective memory leaves memory B cells as the basis of reactive memory. However, a detailed analysis of the biology of long-lived plasma cells is required if we are to understand the quality of humoral immunity, the effectiveness of vaccination protocols, and the significance of stable titers of autoantibodies and allergen-specific antibodies in autoimmunity and allergy.

First, how is longevity of plasma cells regulated? Several experimental observations indicate that longevity is not an intrinsic capacity of plasma cells, but depends on their microenvironment.[8] In cultures of isolated bone marrow plasma cells, we could demonstrate synergistic effects of several factors in sustaining the survival of murine bone marrow plasma cells. Although long-lived *in vivo*, isolated plasma cells die within three days *in vitro*, if cultured in regular medium. Supernatant of total bone marrow cell cultures, but also individual cytokines, such as interleukin (IL)-5, IL-6, SDF-1α, and TNFα, prolong the survival of bone marrow plasma cells, rescuing 30–70% of these cells for more than three days. Moreover, contact to stroma cells or tickling of the matrix receptor CD44 provides additional support for plasma cell survival, such as that observed with the tickling of CD44 in the presence of IL-6 (Cassese *et al.*, submitted for publication).

We propose that these molecules, among other still unidentified factors, define a microenvironment *in vivo* that is able to maintain plasma cells, that is, a plasma cell survival niche. The numbers of survival niches per individual will then determine its capacity for plasma cells and also limit the overall amount of serum immunoglobulins. Competition of plasma cells for survival niches will control establishment of humoral immunity and immunopathology.[9] It has been shown that many more plasma cells are generated during an adaptive immune response than finally survive and become long-lived, suggesting that only a selected fraction of plasma cells enters the pool of long-lived plasma cells.[7,10]

To compete for survival niches, newly generated plasma blasts have to migrate from their site of generation, the secondary lymphoid organs, to the bone marrow. Indeed, between day 4 and day 6 following secondary immunization, antibody-secreting cells migrate via the bloodstream to the bone marrow.[11] Which factors regulate this migration and attract the plasma blasts to the bone marrow? Likely candidates were chemokines, a group of small cytokines that have been shown to play crucial roles in the cellular trafficking of the immune system.[12,13] In chemotaxis experiments, we could show that plasma cells obtained from murine spleen after secondary immunization express functional CXCR3 and CXCR4.[11] Interestingly, the chemotactic responsiveness is restricted to the period of migration of the plasma cells. It has been shown that SDF-1α (CXCL12), the only known ligand for CXCR4 so far, is expressed in high amounts by bone marrow stromal cells.[14] It also has been shown that this chemokine is required for normal accumulation of plasma cells in the bone marrow.[15] Plasma cells from the bone marrow still express CXCR4; however, they no longer respond to a gradient of its ligand SDF-1α by migration. Instead, they respond by increased survival *in vitro*, defining SDF-1α as a constitutive component of the plasma cell survival niche.

The ligands for CXCR3 are MIG (CXCL9), IP-10 (CXCL10), and I-TAC (CXCL11), which share the property that their expression is up-regulated in inflamed tissues to attract a subset of lymphocytes.[16] Plasma cells are often found at sites of inflammation, and in many autoimmune diseases plasma cells can produce autoreactive antibodies and contribute to the pathogenesis of the disease, as is the case for glomerulonephritis of systemic lupus erythematosus. We have shown that plasma cells accumulate and survive in the kidneys of lupus-prone NZW/B mice in numbers comparable to those in the bone marrow.[17] This accumulation is independent of the specificity of those plasma cells and reflects attraction of plasma cells to those inflamed kidneys, probably mediated though CXCR3 and CXCR4. Our results suggest that sites of inflammation may contain considerable numbers of survival niches for plasma cells, providing extra space for plasma cells generated at the time of inflammation and probably secreting antibodies relevant to the pathogens causing that inflammation. A successful immune response and termination of the inflammation will then destroy the survival niches of that tissue. Since they can no longer migrate to other organs, the plasma cells contained will die, leaving humoral protection to those plasma cells that had homed to the bone marrow. In chronic inflammation, however, inflamed tissues would provide stable survival niches for plasma cells, and thus contribute essentially to the stability of pathogenic antibodies, such as persisting autoantibody titers in autoimmunity and titers of allergen-specific antibodies in allergic diseases. And these titers would be refractory to therapies directed against ongoing immune reactions. New therapies will have to be developed to eliminate long-lived plasma cells from their survival niches.

REFERENCES

1. AHMED, R. & D. GRAY. 1996. Immunological memory and protective immunity: understanding their relation. Science **272:** 54–60.
2. STEINHOFF, U. *et al.* 1995. Antiviral protection by vesicular stomatitis virus-specific antibodies in alpha/beta interferon receptor-deficient mice. J. Virol. **69:** 2153–2158.
3. SLIFKA, M.K., M. MATLOUBIAN & R. AHMED. 1995. Bone marrow is a major site of long-term antibody production after acute viral infection. J. Virol. **69:** 1895–1902.
4. SLIFKA, M.K. & R. AHMED. 1996. Long-term humoral immunity against viruses: revisiting the issue of plasma cell longevity. Trends Microbiol. **4:** 394–400.
5. MANZ, R.A., A. THIEL & A. RADBRUCH. 1997. Lifetime of plasma cells in the bone marrow. Nature **388:** 133–134.
6. VIEIRA, P. & K. RAJEWSKY. 1988. The half-lives of serum immunoglobulins in adult mice. Eur. J. Immunol. **18:** 313–316.
7. MANZ, R.A., *et al.* 1998. Survival of long-lived plasma cells is independent of antigen. Int. Immunol. **10:** 1703–1711.
8. SZE, D.M. *et al.* 2000. Intrinsic constraint on plasmablast growth and extrinsic limits of plasma cell survival. J. Exp. Med. **192:** 813–821.
9. MANZ, R.A. *et al.* 2002. Humoral immunity and long-lived plasma cells. Curr. Opin. Immunol. **14:** 517–521.
10. SLIFKA, M.K. & R. AHMED. 1996. Long-term antibody production is sustained by antibody-secreting cells in the bone marrow following acute viral infection. Ann. N.Y. Acad. Sci. **797:** 166–176.
11. HAUSER, A. E. *et al.* 2002. Chemotactic responsiveness toward ligands for CXCR3 and CXCR4 is regulated on plasma blasts during the time course of a memory immune response. J. Immunol. **169:** 1277–1282.
12. BAGGIOLINI, M. 1998. Chemokines and leukocyte traffic. Nature **392:** 565–568.
13. ZLOTNIK, A. & O. YOSHIE. 2000. Chemokines: a new classification system and their role in immunity. Immunity **12:** 121–127.
14. NAGASAWA, T. *et al.* 1996. Defects of B-cell lymphopoiesis and bone-marrow myelopoiesis in mice lacking the CXC chemokine PBSF/SDF-1. Nature **382:** 635–638.

15. HARGREAVES, D.C. *et al.* 2001. A coordinated change in chemokine responsiveness guides plasma cell movements. J. Exp. Med. **194:** 45–56.
16. QIN, S. *et al.* 1998. The chemokine receptors CXCR3 and CCR5 mark subsets of T cells associated with certain inflammatory reactions. J. Clin. Invest. **101:** 746–754.
17. CASSESE, G. *et al.* 2001. Inflamed kidneys of NZB/W mice are a major site for the homeostasis of plasma cells. Eur. J. Immunol. **31:** 2726–2732.

Molecular Mechanism of Serial VH Gene Replacement

ZHIXIN ZHANG,[a] YUI-HSI WANG,[a] MICHAEL ZEMLIN,[a] HARRY W. FINDLEY,[b] S. LOUIS BRIDGES,[c] PETER D. BURROWS,[a] AND MAX D. COOPER[a,d]

[a]*Division of Developmental and Clinical Immunology, University of Alabama at Birmingham, Birmingham, Alabama 35294, USA*

[b]*Division of Pediatric Hematology/Oncology/Bone Marrow Transplantation, Emory University School of Medicine, Atlanta, Georgia 30322, USA*

[c]*Division of Clinical Immunology and Rheumatology, University of Alabama at Birmingham, Birmingham, Alabama 35294, USA*

[d]*The Howard Hughes Medical Institute, Birmingham, Alabama 35294, USA*

ABSTRACT: The molecular mechanism of serial VH replacement was analyzed using a human B cell line, EU12, that undergoes continuous spontaneous differentiation from pro-B to pre-B and then to B cell stage. In earlier studies, we found that this cell line undergoes intraclonal V(D)J diversification. Analysis of the IgH gene sequences in EU12 cells predicted the occurrence of serial VH replacement involving the cryptic recombination signal sequences (cRSS) embedded within framework 3 regions and concurrent extension of the CDR3 region. Detection of double-stranded DNA breaks at the cRSS site and different VH replacement excision circles confirm the ongoing nature of this diversification process in the EU12 cells. *In vitro* binding and cleavage assays using recombinant RAG-1 and RAG-2 proteins further validated the cRSS participation in this RAG-mediated recombination process. Serial VH replacements may represent an additional mechanism for diversification of the primary B cell repertoire.

KEYWORDS: molecular mechanism; serial VH replacement; human B cell; VH gene; VL gene; double-stranded DNA

Virtually unlimited specificities of immunoglobulin are generated by somatic recombination of previously separated variable (V), diversity (D) (for heavy chain only), and joining (J) gene segments to form the variable domain exon of immunoglobulin genes to be expressed, a process known as V(D)J recombination.[1,2] V(D)J recombination is mediated by recombination activating gene (*RAG1* and *RAG2*) products that recognize the recombination signal sequences (RSS) flanking V, D,

Address for correspondence: Max D. Cooper, M.D., University of Alabama at Birmingham, WTI378, 1824 6th Avenue South, Birmingham, Alabama 35294-3300. Voice: 205-975-7204; fax: 205-934-1875.
max.cooper@ccc.uab.edu

Ann. N.Y. Acad. Sci. 987: 270–273 (2003). © 2003 New York Academy of Sciences.

and J gene segments.[3–5] A functional RSS site is composed of a heptamer (CACTGTG), a nonamer (GGTTTTTGT), and a non-conserved spacer region that is either 12 or 23 base pairs in length.[1]

Immunoglobulin gene rearrangement proceeds in sequential steps during early B lineage development in hemotopoietic tissues, prominently the fetal liver and adult bone marrow.[2] DH → JH rearrangement normally occurs before the VH → DHJH rearrangement, and these gene rearrangements usually precede the VL → JL rearrangement in the light-chain gene loci.[6,7] The expression of a functional μ heavy chain generated through VDJ recombination is a pre-request for further development along the B lineage pathway. The μ heavy chains initially associate with surrogate light chains composed of VpreB and λ5 elements and with transmembrane Igα and Igβ signaling partners to form pre-B cell receptors (pre-BCR) that are expressed on the cell surface.[8–11] The pre-BCR functions to stimulate pre-B cell proliferation before its expression is extinguished to allow light-chain gene rearrangement and expression of the BCR on the newly formed B cells.[12,13] The pre-BCR and BCR thus function in two early developmental checkpoints in B cell development.

Since joint formation during V(D)J rearrangement is random, two-thirds of the VH → DHJH joints and VL → JL joints that are formed may be out of reading frame and thus unusable for immunoglobulin production.[2] Cells with non-functional IgH or IgL gene rearrangements cannot survive unless they are rescued by a subsequent V(D)J rearrangement. Even after generating a functional VHDHJH reading frame, the expressed μ heavy chains may fail to pair with the surrogate light chains or conventional light chains to form functional pre-BCR or BCR that is needed to promote further differentiation. Finally, B cells that possess self-reactive antigen receptors require alteration of their antigen specificity or must be eliminated before their exit to the periphery.[14] In all of these cases, the early B lineage cells need to retain the capability to edit the preformed V(D)J joints to continue their development.[2]

The genetic organization of the VL and JL gene segments within Ig κ or λ locus allows secondary rearrangement through the joining of an upstream VL with a downstream JL gene segment, as they are flanked by different types of RSS sites. This secondary rearrangement is mechanistically similar to the primary rearrangement. Thus, secondary rearrangement can be performed as long as the light-chain locus remains accessible and the recombination machinery is still active. With each round of rearrangement, a new VL-JL joint is formed, and the previous VL-JL joint is deleted. In addition, a cell with a nonfunctional κ rearrangement may still undergo rearrangement of λ light-chain genes.[2] In contrast to the relative ease of secondary light-chain VJ rearrangement, secondary rearrangement of an upstream VH to a preformed VHDHJH gene is more difficult, since during the primary VH to DHJH rearrangement, all of the intervening D segments are deleted. This leaves no 12-bp spacer RSS tagged gene segment to recombine with the 23-bp spacer RSS flanked upstream VH genes or 23-bp spacer RSS flanked downstream JH genes.[2]

It has been reported, however, that mouse pre-B cell lines carrying non-functional IgH rearrangements can achieve a secondary rearrangement of IgH genes through VH replacement to change a non-functional V(D)J joint into a functional IgH gene.[15–17] Comparison of the functional V(D)J joints with the previous non-functional V(D)J joints in these cases suggested that VH replacement was mediated through the use of cryptic RSS sequences located within the framework 3 region of VH germline genes.[15–17] The potential biological significance of VH replacement was indicated by studies of site-directed transgenic or knock-in mice carrying self-reactive IgH transgenes.[18–20] The self-reactive VDJH genes that were artificially inserted into the JH locus could be deleted by secondary rearrangements. The analysis of these mice suggested that VH replacement was employed to delete the unwanted IgH gene.[18–20] Most of the functional VH germline genes in mice and humans contain the cRSS motifs. These cRSS sites have a hepatmer (TACTGTG), but lack a clearly definable nonamer partner. However, the true function of the cRSS sites in RAG-mediated recombination is still unknown.

The present studies began with the demonstration that a cell line, EU12,[21] which was established from a childhood leukemia patient, undergoes continuous pro-B cell to pre-B and B cell differentiation. *RAG1* and *RAG2* gene expression can be detected at the pro-B and pre-B stages, but *RAG1* expression is extinguished in the EU12 B cells,[22] and terminal deoxynucleotidyl transferase (TdT) gene expression is extinguished at the pre-B cell stage. Analysis of the BCR repertoire in the B cell subpopulation indicated limited intraclonal diversification of VH and VL gene usage.[22] All the IgH sequences were found to include the same DH3-10JH4 joint, but different VH genes were used. Inspection of the VH-DH junction regions in the EU12 IgH sequences suggested that this cell line was undergoing serial VH replacement of DH-proximal genes through the use of their cRSS sites.

This hypothetical model has been tested by experiments aiming to detect the predicted intermediates that would be generated during different VH replacement events in the EU12 cells. Using a ligation-mediated PCR assay, we were able to detect the predicted double-stranded DNA breaks at the VH2-5 cRSS site. In addition, using seminested primer sets, we were able to amplify the different VH replacement excision circles generated that were predicted to be detectable during serial VH replacement events. Thus, the EU12 cell line was caught in the ongoing process of VH replacement. To elucidate the cRSS participation in these RAG-mediated recombinational events, we performed *in vitro* binding and cleavage assays. Purified RAG1 and RAG2 proteins were found to be able to bind to the cRSS sites derived from VH1-8, VH2-5, VH3-7, and VH4-4 germline genes and to cleave these cRSS sites at the end of their heptamers under the same conditions that promote RAG protein binding and cleavage of conventional 12-bp or 23-bp RSS sites.[23]

In conclusion, we have obtained *in vivo* and *in vitro* evidence to show the participation of the cRSS sites in RAG-mediated recombination as the molecular mechanism of VH replacement in a human cell line. Serial VH replacements may represent a physiological mechanism for diversification of the primary B cell repertoire.

ACKNOWLEDGMENTS

We thank Dr. David Roth and Leslie Huye for providing RAG1, RAG2, and HMG1 expression vectors, initial aliquots of purified proteins, and discussion of experimental conditions for protein purification and coupled cleavage reaction. We thank Yoshiki Kubagawa and Yuanqing Zhu for DNA sequencing, Dr. Larry Gartland for help with FACS sorting, and Drs. Harry Schroeder, Chen-lo Chen, Hiromi Kubagawa, and Robert Stephan for helpful discussions.

REFERENCES

1. TONEGAWA, S. 1983. Somatic generation of antibody diversity. Nature **302:** 575–581.
2. RAJEWSKY, K. 1996. Clonal selection and learning in the antibody system. Nature **381:** 751–758.
3. SCHATZ, D.G. & D. BALTIMORE. 1988. Stable expression of immunoglobulin gene V(D)J recombinase activity by gene transfer into 3T3 fibroblasts. Cell **53:** 107–115.
4. SCHATZ, D.G., M.A. OETTINGER & D. BALTIMORE. 1989. The V(D)J recombination activating gene, RAG-1. Cell **59:** 1035–1048.
5. OETTINGER, M.A., D.G. SCHATZ, C. GORKA & D. BALTIMORE. 1990. RAG-1 and RAG-2, adjacent genes that synergistically activate V(D)J recombination. Science **248:** 1517–1523.
6. ALT, F.W., G.D. YANCOPOULOS, T.K. BLACKWELL, *et al.* 1984. Ordered rearrangement of immunoglobulin heavy chain variable region segments. EMBO J. **3:** 1209–1219.
7. BURROWS, P., M. LEJEUNE & J.F. KEARNEY. 1979. Evidence that murine pre-B cells synthesise mu heavy chains but no light chains. Nature **280:** 838–840.
8. SAKAGUCHI, N. & F. MELCHERS. 1986. Lambda 5, a new light-chain-related locus selectively expressed in pre-B lymphocytes. Nature **324:** 579–582.
9. KUDO, A. & F. MELCHERS. 1987. A second gene, VpreB in the lambda 5 locus of the mouse, which appears to be selectively expressed in pre-B lymphocytes. EMBO J. **6:** 2267–2272.
10. LASSOUED, K., C.A. NUNEZ, L. BILLIPS, *et al.* 1993. Expression of surrogate light chain receptors is restricted to a late stage in pre-B cell differentiation. Cell **73:** 73–86.
11. KARASUYAMA, H., A. ROLINK, Y. SHINKAI, *et al.* 1994. The expression of Vpre-B/lambda 5 surrogate light chain in early bone marrow precursor B cells of normal and B cell-deficient mutant mice. Cell **77:** 133–143.
12. TSUBATA, T., R. TSUBATA & M. RETH. 1992. Crosslinking of the cell surface immunoglobulin (mu-surrogate light chains complex) on pre-B cells induces activation of V gene rearrangements at the immunoglobulin kappa locus. Int. Immunol. **4:** 637–641.
13. KATO, I., T. MIYAZAKI, T. NAKAMURA & A. KUDO. 2000. Inducible differentiation and apoptosis of the pre-B cell receptor-positive pre-B cell line. Int. Immunol. **12:** 325–334.
14. GOODNOW, C.C., J.G. CYSTER, S.B. HARTLEY, *et al.* 1995. Self-tolerance checkpoints in B lymphocyte development. Adv. Immunol. **1995:** 279–368.
15. COVEY, L.R., P. FERRIER & F.W. ALT. 1990. VH to VHDJH rearrangement is mediated by the internal VH heptamer. Int. Immunol. **2:** 579–583.
16. KLEINFIELD, R., R.R. HARDY, D. TARLINTON, *et al.* 1986. Recombination between an expressed immunoglobulin heavy-chain gene and a germline variable gene segment in a Ly 1^+ B-cell lymphoma. Nature **322:** 843–846.
17. RETH, M., P. GEHRMANN, E. PETRAC & P. WIESE. 1986. A novel VH to VHDJH joining mechanism in heavy-chain-negative (null) pre-B cells results in heavy-chain production. Nature **322:** 840–842.
18. CHEN, C., Z. NAGY, E.L. PRAK & M. WEIGERT. 1995. Immunoglobulin heavy chain gene replacement: a mechanism of receptor editing. Immunity **3:** 747–755.
19. CHEN, C., Z. NAGY, M.Z. RADIC, *et al.* 1995. The site and stage of anti-DNA B-cell deletion. Nature **373:** 252–255.
20. CHEN, C., E.L. PRAK & M. WEIGERT. 1997. Editing disease-associated autoantibodies. Immunity **6:** 97–105.
21. ZHOU, M., A.M. YEAGER, S.D. SMITH & H.W. FINDLEY. 1995. Overexpression of the MDM2 gene by childhood acute lymphoblastic leukemia cells expressing the wild-type p53 gene. Blood **85:** 1608–1614.
22. WANG, Y., Z. ZHANG, H.W. FINDLEY, *et al.* 2003. Intraclonal V(D)J diversification during pro-B to B cell differentiation in EU12 cells. Blood **101:** 1030–1037.
23. ZHANG, Z., Y. WANG, M. ZEMLIN, *et al.* 2003. Usage of VH replacement mechanism to generate the B cell repertoire in humans. Submitted for publication.

Oligoclonal Expansions of Antigen-Specific $CD8^+$ T Cells in Aged Mice

DAVID N. POSNETT,[a,b] DMITRY YARILIN,[a,b] JENNIFER R. VALIANDO,[b] FANG LI,[a,c] FOO Y LIEW,[d] MARC E. WEKSLER,[a,c] AND PAUL SZABO[a,c]

Weill Medical College, Cornell University, [a]Immunology Program, Graduate School of Medical Sciences; Department of Medicine [b]Division of Hematology-Oncology, and [c]Division of Geriatrics, 1300 York Avenue, New York, New York 10021, USA

[d]Division of Immunology, University of Glasgow, Glasgow, 911 6NT, Scotland

ABSTRACT: Oligoclonal T cell expansions (TCE) are common in old humans and mice. It is not known whether an Ag-specific response becomes more oligoclonal with age, and, if so, how this might alter biological responses or compromise the immune response, thus contributing to the immunodeficiency of aging. We used a tumor antigen response to study these questions. Early on, antigen reactive T cell numbers at the site of tumor injection were lower and clonally more restricted in old mice. Subsequently, long-term oligoclonal TCE emerged in the blood and spleen of old mice. IL-15 was not necessary for development of TCE in the blood. Overall, the data pointed to a dysregulated immune response in old mice, perhaps due to lack of optimal IL-2 and CD4 help at the earliest stages and a lack of an efficient local peritoneal CTL response. This was associated with a deficient humoral response and, likely, persistence of tumor cells or tumor antigens. Perhaps the spleen is the site of persistence which explains clonal TCE observed primarily in PBL and spleen. The TCE appear to be inefficient as they are often anergic. As a result an occasional peritoneal or splenic tumor may arise in old mice.

KEYWORDS: T lymphocytes; clonal expansion; T cell receptors; *in vivo* animal models

RESPONSE TO Cw3 TUMOR ANTIGEN IN OLD VS. YOUNG MICE

The CD8 response in DBA-2 mice to syngeneic P815-Cw3 tumor cells is directed to an immunodominant peptide of the transfected human HLA-Cw3 gene: $Cw3_{170-179}$ (RYLKNGKETL) presented by K^d.[1,2] Within two weeks after intraperitoneal (i.p.) tumor injection, up to 90% of CD8 cells are tumor antigen-specific CTL,[2] and the tumor cells themselves are rapidly eliminated. In an average mouse, approximately 15–20 clones participate in the response to tumor antigen with TCRs composed of BV10, often BJ1S2, and invariably a 6-amino acid CDR3 β chain.[1,3] Most of the cells with

Address for correspondence: David N. Posnett, Immunology Program, Graduate School of Medical Sciences, Cornell University, 1300 York Avenue, New York, NY 10021. Voice: 212-746-6488; fax: 212-746-8866.
dposnett@med.cornell.edu

Ann. N.Y. Acad. Sci. 987: 274–279 (2003).

this TCR signature bind K^dCw3 tetramers.[4] The TCR α chains of the tumor-specific CD8 cells also have limited diversity: AV8, JV pHDS58, and a CDR3 of 9 aa.

By tracking this response over time, we found[5] that the percentages of Vβ10/CD8 cells in the peritoneum were similar in young and old animals. In PBL and spleen, however, much higher percentages were observed in old mice (on D15, approx. 60% in old vs. 25% Vβ10/CD8 in young mice). The same results were obtained using $Cw3/K^d$ tetramers to identify the tumor Ag-specific CD8 cells. Total numbers of Ag-specific cells were increased (not just percentages). Thus, these cells represented true clonal expansions.

The response was also more oligoclonal in old mice. This was demonstrated by CDR3 spectratyping and by TCR β-chain CDR3 sequences of Vβ10 CD8 cells. On average there were 6.14 ± 1.3 different 6 aa CDR3 sequences per sample in young mice, and 3.08 ± 0.4 in old mice. The differences in diversity of the response between young and old mice were most marked in the early two-week samples obtained from the peritoneum, that is, the site of tumor injection.[5] The clonal expansions in the blood of old mice persisted longer, resulting in a higher proportion of the CD8 subset being occupied by these clones. These cells had an activation/memory phenotype ($CD62L^-$, $CD45RB^+$). Occasionally, they expressed CD69 and showed evidence of CD8 and TCR downmodulation, even months after tumor injection.

These persistent tumor antigen-specific clones were probably non-functional because they were unable to respond to peptide presented by splenic APCs by mounting an IFN–γ response as detected by intracellular staining after 12 h of peptide stimulation. By contrast, CD8 cells from young mice that had previously been injected with tumor cells mounted a vigorous memory response.

Moreover, the antibody levels to P815-Cw3 tumor cells were 10-fold lower in old mice.[5] In addition, old mice tended to develop tumors: 3 of 10 old mice (but 0 of 10 young mice) developed tumors when injected with a subclone of tumor cells (named P815-Cw3.T1) that had been adapted for *in vivo* growth. Thus, overall, the antitumor response in old mice was functionally deficient.

Based on analysis of peritoneal cavity (PC) T cells at the earliest technically feasible time point (day 7), we found the early response was numerically deficient in old mice. Percent Vβ10/CD8 was greater in young mice: 22.9 ± 2.8 (mean ± SEM) versus 9.0 ± 0.9 ($P < 0.004$). On day 14, total CD8 cell numbers in the PC were higher in young mice: 4.48 ± 0.66 million cells/PC versus 1.28 ± 0.73 million cells/PC ($P < 0.01$). Total CD4 cells/PC were twofold higher in young mice (NS), and total $V\beta10^+$ CD8 cells were also higher: 2.74 ± 0.63 million cells/PC versus 0.66 ± 0.37 million cells/PC ($P < 0.03$). Finally, we found that peritoneal macrophages were present in higher numbers in young vs. old mice, and this correlated with lower average tumor cell numbers as measured 5 days after tumor injection. Age-associated deficiency in peritoneal macrophages has been previously noted.[6] In old mice, however, there was considerable heterogeneity in the numbers of peritoneal macrophages and of tumor cells that could be retrieved and counted in these peritoneal washes.

In other experiments, we injected CFSE-labeled P815-Cw3 tumor cells i.p. *In vitro*, 95% remained brightly CFSE stained after three days as a measure of minimal cell proliferation. *Ex vivo*, tumor proliferation was assessed on day 3 in peritoneal lavage samples. In young mice, 19 ± 3% remained brightly CFSE stained. In old mice, only 3.6 ± 2% remained brightly stained, indicating greater proliferation of tumor cells in the PC.

In summary, we suspected that old mice may have a deficient early tumor response at the site of tumor injection. Similar findings were reported by Shrikant *et al.*, who used EL-4 thymoma cells expressing an OVA peptide as the tumor antigen that were injected i.p.[7] Growth of the tumor was critically dependent on the early local response. The tumor antigen OVA was recognized by transgenic CD8 T cells using the OT-1 TCR. When locally infiltrating CD4 Th cells and IL-2 production were boosted by anti-CTLA4 treatment or exogenous IL-2, local expansion of OT-1 CD8 effector cells and control of tumor growth were observed. In the absence of CTLA-4 blockade or exogenous IL-2, OT-1 CD8 cells migrated out of the peritoneum only days after an initial influx and then accumulated in the spleen and draining lymph nodes as anergic effector CD8 cells. Lack of sufficient IL-2 or an adequate response to IL-2 is a prominent feature of the immune response in old mice. Overall, our observations seemed to fit with these earlier findings and with those of Shrikant *et al.*

Another cytokine important for CD8 effector maturation is IL-15.[8,9] We therefore examined the effect of daily injections of recombinant soluble IL-15Rα, an inhibitor of endogenous IL-15. As a control, we used a mutant sIL-15Rα lacking IL-15 binding activity. These recombinant proteins were administered daily at 40 μg/mL i.p. over the first 19 days post tumor injection. The tetramer-positive response in the blood was even higher in mice treated with this IL-15 blocking reagent, regardless of age. We conclude that IL-15 deficiency does not affect the emergence of clonal expansions in old mice.

WHAT CELL TYPE CONSTITUTES PERSISTENT CD8 T CELL CLONAL EXPANSIONS?

A recurrent question is whether the TCE are composed of cells belonging to a normal CD8 effector/memory subset. For instance, these clones could simply be the result of extended antigenic stimulation and prolonged exposure to antigen-loaded DC, known to favor full effector memory maturation.[10,11] Alternatively, the TCE cells may be defective in some way perhaps due to clonal exhaustion, or due to terminally shortened telomeres.[12]

P815 cells do not express CD80/86 and are inefficient APCs for CD8 cells.[13] Therefore, tumor antigen is probably presented by professional APCs by cross presentation. Either peritoneal macrophages or dendritic cells might then migrate to the spleen loaded with tumor antigens. The balance of cytokines, such as IL-2/IL-15 or DC-derived cytokines, may determine the relative predominance of central (c-mem) versus effector (e-mem) memory CD8 cells that emerge from the spleen (FIG. 1). Prolonged IL-15 exposure generates CD8 c-mem while IL-2 exposure generates predominantly CD8 e-mem.[14] In our studies, blocking of IL-15 signaling did not ablate the TCE observed after tumor injection, but resulted in increased TCE instead. Ku *et al.*[9] used adoptive transfer of whole spleen or lymph node cells into β2M knockout mice to show that spontaneous CD8 TCE from old mice proliferated after transfer, apparently without the need for continued antigen presentation. They suggested that proliferation was dependent on IL-15. Perhaps the different results are due to different techniques used and different subsets within a TCE. Indeed, individual TCE clones probably consist of both progenitor $T_{e\text{-mem}}$ cells with proliferative capacity and derived CD8 effector cells. In humans, these cells differ in CD28 expression.

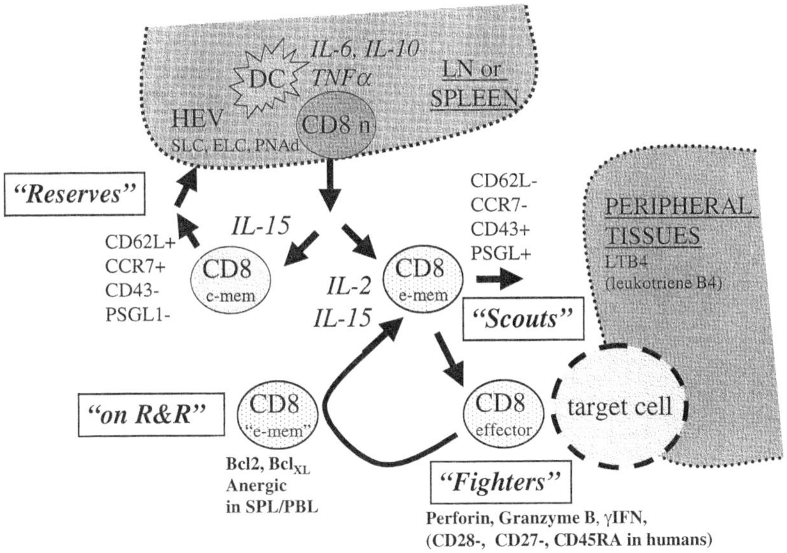

FIGURE 1. Naïve CD8 cells mature in the context of antigen presentation by dendritic cells within the lymph node (LN) or spleen, where they are exposed to DC-derived cytokines such as IL-5, IL-10, and TNFα. They mature into central memory cells, $CD8_{c\text{-}mem}$, possibly under the influence of IL-15. $CD8_{c\text{-}mem}$ circulate in the blood and then find their way back to the LN due to expression of CCR7, the chemokine receptor for SLC and ELC chemokines produced in the high endothelial venules (HEV). CD8 effector memory cells $CD8_{e\text{-}mem}$, by contrast, may be preferentially produced under conditions of IL-2 excess. They lack CCR7 but express other molecules that help them home to peripheral tissues, for example, in response to LTB4. CD8 effectors derive directly from these $CD8_{e\text{-}mem}$ and are responsible for cytolysis of target cells. A new subset is proposed to identify cells that are post-effector stage, have high levels of Bcl2, and BclXL are usually anergic and circulate in non-effector tissues such as the blood and the spleen.

$CD28^+$ $CD8^+$ cells can give rise to $CD28^-CD8^+$ progeny,[15] and many of these cells also lack CD27 expression.[16] As they lose CD28 expression, these cells gain CD57 expression. Individual TCE identified, based on TCR sequences, are often distributed in both subsets.[17]

In man, CD8 TCE have the hallmarks of effector CD8 cells with abundant intracellular perforin, sometimes spontaneous cytolytic effector function, shortened telomers, and the phenotype of fully matured effector CD8 cells. This puts them squarely in the effector arm of CD8 maturation (FIG. 1). One may still debate whether they are pre-effectors or post-effectors. We favor the latter as it is difficult to obtain any kind of proliferative response with these cells, and they appear to be quite anergic. These cells have been described as "terminally differentiated," implying that they are simply awaiting cell death. It makes little sense, however, to accumulate cells that will serve no future function. In addition, these cells have up-regulated Bcl-2 and Bcl-XL and are resistant to apoptotic stimuli in several systems.[15,18] Thus, we propose that they may have potential utility.

An analogy with military functions may be useful. CD8 c-mem can be viewed as "reserves." They return to the lymph nodes and are available for future memory responses. CD8 e-mem correspond to "scouts" as they patrol the peripheral tissues. They can mount rapid and fatal responses by generating effector cells or "fighters" that lyse target cells such as tumors. Surviving fighters are put on "rest and relaxation" in the military and have future usefulness. We speculate that the cells of TCE are in a similar position.

ACKNOWLEDGMENTS

This work was supported in part by PHS grants AG14669 and AI22333 and an NIDA supplement, and a grant from the Harriman Foundation.

REFERENCES

1. Casanova, J-L., J-C. Cerottini, M. Matthes, *et al.* 1992. H-2-restricted cytolytic T lymphocytes specific for HLA display T cell receptors of limited diversity. J. Exp. Med. **176:** 439–447.
2. MacDonald, H.R., J-L. Casanova, J.L. Maryanski & J.-C. Cerottini. 1993. Oligoclonal expansion of major histocompatibility complex class I-restricted cytolytic T lymphocytes during a primary immune response in vivo: direct monitoring by flow cytometry and polymerase chain reaction. J. Exp. Med. **177:** 1487–1492.
3. Maryanski, J.L., C.V. Jongeneel, P. Bucher, *et al.* 1996. Single-cell PCR analysis of TCR repertoires selected by antigen in vivo: a high magnitude CD8 response is comprised of very few clones. Immunity **4:** 47–55.
4. Bousso, P., A. Casrouge, J. D. Altman, *et al.* 1998. Individual variations in the murine T cell response to a specific peptide reflect variability in naive repertoires. Immunity **9:** p169–p178.
5. Li, F., D.A. Yarilin, J. Valiando, *et al.* 2002. Tumor antigen drives a persistent oligoclonal expansion of CD8+ T cells in aged mice. Eur. J. Immunol. **32:** 1650–1658.
6. Weksler, M.E., R. Schwab, F. Huetz, *et al.* 1990. Cellular basis for the age-associated increase in autoimmune reactions. Int. Immunol. **2:** p329–p335.
7. Shrikant, P., A. Khoruts & M.F. Mescher. 1999. CTLA-4 blockade reverses CD8+ T cell tolerance to tumor by a CD4+ T cell- and IL-2-dependent mechanism. Immunity **11:** p483–p493.
8. Ku, C.C., M. Murakami, A. Sakamoto, *et al.* 2000. Control of homeostasis of CD8+ memory T cells by opposing cytokines. Science **288:** 675–678.
9. Ku, C.C., J. Kappler & P. Marrack. 2001. The growth of the very large CD8+ T cell clones in older mice is controlled by cytokines. J. Immunol. **166:** 2186–2193.
10. Lanzavecchia, A. & F. Sallusto. 2000. Dynamics of T lymphocyte responses: intermediates, effectors, and memory cells. Science Oct. 6. **290:** 92–97.
11. Lanzavecchia, A. & F. Sallusto. 2001. Antigen decoding by T lymphocytes: from synapses to fate determination. Nat. Immunol. **2:** 487–492.
12. Monteiro, J., F. Batliwalla, H. Ostrer & P.K. Gregersen. 1996. Shortened telomeres in clonally expanded CD28-CD8+ T cells imply a replicative history that is distinct from their CD28+CD8+ counterparts. J. Immunol. **156:** 3587–3590.
13. Azuma, M. & L.L. Lanier. 1995. The role of CD28 costimulation in the generation of cytotoxic T lymphocytes. Curr. Top. Microbiol. Immunol. **198:** 59–74.
14. Weninger, W., M.A. Crowley, N. Manjunath & U.H. von Andrian. 2001. Migratory properties of naive, effector, and memory CD8(+) T cells. J. Exp. Med. **194:** 953–966.

15. POSNETT, D.N., C. IRWIN, J. EDINGER, *et al.* 1999. Differentiation of human CD8 T cells: implications for in vivo persistence of oligoclonal CD8+28 cytotoxic effector clones. Int. Immunol. **11:** 229–241.
16. HAMANN, D., P.A. BAARS, M.H.G. REP, *et al.* 1997. Phenotypical and functional separation of memory and effector human CD8+ T cells. J. Exp. Med. **186:** 1407–1418.
17. MORLEY, J.K., F.M. BATLIWALLA, R. HINGORANI & P. K. GREGERSEN. 1995. Oligoclonal CD8+ T cells are preferentially expanded in the CD57+ subset. J. Immunol. **154:** 6182–6190.
18. KANEKO, H., K. SAITO, H. HASHIMOTO, *et al.* 1996. Preferential elimination of CD28+ T cells in systemic lupus erythematosus (SLE) and the relation with activation-induced apoptosis. Clin. Exp. Immunol. **106:** 218–229.

A Genome-Wide Analysis of the Acute-Phase Response and Its Regulation by Stat3β

STEPHEN DESIDERIO AND JOO-YEON YOO

Department of Molecular Biology and Genetics and Howard Hughes Medical Institute, Johns Hopkins University School of Medicine, Baltimore, Maryland 21205, USA

ABSTRACT: The acute-phase response (APR) is the systemic inflammatory component of innate immunity. A global assessment of hepatic gene expression during an APR has been undertaken. In response to endotoxin, an inducer of the APR, about 7% of mouse genes exhibited significant changes in expression. Genes for cholesterol, fatty acid, and phospholipid synthesis were suppressed, while genes participating in innate defense and antigen presentation were induced. Upon challenge with endotoxin, mice deficient in Stat3β, a dominant-negative variant of Stat3, exhibited impaired recovery and susceptibility to protracted shock. These findings are accompanied by overexpression and hyperresponsiveness of a subset of lipopolysaccharide (LPS)-inducible genes in liver, suggesting a critical role for Stat3β in the control of systemic inflammation.

KEYWORDS: acute-phase response; transcriptional control; innate immunity; inflammation

A GENOME-WIDE SURVEY OF THE ACUTE-PHASE RESPONSE

Innate immune responses are initiated by chemical patterns that are presented by invading microorganisms or revealed by tissue damage. These stereotypic structures are detected by pattern-recognition receptors, whose engagement can result in the production of proinflammatory cytokines that act on target organs such as the liver.[1] The actions of proinflammatory cytokines, in turn, modulate innate immunity by inducing changes in the expression of pattern-recognition receptors and their downstream effectors.

The systemic inflammatory component of innate immunity, or the acute-phase response (APR), is invoked when organismic integrity is breached. The APR involves diverse extracellular and intracellular proteins whose levels change in response to tissue damage or infection.[2–7] Systemic inflammation is accompanied by an increase in triglyceride-rich lipoproteins, a reduction in high-density lipoprotein (HDL) cholesterol, and impairment of cholesterol transport.[5] These metabolic alterations,

Address for correspondence: Stephen Desiderio, M.D., Ph.D., Johns Hopkins University School of Medicine, 725 North Wolfe Street, Baltimore, MD 21205. Voice: 410-955-4735; fax: 410-955-9124.

sdesider@jhmi.edu

Ann. N.Y. Acad. Sci. 987: 280–284 (2003).

which promote atherosclerosis, may explain the epidemiologic link between chronic inflammation and cardiovascular disease.[8]

A genome-wide analysis of the hepatic inflammatory response to lipopolysaccharide (LPS) in the mouse documents an adaptation to endotoxic challenge involving about 7% of known genes.[9] Metabolic adjustments include suppression of pathways for cholesterol, fatty acid, and phospholipid synthesis. While many of the genes induced by LPS function in innate immunity, the major histocompatibility complex (MHC) class I antigen presentation machinery is also coordinately induced, illustrating an intersection between innate and adaptive immunity.

To provoke a systemic inflammatory response, 129/SvJ X C57BL/6 mice were injected intraperitoneally with a sublethal dose of LPS, and liver RNA was harvested at various times up to 48 h. These RNA samples were used to interrogate oligonucleotide arrays containing 12,488 probe sets, or roughly one-third of the protein coding capacity of the mouse genome. Expression of about 900 genes, or roughly 7% of the markers assayed, exhibited alterations in expression in response to LPS. This observation suggests that 5–10% of the protein-encoding portion of the mouse genome is mobilized in response to a single inflammatory stimulus. Of the LPS-responsive genes, about 500 encode proteins of known function; these were sorted into functional categories, as defined by the Gene Ontology Consortium (www.geneontology.org). Three categories — defense and immunity, intracellular signaling, and metabolism — account for about 70% of the LPS-responsive genes. A reciprocal relationship is observed between the responses of metabolic genes and host-defense genes. Genes with direct roles in defense and immunity or intracellular signaling constitute about 60% of the LPS-induced set, but only 20% of the down-regulated

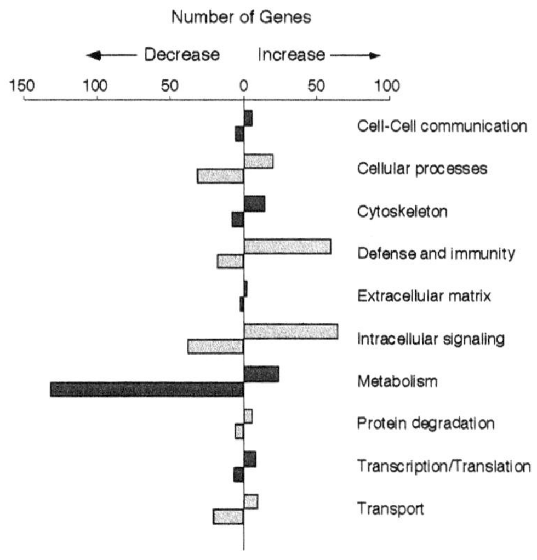

FIGURE 1. Functional profiles of LPS-responsive genes expressed in liver. Genes were assigned to single functional categories, as defined by the Gene Ontogeny Consortium, and determined by interrogation of PubMed, Unigene, and SwissProt databases. For each category, listed at the right, the number of genes whose expression increased or decreased in response to LPS is plotted.

group. In contrast, genes with functions in intermediary metabolism make up half of the set whose expression is suppressed by LPS, but only one-tenth of the induced group (FIG. 1).

EFFECTS ON GENES OF LIPID AND CHOLESTEROL METABOLISM

At least 50 effectors of lipid metabolism were observed to undergo significant alterations in expression. Fatty acid synthesis, phospholipid synthesis, fatty acid oxidation, bile acid synthesis, and later stages of cholesterol synthesis were coordinately down-regulated. Although the APR is accompanied by a transient increase in the level of LDL cholesterol,[10] we observed coordinate decreases in the levels of mRNA for cholesterol biosynthetic enzymes downstream of mevalonate. Our observations support the view that the inflammatory increase in serum cholesterol reflects diminished cholesterol uptake and decreased conversion of cholesterol to bile acids. LDL receptor mRNA was reduced at least 4-fold in response to LPS; a reduction in LDL receptor expression is predicted to result in diminished serum cholesterol clearance.[11] The expression of enzymes responsible for the synthesis of bile acids was also reduced, consistent with a decrease in the consumption of cholesterol that could also contribute to increased serum cholesterol levels.[12]

An elevation in serum triglyceride levels is characteristic of the APR.[5] Our observations support decreased clearance as an underlying mechanism, as mRNA for lipoprotein lipase (*lpl*), which plays a critical role in the cellular uptake of triglycerides, was reduced to undetectable levels by 6 h after LPS administration; reductions in lipoprotein lipase levels during an APR have been previously documented.[13]

AN INTERFACE BETWEEN INNATE AND ADAPTIVE IMMUNITY

At least 58 genes with direct roles in host defense or immunity showed significant changes in expression—usually increased—in response to LPS. Several classic APR markers were classed in this group, including serum amyloid A proteins 1 through 4 and other lipocalins. In livers of LPS-treated mice, we observed coordinate induction of genes whose products act in concert to present endogenous antigens. This wave of increased expression, sustained 3–12 h after LPS administration, included transcripts for all major components of the MHC class-I-specific antigen presentation pathway. Taken together, our results suggest that mobilization of antigen presentation by hepatocytes through MHC class I is an intrinsic feature of the APR.

REGULATION OF SYSTEMIC INFLAMMATION BY STAT3β ISOFORMS

Several cytokines, including IL-1β, tumor necrosis factor α (TNFα), and IL-6, are produced at a site of local inflammation and act on hepatocytes to stimulate or suppress the synthesis of acute-phase proteins.[3] IL-6 exerts its effects on APR genes principally through the latent transcription factor signal transducer and activator of transcription 3 (Stat3).[14–17]

Alternative splicing of the gene for Stat3 produces two isoforms: Stat3α and a dominant-negative variant, Stat3β. Stat3β-deficient mice were generated by gene-targeted ablation of a cryptic splice-acceptor site in exon 23 of the *Stat3*β gene.[18] Despite intact expression and phosphorylation of Stat3α, overall Stat3 activity was impaired in *Stat3*$\beta^{-/-}$ cells. Global comparison of transcription in *Stat3*$\beta^{+/+}$ and *Stat3*$\beta^{-/-}$ cells revealed stable differences. Stat3β-deficient mice exhibit diminished recovery from endotoxic shock.[18] Stat3β-deficient mice that had died at about 96 h after administration of LPS consistently exhibited severe acute tubular necrosis (ATN), suggesting that they may have suffered a more severe or protracted period of shock than their wild-type littermates. Activated macrophages from Stat3β-deficient mice and wild-type littermates produced similar levels of TNFα, IL-1β, IL-6, and IL-10 in response to LPS, suggesting that the detection of LPS (and the subsequent production of proinflammatory cytokines by macrophages) is intact in the absence of Stat3β. We therefore considered that some detrimental effects of bacterial endotoxin might be sustained by the hepatic APR itself, and that Stat3β might play a role in opposing these effects.

The effect of Stat3β-deficiency on the hepatic APR was examined using oligonucleotide microarrays. Of 776 LPS-responsive genes, 128 were differentially expressed in wild-type and Stat3β-deficient mice ($P < 0.01$). A striking bias in the expression of these genes was observed, with the overwhelming majority (124/128) exhibiting increased accumulation in the Stat3β-deficient animals, relative to wild-type. In most instances, the transcripts were overexpressed at baseline and overinduced by LPS. The bias toward overexpression of LPS-responsive genes in the mutant mice supports the interpretation that Stat3β serves negative regulatory functions in the APR. Consistent with this proposal, the hepatic response to endotoxin in wild-type mice is accompanied by a transient increase in the ratio of Stat3β to Stat3α. Taken together, our observations[18] indicate that Stat3β negatively regulates a subset of LPS-responsive genes in liver, revealing a critical role for Stat3β in the control of systemic inflammation.

ACKNOWLEDGMENTS

This work was supported by the Howard Hughes Medical Institute. Microarray hybridization and scanning were performed by the Center for Cancer Research–Howard Hughes Medical Institute Biopolymers Laboratory at M.I.T.

REFERENCES

1. MEDZHITOV, R. 2001. Toll-like receptors and innate immunity. Nat. Rev. Immunol. **1:** 135–145.
2. BAYNE, C.J. & L. GERWICK. 2001. The acute phase response and innate immunity of fish. Dev. Comp. Immunol. **25:** 725–743.
3. GABAY, C. & I. KUSHNER. 1999. Acute-phase proteins and other systemic responses to inflammation. N. Engl. J. Med. **340:** 448–454.
4. RAMADORI, G. & B. CHRIST. 1999. Cytokines and the hepatic acute-phase response. Semin. Liver Dis. **19:** 141–155.
5. KHOVIDHUNKIT, W., R.A. MEMON, K.R. FEINGOLD & C. GRUNFELD. 2000. Infection and inflammation-induced proatherogenic changes of lipoproteins. J. Infect. Dis. **181:** S462–S472.

6. DHAINAUT, J.F., N. MARIN, A. MIGNON & C. VINSONNEAU. 2001. Hepatic response to sepsis: interaction between coagulation and inflammatory processes. Crit. Care Med. **29:** S42–S47.
7. OPAL, S.M. 2000. Phylogenetic and functional relationships between coagulation and the innate immune response. Crit. Care Med. **28:** S77–S80.
8. ROSS, R. 1999. Atherosclerosis—an inflammatory disease. N. Engl. J. Med. **340:** 115–126.
9. YOO, J.-Y. & S. DESIDERIO. 2003. Innate and acquired immunity intersect in a global view of the acute-phase response. Proc. Natl. Acad. Sci. USA **100:** 1157–1162.
10. FEINGOLD, K.R., I. HARDARDOTTIR, R. MEMON, *et al.* 1993. Effect of endotoxin on cholesterol biosynthesis and distribution in serum lipoproteins in Syrian hamsters. J. Lipid Res. **34:** 2147–2158.
11. GOLDSTEIN, J.L., H.H. HOBBS & M.S. BROWN. 1995. *In* The Metabolic and Molecular Bases of Inherited Disease, Vol. 2. C.R. Scriver, A.L. Beaudet, W.S. Sly, and D. Valle, Eds.: 1981–2030. McGraw Hill. New York.
12. MEMON, R.A., A.H. MOSER, J.K. SHIGENAGA, *et al.* 2001. In vivo and in vitro regulation of sterol 27-hydroxylase in the liver during the acute phase response. J. Biol. Chem. **276:** 30118–30126.
13. FEINGOLD, K.R., R.A. MEMON, A.H. MOSER, *et al.* 1999. Endotoxin and interleukin-1 decrease hepatic lipase mRNA levels. Atherosclerosis **142:** 379–387.
14. AKIRA, S., Y. NISHIO, M. INOUE, *et al.* 1994. Molecular cloning of APRF, a novel IFN-stimulated gene factor 3 p91-related transcription factor involved in the gp130-mediated signaling pathway. Cell **77:** 63–71.
15. ZHONG, Z., Z. WEN & J.E. DARNELL, JR. 1994. Stat3: a STAT family member activated by tyrosine phosphorylation in response to epidermal growth factor and interleukin-6. Science **264:** 95–98.
16. IHLE, J.N., W. THIERFELDER, S. TEGLUND, *et al.* 1998. Signaling by the cytokine receptor superfamily. Ann. N.Y. Acad. Sci. **865:** 1–9.
17. SCHINDLER, C. & J.E. DARNELL, JR. 1995. Transcriptional responses to polypeptide ligands: the JAK-STAT pathway. Annu. Rev. Biochem. **64:** 621–651.
18. YOO, J.-Y., D.L. HUSO, D. NATHANS & S. DESIDERIO. 2002. Specific ablation of Stat3beta distorts the pattern of Stat3-responsive gene expression and impairs recovery from endotoxic shock. Cell **108:** 331–344.

Ii-CS on Gastric Epithelial Cells Interacts with CD44 on T Cells and Induces Their Proliferation

C.A. BARRERA,[a] T. CHAN,[a] S.E. CROWE,[b] P.B. ERNST,[b] AND V.E. REYES[a]

[a]*University of Texas Medical Branch, Galveston, Texas, USA*

[b]*University of Virginia, Charlottesville, Virginia, USA*

KEYWORDS: gastric epithelial cell; *H. pylori*; CD44

Helicobacter pylori is an important pathogen that is implicated in peptic ulceration and gastric cancer. The infected gastric mucosa has a substantial increase of $CD4^+$ T cells with markers of activation, including CD44. Although human gastric epithelial cells express class II major histocompatibility complex (MHC) molecules, their role in activating these T cells is undefined. Our studies have documented that gastric epithelial cells represent a novel antigen-presenting cell (APC) phenotype because they express elements required during antigen processing and presentation.[1–3] These cells express on the surface the class II MHC associated invariant chain (Ii). Since the interaction between CD44 and the chondroitin sulfate form of Ii (Ii-CS) on conventional APCs leads to enhanced $CD4^+$ T cell proliferation,[4] we tested the hypothesis that human gastric epithelial cells express Ii-CS that binds to CD44 on T cells, and further implicates the epithelium in mucosal immunity to *H. pylori*.

Gastric epithelial cells were found to express Ii-CS by immunoprecipitation, from $^{35}SO_4$-labeled cells, as a heterogeneous band in the range of 40–120 kDa on B cells and 69–190 kDa on Kato III cells (FIG. 1). The size differences are due to variations in the glycosylation within those cell types. To examine whether Ii-CS from gastric epithelial cells acts as co-receptor for CD44 on T cells, we performed binding studies in a competition-binding assay. A recombinant soluble form of CD44 (CD44HIg) was immobilized on 96-well plates. Affinity-purified Ii-CS from $^{35}SO_4$-labeled Kato III cells was serially diluted in binding buffer and added to the wells. The samples were added in triplicate to CD44HIg-coated wells and incubated for 4 h at 2°C. The bound proteins were eluted with SDS sample buffer and counted in a scintillation counter. Ii-CS bound in a dose-dependent fashion to CD44HIg-coated wells. The specificity of the binding was confirmed by blocking with cold Ii-CS or with anti-CD44. To confirm that Ii-CS from gastric epithelial cells mediates the ad-

Address for correspondence: V.E. Reyes, Children's Hospital, UTMB, 301 University Boulevard, Rm. 2.300, Galveston, TX 77555-0366. Voice: 409-772-3824; fax: 409-772-1761.
Vreyes@utmb.edu

Ann. N.Y. Acad. Sci. 987: 285–287 (2003). © 2003 New York Academy of Sciences.

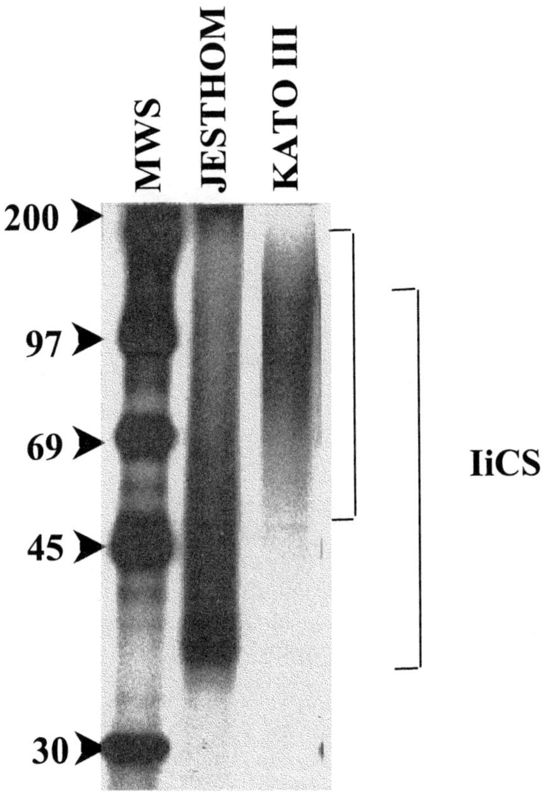

FIGURE 1. Characterization of Ii-CS in gastric epithelial cells. Ii-CS was enriched through a DEAE-Sephacel column; Ii-CS was then run on SDS/PAGE gel. One-dimensional gel electrophoresis showed expression of Ii-CS in KATO III gastric epithelial cells and in control Jesthom B cell lines.

herence of $CD44^+$ T cells, we determined the binding of $CD44^+$ T cells as well as $CD44^-$ T cells to wells coated with Ii-CS. ^{51}Cr-labeled T cells were then added to the coated wells in triplicate. After lysis of the bound cells and determination of the recovered γ counts, it was noted that $CD44^+$ T cells bound in a dose-dependent manner to the Ii-CS-containing wells, while the $CD44^-$ T cells did not bind. The role of the Ii-CS/CD44 interaction on T cell proliferation was assessed by [^{3}H]thymidine uptake, which showed that Ii-CS from Kato III gastric epithelial cells stimulated the proliferation of $CD44^+$ T cells, but not the $CD44^-$ T cells, and was blocked by soluble CD44HIg (FIG. 2).

The results further support the role of the gastric epithelial cell as a novel APC and specifically suggest that the interaction of Ii-CS and CD44 at the mucosal level may play a role in the modulation of gastric T cell functions. These studies enhance our understanding of cell-mediated immunity in the gastric mucosa in response to *H. pylori* infection.

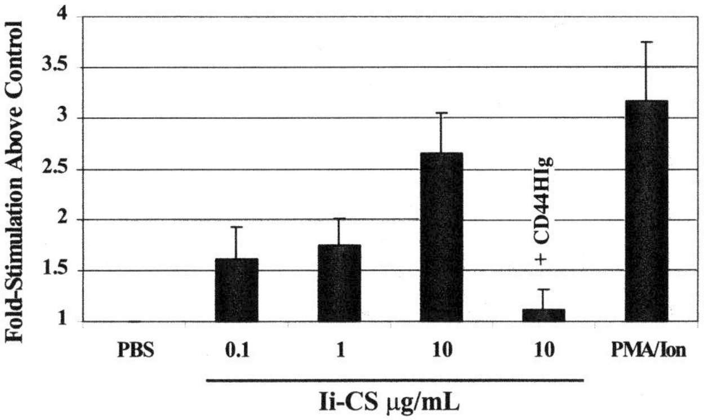

FIGURE 2. Proliferation of CD4$^+$ CD44$^+$ T cells induced by Ii-CS. CEM/C7 CD4$^+$ CD44$^+$ T cells proliferated in a dose-dependent manner to Ii-CS coated to microtiter wells, and the response was blocked by soluble CD44HIg. Proliferation was assessed by thymidine incorporation.

ACKNOWLEDGMENTS

This work was supported by grant DK 50669 from the National Institutes of Health.

REFERENCES

1. YE, G., C. BARRERA, X.J. FAN, *et al.* 1997. Expression of B7-1 and B7-2 costimulatory molecules by human gastric epithelial cells: potential role in CD4$^+$ T cell activation during Helicobacter pylori infection. Clin. Invest. **99:** 1628–1636.
2. BARRERA, C., G. YE, R. ESPEJO, *et al.* 2001. Expression of cathepsins B, L, S, and D by gastric epithelial cells implicates them as antigen presenting cells in local immune responses. Hum. Immunol. **62:** 1081–1091.
3. BARRERA, C., R. ESPEJO & V.E. REYES. 2002. Differential glycosylation of MHC class II molecules on gastric epithelial cells: implications in local immune responses. Hum. Immunol. **63:** 384–393.
4. NAUJOKAS, M.F., M. MORIN, M.S. ANDERSON, *et al.* 1993. The chondroitin sulfate form of invariant chain can enhance stimulation of T cell responses through interaction with CD44. Cell **74:** 257–268.

Tracking CD40 Signaling during Normal Germinal Center Development by Gene Expression Profiling

KATIA BASSO,[a] ULF KLEIN,[a] HUIFENG NIU,[a] GUSTAVO A. STOLOVITZKY,[b] YUHAI TU,[b] ANDREA CALIFANO,[c] GIORGIO CATTORETTI,[a] AND RICCARDO DALLA-FAVERA[a]

[a]*Institute for Cancer Genetics, Columbia University, New York, New York 10032, USA*

[b]*IBM T.J. Watson Research Center, Yorktown Heights, New York 10598, USA*

[c]*First Genetic Trust, Inc., Lyndhurst, New Jersey 07071, USA*

KEYWORDS: CD40 signaling; germinal center; DNA microarray

Signaling through the CD40 receptor plays an important role in multiple events in T-cell dependent immune responses, including germinal center (GC) and memory B-cell formation and immunoglobulin isotype switching.[1] CD40 signaling occurs in B cells following the interaction between the CD40 receptor on B cells and its ligand (CD40L) on T cells. The NF-κB transcription complex is the main nuclear mediator of CD40 signaling and induces the transcription of CD40 target genes. While the CD40–CD40L interaction has been extensively studied in the past years, it is not yet fully understood at which stages of B cell development signaling through CD40 occurs. In particular, while it has been shown that CD40 signaling is required for GC development, it remains to be elucidated whether this signal occurs continuously or at particular stages of the development of this structure. To address these questions, we first used gene expression profiling to identify the genes whose expression is induced or repressed upon CD40 signaling of B cells *in vitro* (CD40 signature). This signature was then used to track CD40 signaling in human B cells corresponding to the main developmental stages of GC—namely, naïve, centroblasts (CB), centrocytes (CC), and memory B cells.

CD40 signaling was induced in a Burkitt lymphoma cell line (Ramos) by co-cultivation with fibroblasts that were stably transfected to express the CD40 ligand. After co-cultivation for 24 h, B cells were purified from fibroblasts by magnetic cell separation using anti-CD19 antibodies. The expression of the known CD40-responsive genes bfl-1, BCL-6, CD95/FAS, and CD80 was used as internal

Address for correspondence: Dr. Riccardo Dalla-Favera, Institute for Cancer Genetics, Columbia University, 1150 St. Nicholas Avenue, New York, NY 10032. Voice: 212-851-5274; fax: 212-851-5256.

rd10@columbia.edu

Ann. N.Y. Acad. Sci. 987: 288–290 (2003). © 2003 New York Academy of Sciences.

control for CD40 signaling, and verified by Northern blot or flow cytometric analysis. Labeled cRNAs were then generated from total RNA isolated from six samples each of CD40-stimulated and unstimulated Ramos cells, and hybridized to Affymetrix U95A Gene Chips containing oligonucleotides corresponding to ~12,000 genes. The gene expression data were analyzed by the Genes@Work software platform that is based on the pattern discovery algorithm SPLASH.[2,3] Genes differentially expressed between the two groups of CD40-stimulated and unstimulated samples were identified by supervised cluster analysis using highly stringent criteria (requiring consistency in the expression values across all six samples of each group).

The resulting CD40 gene expression signature comprised 100 up-regulated and 58 down-regulated genes. The signature contained several genes involved in apoptosis: anti-apoptotic genes were up-regulated, while pro-apoptotic genes were down-regulated. RNAs for the adhesion molecules ICAM-1 and LFA3, as well as MHC class II genes, were strongly up-regulated consistent with the induction of these genes in T cell–B cell interaction. Several genes encoding chemokine receptors were up-regulated, including CXCR5 and CCR7, that are important for lymphocyte homing in secondary lymphoid tissues. The mRNAs encoding lymphotoxin-α and -β, which play a role in dendritic cell recruitment and GC formation, were up-regulated following CD40 stimulation. The CD40 receptor gene was also induced as well as the genes coding for proteins involved in CD40 signal transduction (TRAF3, PI-3 kinase, Lyn tyrosine kinase, NF-κB subunits p50 and NF-κB2). Overall, the most relevant aspects of the identified CD40 signature were the up-regulation of anti-apoptotic and adhesion-related genes.

The CD40 gene expression signature was then used to track the activity of CD40 signaling in normal B cell subsets using the gene expression profiles generated from pre-GC naïve B cells, GC B-cell populations (centroblasts and centrocytes), and post-GC memory B cells.[4,5] Unsupervised hierarchical clustering showed that CB and CC lacked CD40 signaling, while 71 and 74 of the 100 genes up-regulated in CD40-stimulated cells were found up-regulated also in naïve and memory B cells, respectively, suggesting that these cells may be subjected to CD40 signaling *in vivo*. To further quantify the presence of the CD40 signature, we used a binary scoring approach showing that the CD40 signature was detectable in naïve and memory B cells at P values of 0.15×10^{-11} and 0.72×10^{-13}, respectively. These results suggest that CD40 signaling is not activated in most of the GC B cells, while naïve and memory B cells are subjected to CD40 stimulation or to other signals that can induce a similar pattern of gene expression.

To validate the gene expression results, we investigated the status of the NF-κB transcriptional complex, the main nuclear mediator of CD40 signaling, in the normal B cell subpopulations. The presence of cytoplasmic NF-κB molecules would imply inactivity of NF-κB and the absence of CD40 signaling, while nuclear NF-κB would imply activation of NF-κB by CD40 or other signals. Immunohistochemical analysis of tonsil sections using anti-cRel and anti-p65 antibodies showed that NF-κB was found to localize in the cytoplasm in the majority of GC B cells, while a subset of GC centrocytes in the light zone showed nuclear staining for c-Rel and p65. In addition, nuclear NF-κB was present in a large fraction of B cells in the marginal zone and in the tonsillar subepithelium where memory B cells localize. These results were confirmed by electrophoretic mobility gel shift analysis (EMSA), which showed the presence of DNA-binding NF-κB complex as in memory B cells, but not in CB and naïve B cells. Together, these findings suggest that CD40 signaling does not occur in CB, while it can be tracked in naïve and memory B cells, as well as in a small subset of GC centrocytes that might be committed to differentiate into memory B cells.

In conclusion, these results have implications for the role of CD40 signaling during GC development. Although *in vitro* studies have demonstrated that CD40 activation rescues isolated GC B cells from apoptosis and induces naïve B cells to differentiate into memory B cells, the data herein suggest that CD40 signaling is not active in the majority of GC B cells, including most CB and CC *in vivo.* Conversely, evidence of CD40 signaling (nuclear NF-κB) and response (specific gene expression profile) is found in naïve B cells and a subset of CC as well as in memory B cells *in vivo.* Overall, these data support a role of CD40 in the final steps (CC to memory) and perhaps the initial ones (naïve to CB) of the GC reaction, but not in its proliferative phase represented by CB.

REFERENCES

1. VAN KOOTEN, C. & J. BANCHEREAU. 2000. CD40-CD40 ligand. J. Leukocyte Biol. **67:** 2–17.
2. CALIFANO, A., G. STOLOVITZKY & Y. TU. 2000. Analysis of gene expression microarrays for phenotype classification. Proc. Int. Conf. Intell. Syst. Mol. Biol. **8:** 75–85.
3. KLEIN, U., Y. TU, G. STOLOVITZKY, *et al.* 2001. Gene expression profiling of B-cell chronic lymphocytic leukemia reveals a homogeneous phenotype related to memory B-cells. J. Exp. Med. **194:** 1625–1638.
4. KLEIN, U. *et al.* 2003. Gene expression dynamics during germinal center transit in B cells. Ann. N.Y. Acad. Sci. **987:** this volume.
5. KLEIN, R. *et al.* 2003. Transcriptional analysis of the B cell germinal center reaction. Proc. Natl. Acad. Sci. USA **100:** 2639–2644.

Elevated Levels of Soluble CD27 in Sera of Primary Sjögren's Syndrome Patients Correlate Positively with Serum Concentration of IgG: A Result of Abnormal B Cell Differentiation?

JANNE Ø. BOHNHORST,[a] JØRN E. THOEN,[b] ROLAND JONSSON,[c] JACOB B. NATVIG,[a] AND KEITH M. THOMPSON[a]

[a]*Institute of Immunology and* [b]*Centre for Rheumatic Disease, Rikshospitalet University Hospital, University of Oslo, 0027 Oslo, Norway*

[c]*Broegelmann Research Laboratory, University of Bergen, 5021 Bergen, Norway*

KEYWORDS: Sjögren's syndrome; IgG; soluble CD27

Primary Sjögren's syndrome (pSS) is an autoimmune rheumatic disease with demonstrably strong abnormalities in B cell development and immunoglobulin production. It is characterized by a chronic inflammation of affected salivary glands, autoantibody production, and hypergammaglobulinemia. We have previously shown that patients with pSS have a significantly reduced percentage of CD27$^+$ (memory) B cells in peripheral blood (PB) (15 ± 2%) compared to healthy donors (31 ± 5%, $P = 0.005$).[1] CD27 is a marker for somatically mutated B cells and is expressed on PB memory B cells.[2] CD27 can also occur in a soluble form (sCD27, 32-kDa), which is most likely generated by proteolytic cleavage of the membrane-bound CD27 molecule.[3] Although elevated levels of sCD27 have been related to activated T cells, CD27 is also expressed on memory B cells and is highly expressed on plasma cells, and thus these are potentially significant sources of sCD27.[4] It is possible that the low percentage of CD27$^+$ PB B cells in pSS may merely be due to a loss of surface CD27. To determine if CD27 expression distinguishes between memory and naive B cells in pSS (as in healthy donors), V_H region genes of CD27$^+$ and CD27$^-$ sorted PB B cells from pSS patients were sequenced. The low percentage of CD27$^+$ PB B cells in pSS may be caused by a bias towards plasma cell differentiation. CD27 is highly expressed on plasma cells, and consequently they may be the major source of sCD27 in serum. Thus, we compared serum levels of IgG with that of sCD27 in patients with pSS.

Address for correspondence: Janne O. Bohnhorst, Institute of Immunology, Rikshospitalet University Hospital, University of Oslo, 0027 Oslo, Norway. Voice: +472307000.
janne.bohnhorst@labmed.uio.no

Ann. N.Y. Acad. Sci. 987: 291–294 (2003). © 2003 New York Academy of Sciences.

In the present study, we demonstrate that the mutation frequencies of $CD27^+$ and $CD27^-$ PB B cells in pSS correspond to those of healthy donors. In addition, serum levels of sCD27 are significantly increased in pSS patients compared to healthy donors, and in pSS the serum concentration of sCD27 correlates with the serum level of IgG. Our data indicate that in pSS, there is an abnormal differentiation of B cells towards plasma cells resulting in a depression of the circulating memory B cell pool and the release of significant amounts of sCD27.

MATERIALS AND METHODS

PB samples were collected from three patients with pSS for sorting and sequencing of $CD27^-$ and $CD27^+$ ($CD19^+$) B cells. Sera were collected from 24 pSS patients for analyses of sCD27 and IgG concentrations, and from 11 healthy donors for sCD27 analysis.

Sequencing of V_H genes. Mononuclear cells from pSS patients were isolated, and T cells were depleted by anti-CD3-coated Dynabeads (Dynal). Cells were stained for CD19 and CD27 and sorted into $CD27^+$ and $CD27^-$ B cells by FACSVantage (B&D). V_H genes from the sorted cells were amplified by PCR, cloned by using a TOPO cloning kit (Invitrogen), and sequenced by BigDye ready reaction kit and a DNA sequencer (PE Biosystems).

sCD27. The serum concentrations of sCD27 were measured by PeliKine compact human-soluble CD27 ELISA kit (Research Diagnostics).

Serum Ig. Levels of IgG, IgM, IgA, and IgG subclasses were measured by a standard nephelometry assay.

Statistics. Comparison of the serum concentrations of sCD27 between healthy donors and pSS patients was performed by an unpaired two-tailed t test, and the correlations between sCD27 and major Ig isotypes and the IgG subclasses were performed by multiple regression analysis.

RESULTS AND DISCUSSION

The serum concentrations of sCD27 were significantly higher in pSS patients (376.5 ± 39.7 U/mL) compared to healthy donors (135.8 ± 12.2 U/mL, $P < 0.0001$) (FIG. 1A). Analyses of V_H region genes from sorted $CD27^+$ and $CD27^-$ B cells from three pSS patients demonstrated that all the sequences obtained from the $CD27^+$ PB B cells were highly mutated (mean mutation rate: 4.5%). In contrast, there was an almost complete absence of mutations in the $CD27^-$ B cells (mean mutation rate: 0.7%) (FIG. 2). These data confirmed that the $CD27^+$ population in pSS corresponds to the somatically mutated (memory) B cell compartment, as in healthy donors (mean mutation rate: 3.1%).[5] Thus, abnormal absence of CD27 from the membrane of memory B cells is likely not to explain the reduced percentage of $CD27^+$ B cells in PB of pSS. The serum IgG concentrations were above normal in the majority of the pSS patients, and the IgG concentration correlated positively with sCD27 ($r = 0.6797$, $P = 0.0007$) (FIG. 1B), suggesting that plasma cells or B cells undergoing plasma cell differentiation may be responsible for the production of sCD27 in addition to IgG. Interestingly, significant positive correlations were seen between sCD27

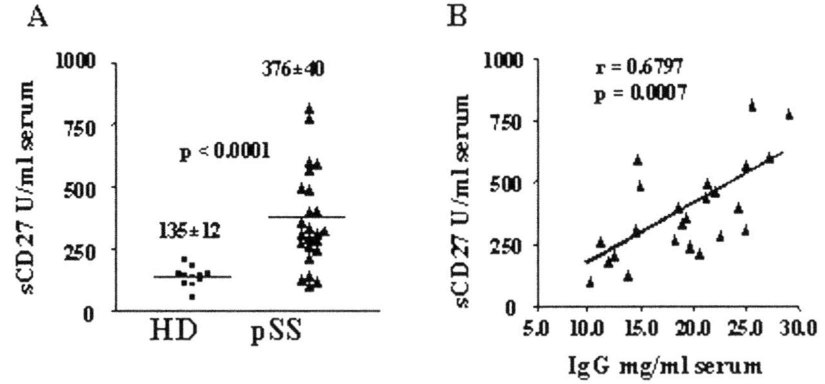

FIGURE 1. (**A**) Serum concentration of sCD27 in patients with pSS and healthy donors (HD). The concentrations of sCD27 (U/mL serum) for 24 pSS patients (*squares*) and 11 healthy donors (*triangles*) are illustrated in this figure. The lines represent the mean concentrations of sCD27 in each group. The mean of sCD27 for pSS was significantly higher compared to healthy donors ($P < 0.0001$). (**B**) Correlation analysis of IgG and sCD27 serum concentration in pSS patients. The positive correlation between IgG and sCD27 was significant, with $P = 0.0007$ and $r = 0.6797$.

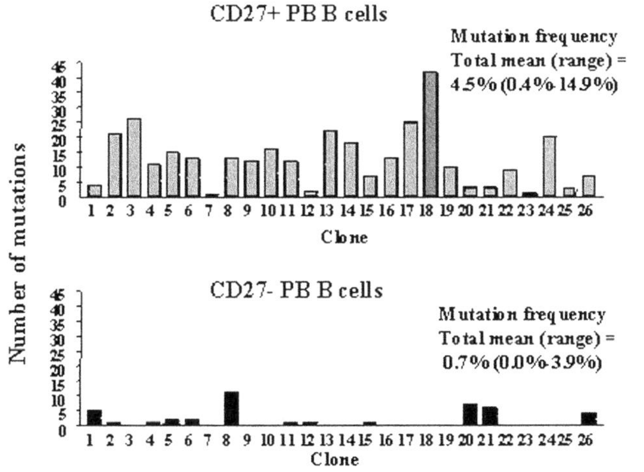

FIGURE 2. Somatic mutation in V^H region genes from $CD27^+$ and $CD27^-$PB ($CD19^+$) B cells from three patients with pSS. *Upper panel*: A histogram illustrating the number of mutations in each clone from $CD27^+$ PB B cells. *Lower panel*: A histogram illustrating the number of mutations in each clone from $CD27^-$ PB B cells. The replacement-to-silent mutation (R:S) ratio in the complementarity-determining region (CDR) 1 + 2 was 3.38 (81:24) of $CD27^+$ B cells and 2.25 (9:4) for the $CD27^-$ PB B cells.

and IgG1 ($r = 0.6097$, $P = 0.0496$) and IgG3 ($r = 0.4438$, $P = 0.0091$), but there was no correlation between IgG2 and IgG4 (data not shown). Increased frequency of cells producing IL-10 has been found in pSS patients, where IL-10 may function as a switching factor that stimulates IgG1 and IgG3 production.[6] In addition, IL-10 promotes plasma cell differentiation,[7] and thus it is likely to be indirectly responsible for the correlation between IgG1, IgG3, and sCD27. There were no significant correlations between serum levels of sCD27 and IgM and IgA, and the concentration of both IgM (mean: 1.43 mg/mL) and IgA (mean: 2.56 mg/mL) were within the normal range (data not shown). By *in vitro* experiments, we have previously demonstrated that normal B cells cultured under conditions driving plasma cell differentiation release a substantial amount of sCD27 compared to *in vitro* activated T cells.[8] Since it has been established that B cells differentiate either to plasma cells or memory B cells,[9] our data indicate that in pSS, there is an abnormal differentiation of B cells to plasma cells resulting in a depression of the circulating memory B cell pool and the release of significant amounts of sCD27 and IgG.

REFERENCES

1. Bohnhorst, J., J.E. Thoen, J.B. Natvig & K.M. Thompson. 2001. Significantly depressed percentage of $CD27^+$ (memory) B cells among peripheral blood B cells in patients with primary Sjögren's syndrome. Scand. J. Immunol. **54:** 421–427.
2. Klein, U., K. Rajewsky & R. Kuppers. 1998. Human immunoglobulin $(Ig)M^+IgD^+$ peripheral blood B cells expressing the CD27 cell surface antigen carry somatically mutated variable region genes: CD27 as a general marker for somatically mutated (memory) B cells. J. Exp. Med. **188:** 1679–1689.
3. Loenen, W.A., E. de Vries, L.A. Gravestein, *et al.* 1992 The CD27 membrane receptor, a lymphocyte-specific member of the nerve growth factor receptor family, gives rise to a soluble form by protein processing that does not involve receptor endocytosis. Eur. J. Immunol. **22:** 447–455.
4. Odendahl, M., A. Jacobi, A. Hansen, *et al.* 2000. Disturbed peripheral B lymphocyte homeostasis in systemic lupus erythematosus. J. Immunol. **165:** 5970–5979.
5. Bohnhorst, J., M.B. Bjørgan, J.E. Thoen, *et al.* 2001. Bm1-Bm5 classification of peripheral blood B cells reveals circulating germinal center founder cells in healthy individuals and disturbance in the B cell subpopulations in patients with primary Sjögren's syndrome. J. Immunol. **167:** 3610–3618.
6. Briere, F., C. Servet-Delprat, J.M. Bridon, *et al.* 1994. Human interleukin-10 induces naive surface immunoglobulin D^+ ($sIgD^+$) B cells to produce IgG1 and IgG3. J. Exp. Med. **179:** 757–762.
7. Agematsu, K., H. Nagumo, Y. Aguchi, *et al.* 1998 Generation of plasma cells from peripheral blood memory B cells: synergistic effect of interleukin-10 and CD27/CD70 interaction. Blood **91:** 173–180.
8. Bohnhorst, J., M.B. Bjørgan, J.E. Thoen, *et al.* 2002. Abnormal B cell differentiation in primary Sjögren's syndrome results in a depressed percentage of circulating memory B cells and elevated levels of soluble CD27 that correlate with serum IgG concentration. Clin. Immunol. **103:** 79–88.
9. Morimoto, S., Y. Kanno, Y. Tanaka, *et al.* 2000. CD134L engagement enhances human B cell Ig production: CD154/CD40, CD70/CD27, and CD134/CD134L interactions coordinately regulate T cell-dependent B cell responses. J. Immunol. **164:** 4097–4104.

IL-1β and TNF-α Produce Divergent Acute Inflammatory and Skeletal Lesions in the Knees of Lewis Rats

GIUSEPPE CAMPAGNUOLO,[a,c] BRAD BOLON,[b,c] LI ZHU,[a] DIANE DURYEA,[b] ULRICH FEIGE,[a] AND DEBRA ZACK[a]

Departments of [a]Inflammation and [b]Pathology, Amgen Inc., Thousand Oaks, California 91320-1789, USA

KEYWORDS: pro-inflammatory; cytokine; osteoclast

INTRODUCTION

Arthritis is initiated and sustained by the release of myriad pro-inflammatory cytokines.[1] In patients with rheumatoid arthritis, the dominant pro-inflammatory cytokines are interleukin-1 (IL-1) and tumor necrosis factor-α (TNF-α).[2] Inhibition of IL-1β and/or TNF-α reduces the extent of inflammation in rheumatoid arthritis[3,4] and in animal models of arthritis.[5–7] Therefore, biologic agents that inhibit the action of these two cytokines are gaining rapid acceptance as early, aggressive treatments for rheumatoid arthritis.

IL-1β and TNF-α act synergistically to induce and maintain inflammation and skeletal erosion in arthritis.[1] The present study was designed to explore the early effects (48 h) by direct intra-articular injection of IL-1β or TNF-α. Our data show that both cytokines can initiate disease, but that IL-1β appears to play a more substantial role than TNF-α in the early stages of arthritis, particularly with respect to the induction of osseous lesions.

MATERIALS AND METHODS

Animals

Young adult, male Lewis rats (6–7 weeks and 175–200 g; Charles River, Wilmington, MA) were used. At necropsy, rats were sacrificed by carbon dioxide inhalation. These studies were conducted in accordance with federal animal care guidelines and were pre-approved by the Amgen Institutional Animal Care and Use Committee (AIACUC).

Address for correspondence: Dr. Debra Zack, Amgen, One Amgen Center Drive, M/S 29-M-B, Thousand Oaks, CA 91320-1799. Voice: 805-447-6684; fax: 805-480-1329.
dzack@amgen.com

[c]G.C. and B.B. are equal contributors.

Ann. N.Y. Acad. Sci. 987: 295–298 (2003). © 2003 New York Academy of Sciences.

TABLE 1. Semiquantitative grading criteria for histopathologic lesion scores[a]

Lesion/Grades	Location of assessment/Criteria for grades
Inflammation	Assessment made for intra-articular and peri-articular soft tissues
0	Normal
1	Few inflammatory cells
2	Mild inflammation
3	Moderate inflammation
4	Marked inflammation (generally diffuse)
Bone Erosions	Assessment made for subchondral bone on distal femur and proximal tibia
0	Normal
1	Minimal increase in the depth and/or number of resorption lacunae (scallops) in subchondral bone
2	Mild increase in the depth and/or number of resorption lacunae in subchondral bone
3	Moderate increase in the depth and/or number of resorption lacunae in subchondral bone
Osteoclasts	Assessment made for distal femur and proximal tibia
0	Normal
1	Minimal increase in osteoclasts (crescentic) lining subchondral bone
2	Mild increase in osteoclasts (oval < crescentic) lining subchondral bone
3	Moderate increase in osteoclasts (crescentic < oval) lining subchondral bone
4	Marked increase in osteoclasts (oval) lining subchondral bone

[a]Osteoclast morphology: crescentic = resting; oval = activated.

Cytokine Instillation

Each rat was anesthetized with isoflurane (3–5%) in oxygen (4 L/min), the left knee was shaved with electric clippers, and the skin was disinfected with isopropanol. A sterile 28-gauge needle was used to penetrate through the skin and subjacent infrapatellar ligament into the joint space of the left knee, and 50 µL of either IL-1β or TNF-α (0, 3, 10, or 30 µg of purified recombinant human protein in phosphate buffered saline; Amgen Inc., Thousand Oaks, CA) was injected.

Histopathologic Analysis

The left knee of each rat was removed, fixed in 70% ethanol, decalcified, trimmed in the frontal plane (to expose the center of the femorotibial joint, including cruciate ligaments and menisci), and processed into paraffin. Four-µm-thick sections were cut and stained to show osteoclasts (visualized using a monoclonal anti-cathepsin K antibody and indirect immunoperoxidase) as well as architectural features of the bone and soft tissues (via counterstaining with hematoxylin and eosin). Histopathologic lesions for each knee were scored using tiered, semi-quantitative grading criteria and a "blinded" analytical paradigm. Data were compared using the χ-square test (a conservative non-parametric method) using a P value ≤ 0.05 to define significant differences between groups.

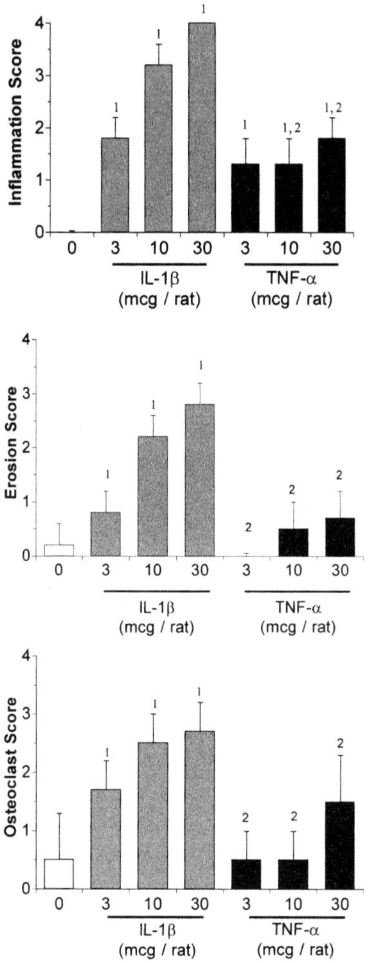

FIGURE 1. Intra-articular instillation of IL-1β or TNF-α (0, 3, 10, or 30 μg of purified protein) into the left knee of male Lewis rats (n = 6/group) induced significant, dose-dependent increases in inflammation, bone erosion, and osteoclasts after 48 h relative to normal animals (denoted by "1"). At each dose, the elevations produced by IL-1β were greater than those provided by TNF-α (significant differences are denoted by "2"). mcg = μg.

RESULTS

Instillation of either IL-1β or TNF-α into the femorotibial joint cavity resulted (after 48 h) in substantial and dose-dependent intra-articular and peri-articular inflammation as well as erosion of subchondral bone in association with local expansion of the osteoclast population (FIG. 1). Injection of IL-1β, however, significantly augmented erosion of subchondral bone and amplified subchondral osteoclasts, while introduction of TNF-α did not (FIG. 1). At each dose, the response for all three parameters elicited by IL-1β was greater than that obtained with TNF-α. A finding of particular note is that IL-1β injection resulted in substantially more activated osteoclasts (very large, cathepsin K-labeled cells) than did TNF-α.

DISCUSSION

Inflammation and skeletal destruction in arthritis are induced and maintained by the cooperative interaction between IL-1β and TNF-α. Numerous studies indicate that these two cytokines act synergistically to promote inflammation.[8–11] *In vivo*, IL-1 and TNF-α act synergistically to recruit leukocytes into rabbit joints.[12] In like manner, biomolecules that bind IL-1 or TNF-α, thereby blocking signal transduction through these two pathways, exhibit synergistic efficacy in decreasing the severity of inflammation and bone destruction in the Lewis rat model of adjuvant arthritis.[7]

We undertook the present work to compare the effects elicited by IL-1β and TNF-α in the earliest stages of joint inflammation (48 h). In particular, we sought evidence to select between two contending postulates: preferential control of inflammation by TNF-α and bone erosion by IL-1β,[13] or co-regulation of inflammation and skeletal destruction by the actions of both cytokines.[7] Our data show that IL-1β drives joint lesions to a substantially greater degree than does TNF-α. Intriguingly, IL-1β was a more potent stimulus for bone erosion and osteoclast expansion than was TNF-α. Further work with this intra-articular instillation model offers considerable promise for elucidating specific *in vivo* functions of IL-1β vs. TNF-α, as well as for comparing the effects of additional pro- and anti-inflammatory cytokines in the onset and progression of arthritis.

REFERENCES

1. BRENNAN, F.M. *et al.* 1998. Springer Sem. Immunopathol. **20:** 133–147.
2. BRENNAN, F.M. *et al.* 1991. Br. J. Rheumatol. **30:** 76–80.
3. BRESNIHAN, B. *et al.* 1998. Arthritis Rheum. **41:** 2196–2204.
4. RICHARD-MICELI, C. *et al.* 2001. BioDrugs **15:** 251–259.
5. JOOSTEN, L.A. *et al.* 1996. Arthritis Rheum. **39:** 797–809.
6. KUIPER, S. *et al.* 1998. Cytokine **10:** 690–702.
7. FEIGE, U. *et al.* 2000. Cell. Mol. Life Sci. **57:** 1457–1470.
8. CAMPBELL, I.K. *et al.* 1990. Arthritis Rheum. **33:** 542–552.
9. MEYER, F.A. *et al.* 1990. Arthritis Rheum. **33:** 1518–1525.
10. HARIGAI, M. *et al.* 1991. J. Clin. Lab. Immunol. **34:** 107–113.
11. RATHANASWAMI, P. *et al.* 1993. Arthritis Rheum. **36:** 1295–1304
12. HENDERSON, B. *et al.* 1989. Clin. Exp. Immunol. **75:** 306–310.
13. VAN DEN BERG, W.B. 1998. Springer Semin. Immunopathol. **20:** 149–164.

Correlation of Innate Immune Response with IgA against *Gardnerella vaginalis* Cytolysin in Women with Bacterial Vaginosis

SABINA CAUCI,[a] SECONDO GUASCHINO,[b] DOMENICO DE ALOYSIO,[c] SILVIA DRIUSSI,[a,d] DAVIDE DE SANTO,[b] PAOLA PENACCHIONI,[c] ALINE BELLONI,[a] PAOLO LANZAFAME,[e] AND FRANCO QUADRIFOGLIO[a]

[a]*Department of Biomedical Sciences and Technologies, School of Medicine, University of Udine, Udine, Italy*

[b]*Obstetric and Gynecologic Unit, Department of Reproductive and Development Sciences, IRCCS Burlo Garofolo Hospital, School of Medicine, University of Trieste, Trieste, Italy*

[c]*Department of Obstetrics and Gynecology, Sant'Orsola Hospital, School of Medicine, University of Bologna, Bologna, Italy*

[d]*Azienda Servizi Sanitari N. 4 Medio Friuli, Udine, Italy*

[e]*Microbiology Unit, Santa Maria della Misericordia Hospital, Udine, Italy*

KEYWORDS: **interleukin-8; interleukin-1β; IgA; *Gardnerella vaginalis;* hemolysin; bacterial vaginosis; vaginal flora; vaginal mucosa; mucosal immunity; innate immunity; local adaptive immunity**

INTRODUCTION

Bacterial vaginosis (BV) is the main vaginal disorder in non-pregnant and pregnant women and is associated with several adverse outcomes, such as increased susceptibility to HIV sexual transmission, upper genital tract infections, post-surgical infections, and adverse pregnancy outcomes.[1,2] BV is a complex polymicrobial alteration of the vaginal ecology, characterized by decreased lactobacilli flora and largely increased colonization of several facultative and strictly anaerobic microorganisms. *Gardnerella vaginalis* is almost always present. Thus far, the only vaginal-specific IgA response characterized in women with BV is the IgA against the hemolysin produced by *G. vaginalis* (anti-Gvh IgA).[3] The anti-Gvh IgA appears to be a critical host response in vaginal fluid of BV-positive women.[3–5] Interestingly, preliminary observations show that high anti-Gvh IgA levels are protective, whereas low levels of anti-Gvh IgA and high levels of microbial hydrolytic enzymes are cor-

Address for correspondence: Dr. Sabina Cauci, Dipartimento di Scienze e Tecnologie Biomediche, Facoltà di Medicina e Chirurgia, Piazzale Kolbe 4, 33100 Udine, Italy. Voice: ++39 0432.494312; fax: ++39 0432.494301.
scauci@mail.dstb.uniud.it

Ann. N.Y. Acad. Sci. 987: 299–301 (2003). © 2003 New York Academy of Sciences.

related with an increased risk of adverse pregnancy outcomes, including early preterm birth (<34 weeks' gestation).[6,7]

The present study was performed to assess the correlation between levels of neutrophils, interleukin (IL)-8, IL-1β, and anti-Gvh IgA in vaginal fluid of healthy and BV-positive women.

PATIENTS AND METHODS

Non-pregnant white women aged 18–45 years were recruited during routine gynecologic examinations to undergo the Papanicolau exam, which was administered in Italy from May 2001 to March 2002. Women were enrolled after informed consent according to local Ethics Committee. Inclusion and exclusion criteria were analogous to those previously adopted.[4] Anti-Gvh IgA, IL-8, IL-1β, and neutrophils were quantified in the vaginal fluid as previously described.[3–5]

Informed Consent

Appropriate informed consent was obtained, and clinical research was conducted in accordance with guidelines for human experimentation that had been adopted by the authors' institutions.

Statistical Analysis

Two-tailed significance of Spearman's ρ coefficient was reported to assess correlation. The Mann-Whitney U-test was used to compare factors levels between groups. A P value <0.05 was considered statistically significant.

RESULTS

Levels of IL-8 and of neutrophils were not statistically increased in 40 BV-positive women with respect to 40 healthy controls, whereas IL-1β levels were elevated 15-fold ($P < 0.001$), and anti-Gvh IgA levels were twofold higher ($P < 0.05$).

The number of neutrophils was strongly associated with IL-8 and IL-1β in all 80 enrolled women ($P < 0.001$), in healthy controls ($P < 0.001$), and in BV-positive women ($P < 0.001$).

Overall, the women's vaginal IL-8 and IL-1β levels were positively correlated ($P < 0.001$). In the group of women positive for BV, the level of anti-Gvh IgA was positively associated with the vaginal IL-8 and IL-1β levels and neutrophil counts ($P < 0.001$).

DISCUSSION

BV status causes a dramatic increase of IL-1β levels (15-fold). This finding shows that the innate immune system is strongly reacting to abnormal microbial colonization, although most BV-positive women do not show any signs of inflammation. In fact, IL-8 levels and neutrophil counts are not statistically increased in BV-positive women. These last findings suggest that the BV microbial consortium produce virulence factors specific to inhibit IL-8 rather than IL-1β. The scarcity of IL-8 may be responsible for the clinically observed absence of inflammatory symptoms and poor counts of vaginal leukocytes in most women with BV. Local adaptive immune levels were correlated with vaginal innate immune factors. Further studies on innate and acquired immunity in BV-positive women are ongoing.

ACKNOWLEDGMENTS

This research has been carried out with the financial support of the "Ministero Istruzione Ricerca e Università" of Italy (COFIN grant); the Regione Friuli Venezia Giulia (year 2000 grant); the University of Udine; the University of Trieste; the IRCCS Burlo Garofolo Hospital, Trieste, Italy; and the University of Bologna, Bologna, Italy.

REFERENCES

1. SEWANKAMBO, N., R.H. GRAY, M.J. WAWER, *et al.* 1997. HIV-1 infection associated with abnormal vaginal flora morphology and bacterial vaginosis. Lancet **350:** 546–550.
2. ESCHENBACH, D.A. 1993. History and review of bacterial vaginosis. Am. J. Obstet. Gynecol. **169:** 441–445.
3. CAUCI, S., F. SCRIMIN, S. DRIUSSI, *et al.* 1996. Specific immune response against *Gardnerella vaginalis* hemolysin in patients with bacterial vaginosis. Am. J. Obstet. Gynecol. **175:** 1601–1605.
4. CAUCI, S., S. DRIUSSI, S. GUASCHINO, *et al.* 2002. Correlation of local interleukin-1β levels with specific IgA response against *Gardnerella vaginalis* cytolysin in women with bacterial vaginosis. Am. J. Reprod. Immunol. **47:** 1–8.
5. CAUCI, S., S. GUASCHINO, S. DRIUSSI, *et al.* 2002. Correlation of local interleukin-8 with IgA against *Gardnerella vaginalis* hemolysin and with prolidase and sialidase levels in women with bacterial vaginosis. J. Infect. Dis. **185:** 1614–1620.
6. CAUCI, S., P. THORSEN, D.E. SCHENDEL, *et al.* 2000. IgA mucosal response, sialidase and prolidase activities as markers for low birth weight in women with BV [abstract no. 28]. BV 2000 Third International Meeting on Bacterial Vaginosis. Ystad, Sweden, Sept. 14–17.
7. HITTI, J., S. CAUCI, C. NOONAN, *et al.* 2001. Vaginal hydrolytic enzyme activity, bacterial vaginosis and risk of early preterm birth among women in preterm labor [abstract]. Am. J. Obstet. Gynecol. **185:** S193.

B Cell Chronic Lymphocytic Leukemia

Distinct Phenotypic Features and Replicative Histories among Subgroups

RAJENDRA N. DAMLE, FABIO GHIOTTO, ANGELO VALETTO, EMILIA ALBESIANO, STEVEN L. ALLEN, PHILIP SCHULMAN, VINCENT P. VINCIGUERRA, KANTI R. RAI, MANLIO FERRARINI,[a] AND NICHOLAS CHIORAZZI

North Shore–LIJ Research Institute, Manhasset, New York, USA

[a]*Istituto per la Ricerca sul Cancro, Genoa, Italy*

KEYWORDS: B-CLL subgroups; phenotype; telomere; telomerase

B cell chronic lymphocytic leukemia (B-CLL) results from the clonal accumulation of $CD5^+$ B lymphocytes. We and others have demonstrated that B-CLL can be divided into at least two clinically distinct subgroups based on the presence of mutations in genes encoding the immunoglobulin V region (Ig V) and on the percentage of the leukemic cells expressing the marker CD38.[1–3] Based on these observations, we speculated that cases with mutated Ig V genes probably arose from the transformation of post-germinal center (GC) memory cells. However, the origin of clonal B-CLL cells with unmutated Ig V genes is unclear. Therefore, it is plausible that these cases may have derived from transformed cells that had not undergone a classical GC reaction.

To characterize the state of activation/maturation of B-CLL cells in detail, peripheral blood mononuclear cells from B-CLL cases and age-matched normal donors were subjected to three-color immunofluorescent staining and analyzed by flow cytometry. A significantly higher percentage of B-CLL cells expressed activation markers such as CD23, CD25, and CD69 compared with $CD5^+$ B cells from normal age-matched controls. When phenotypes of B-CLL cases grouped by V gene mutations were compared, the unmutated cases showed HLA-DR expression at significantly higher densities than the mutated cases. In addition, higher percentages of B-CLL cells from unmutated cases expressed CD38, CD69, and CD40, whereas significantly lower percentages expressed CD39, CD62L, and CD71 compared to the mutated cases (FIG. 1A and B). These data suggest that cells in the subgroups of B-CLL are arrested at different stages of activation.

Address for correspondence: Nicholas Chiorazzi, North Shore–LIJ Research Institute, 350 Community Drive, Manhasset, NY 11030. Voice: 516-562-1085; fax: 516-562-1022.
nchizzi@nshs.edu

Ann. N.Y. Acad. Sci. 987: 302–304 (2003). © 2003 New York Academy of Sciences.

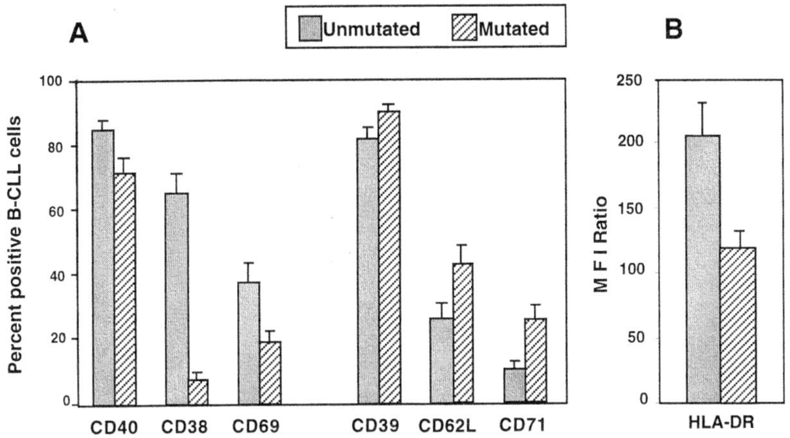

FIGURE 1. (**A**) Percentage of B-CLL cells expressing activation markers. (**B**) Mean fluorescent intensity ratio of HLA-DR expression on B-CLL cells.

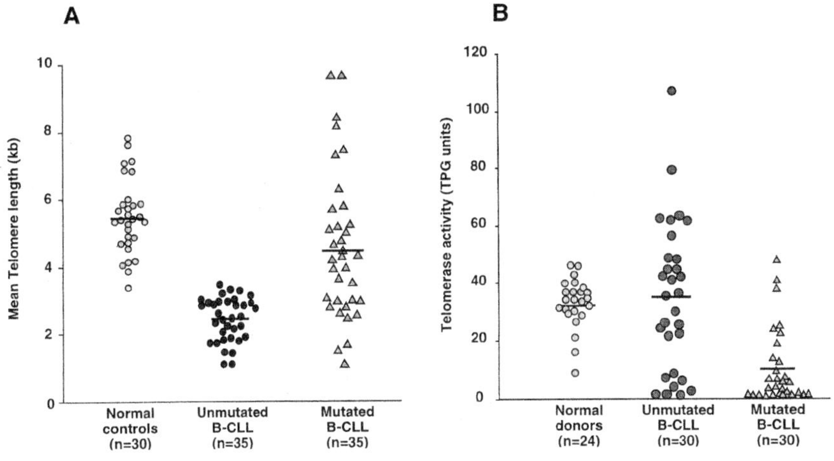

FIGURE 2. (**A**) Telomere length of B cells from age-matched normal donors and unmutated and mutated B-CLL cases. (**B**) Telomerase activity of B cells from age-matched normal donors and unmutated and mutated B-CLL cases.

B cells from all B-CLL cases exhibited features of activated cells compared to CD5 expressing B cells from age-matched normal donors. Telomere lengths of B cells shorten after activation and during differentiation.[4] This effect is also seen as an outcome of the normal "aging phenomenon."[5] To determine possible differences in the replicative histories between B-CLL cells and normal cells, and within B-CLL subgroups, we quantified telomere lengths by Flow-FISH[6] and telomerase activity by the TRAP assay.[7]

B-CLL cells exhibited significantly reduced telomere lengths compared to B cells from age-matched normal controls and autologous polymorphonuclear cells. Although unmutated B-CLL cases exhibited significantly shorter telomeres, they showed significantly elevated telomerase activity compared to mutated cases, the former observation suggesting a more extensive replicative history of the unmutated cases (FIG. 2A and B). B-CLL cells showed a greater rate of decline in telomere lengths compared to B cells from normal individuals. However, the rate of decline in telomere length did not differ between the B-CLL subgroups. In conclusion, these data suggest that (1) all B-CLL cases have been activated *in vivo*, (2) the two subgroups exhibit features of cells arrested at different stages after *in vivo* activation, and (3) the cells from the two subgroups show a distinct replicative history.

REFERENCES

1. FAIS, F., F. GHIOTTO, S. HASHIMOTO, *et al.* 1998. Chronic lymphocytic leukemia B cells express restricted sets of mutated and unmutated antigen receptors. J. Clin. Invest. **102:** 1515–1525.
2. DAMLE, R.N., T. WASIL, F. FAIS, *et al.* 1999. Ig V gene mutation status and CD38 expression as novel prognostic indicators in chronic lymphocytic leukemia. Blood **94:** 1840–1847.
3. HAMBLIN, T.J., Z. DAVIS, A. GARDINER, *et al.* 1999. Unmutated Ig VH genes are associated with a more aggressive form of chronic lymphocytic leukemia. Blood **94:** 1848–1854.
4. WENG, N.P., L. GRANGER & R.J. HODES. 1997. Telomere lengthening and telomerase activation during human B cell differentiation. Proc. Natl. Acad. Sci. USA **94:** 10827–10832.
5. WENG, N.P., L.D. PALMER, B.L. LEVINE, *et al.* 1997. Tales of tails: regulation of telomere length and telomerase activity during lymphocyte development, differentiation, activation, and aging. Immunol. Rev. **160:** 43–54.
6. RUFER, N., W. DRAGOWSKA, G. THORNBURY, *et al.* 1998. Telomere length dynamics in human lymphocyte subpopulations measured by flow cytometry. Nat. Biotechnol. **16:** 743–747.
7. KIM, N.W., M.A. PIATYSZEK, K.R. PROWSE, *et al.* 1994. Specific association of human telomerase activity with immortal cells and cancer. Science **266:** 2011–2015.

New Structural Model for HLA-Restricted Presentation of Nickel to T Cells

KATHARINA GAMERDINGER, CORINNE MOULON, AND
HANS ULRICH WELTZIEN

Max-Planck-Institute for Immunobiology, Freiburg, Germany

KEYWORDS: nickel ions; T cells

Nickel ions are potent inducers of T cell-mediated contact hypersensitivity in humans. Human leukocyte antigen (HLA)-restricted, nickel (Ni)-specific T cells have been repeatedly cloned from Ni-allergic patients.[1] However, unlike the case for peptide- or hapten-reactive T cells, the structural identity of Ni-containing allergenic epitopes remains to be defined. We analyzed the unique T cell receptor (TCR) of the human $CD4^+$ T cell clone SE9, which specifically reacts to Ni in the context of most HLA-DR alleles. Furthermore, its Ni-specificity and DR-restriction are defined largely by the α-chain of its α,β-heterodimer.[2] Ni recognition by this TCR is independent of the nature of HLA-DR-associated peptides, but crucially relies on a conserved, surface-exposed histidine in position 81 of HLA-DR β-chains. A variety of His81 mutations completely abolished nickel presentation, while a mutant replacing His by Tyr retained partial activity. Antigen contact residues within the TCR α-chain were studied by expressing wild-type and mutated TCR α- and β-chains in the TCR-deficient murine hybridoma line 54ζ17,[3] resulting in hybridoma T913. These studies revealed a lack of involvement of the α-chain CDR2 loop, but identified two tyrosines as absolutely essential contact residues of the TCR α-chain: one in position 29 of CDR1, the other in the N-nucleotide-encoded position 94 of CDR3. Exchange of either of the two tyrosines to Ala or Phe extinguished Ni specificity, whereas Tyr29 in CDR1 could be functionally replaced by His. Other mutations in CDR1 or CDR2 had either no or only quantitative effects on Ni recognition. Even an elongation of the CDR3 loop by insertion of Ala between positions 92 and 93 did not alter the hybridoma's sensitivity for nickel. These findings reveal that Ni ions may activate TCRs by a mechanism differing not only from HLA-allele-specific peptide recognition, but also from peptide-independent activation by superantigens. In the case of hybridoma T913, peptide-independent bridging of the TCR and the major histocompatibility complex results from the cross-linking by Ni^{2+} of Tyr or His residues in hypervariable (and particularly in idiotypic) sequences of the TCR with conserved

Address for correspondence: Hans Ulrich Weltzien, Max-Planck-Institute for Immunobiology, Postfach 1169, D-79011 Freiburg, Germany. Voice: +49-761-5108-531; fax: +49-761-5108-534. weltzien@immunbio.mpg.de

Ann. N.Y. Acad. Sci. 987: 305–306 (2003). © 2003 New York Academy of Sciences.

amino acids of the HLA-DR surface. Hence, the T913 receptor will encounter higher numbers of nickel epitopes on antigen-presenting cells than peptide-dependent and/or HLA-allele-restricted TCRs at comparable Ni concentrations. The existence of Ni-specific receptors of the T913-type, representing intermediates between TCR recognition of nominal peptides or of superantigens, may partially explain the extraordinary strong allergenicity of nickel.

ACKNOWLEDGMENTS

This work was supported by the Freiburg Clinical Research Group "Pathomechanisms of Allergic Inflammation" (BMBF FKZ 01GC0102).

REFERENCES

1. MOULON, C., J. VOLLMER & H.U. WELTZIEN. 1995. Characterization of processing requirements and metal cross-reactivities in T-cell clones from patients with allergic contact-dermatitis to nickel. Eur. J. Immunol. **25:** 3308–3315.
2. VOLLMER, J. *et al.* 2000. Antigen contacts by Ni-reactive TCR: typical α,β chain cooperation versus α chain-dominated specificity. Int. Immunol. **12**: 1723–1731.
3. BLANK, U. *et al.* 1993. Analysis of tetanus toxin peptide/DR recognition by human T cell receptors reconstituted into a murine T cell hybridoma. Eur. J. Immunol. **23:** 3057–3065.

Why Are Mice with Targeted Mutation of Co-Stimulatory Molecules Prone to Autoimmune Disease?

YANG LIU, JIAN-XIN GAO, HUIMING ZHANG, XUE-FENG BAI, JING WEN, XINCHENG ZHENG, JINQING LIU, AND PAN ZHENG

Division of Cancer Immunology, Department of Pathology, and Comprehensive Cancer Center, Ohio State University Medical Center, Columbus, Ohio 43210, USA

KEYWORDS: co-stimulatory molecules; autoimmunity; clonal deletion

It is puzzling that mice with a targeted mutation in CD28[1] or CTLA-4[2,3] and B7-2,[4] are more prone to autoimmune diseases. Although many have attributed the autoimmune lymphoproliferative disease in CTLA-4-deficient mice to negative regulatory function of CTLA-4,[2,3] the recent findings of similar, although less severe, predisposition in mice with mutations of other co-stimulatory molecules have raised questions about this interpretation. A largely unexplored possibility is whether blockade of co-stimulatory molecules during T cell development may increase the number and broaden the spectrum of potential autoreactive T cells, and if so, whether this effect contributes, at least in part, to the paradoxical auto-immune phenotypes of these mice.[5]

Using a transgenic model, we have shown that perinatal treatment with anti-B7 monoclonal antibodies prevents clonal deletion of antigen-specific T cells. Moreover, even in the conventional mice with a polyclonal T cell receptor (TCR) repertoire, anti-B7 prevented deletion of T cells specific for VSAg. The results demonstrate that co-stimulatory molecules B7-1 and B7-2 play an important role in T cell clonal deletion. The anti-B7 blockade, in terms of both the spectrums (all four Vβs among CD4 T cells, and three of four Vβs among CD8 T cells) and also the quantity of the effect (3–6-fold increase of VSAg-reactive T cells in the periphery in most cases), argue for a critical role for B7-mediated co-stimulation in the clonal deletion of a significant portion, although certainly not all, of self-reactive T cells, especially in cases where the antigens are limiting or where the affinity of antigen-TCR interaction is low.

We enriched the VSAg-reactive T cells rescued by perinatal treatment of anti-B7-1 and anti-B7-2 blockade and transferred them into irradiated hosts. Our

Address for correspondence: Yang Liu, 185 Hamilton Hall, 1645 Neil Avenue, Columbus, OH 43210-1218. Voice: 614-292-3054; fax: 614-688-8152.
liu-3@medctr.osu.ed

Ann. N.Y. Acad. Sci. 987: 307–308 (2003). © 2003 New York Academy of Sciences.

analysis indicated that these VSAg-reactive T cells expanded substantially, and induced severe inflammation in multiple organs. These results directly demonstrate that the VSAg-reactive T cells that escape clonal deletion as a result of B7 blockade are highly pathogenic. However, the autopathogenic T cells are not limited to VSAg-reactive cells, as anti-B7 also rescues autoreactive T cells in C57BL/6j mice that are pathogenic in the syngeneic VSAg$^-$ host (C57BL/6). In three distinct settings, we have shown that perinatal treatment with both anti-B7-1 and anti-B7-2 mAbs induces pathogenic self-reactive T cells in the thymus and in the spleen. Upon adoptive transfer, both thymocytes and spleen T cells killed the syngeneic young recipients within 1–2 weeks. They died of severe multiple-organ inflammation. While adult recipients survived the acute autoimmune attack, severe chronic autoimmune diseases developed within three months of adoptive transfer. Thus, the adult recipients were less susceptible to the autoreactive T cells. The extreme susceptibility of the young (2–3 weeks old) recipients may explain, at least in part, the lethal lymphoproliferative disease in young CTLA-4(–/–) mice.[2,3]

Taken together, our results demonstrate that co-stimulation by B7-1 and B7-2 plays a critical role in intrathymic deletion of a wide spectrum of autoreactive T cells, and the cells rescued by B7 blockade are highly responsive to antigen *in vivo*. More strikingly, blockade of B7 leads to an accumulation of highly pathogenic autoreactive T cells.[6] The strong effect of the B7 blockade on T cell clonal deletion suggests that the accumulation of pathologic autoreactive T cells results from a blockade of T cell clonal deletion. The fact that the VSAg-specific T cells rescued by anti-B7 are highly pathogenic in syngeneic mice provides strong support for this hypothesis. The significant increase of pathogenic autoreactive T cells may contribute to the paradoxical increase of autoimmunity in mice deficient for B7-2,[4] CD28,[1] or CTLA-4.[2,3]

REFERENCES

1. SALOMON, B., D.J. LENSCHOW, L. RHEE, *et al.* 2000. Immunity **12:** 431–440.
2. TIVOL, E.A., F. BORRIELLO, A.N. SCHWEITZER, *et al.* 1995. Immunity **3:** 541–547.
3. WATERHOUSE, P., J.M. PENNINGER, E. TIMMS, *et al.* 1995. Science **270:** 985–988.
4. SALOMON, B., L. RHEE, H. BOUR-JORDAN, *et al.* 2001. J. Exp. Med. **194:** 677–684.
5. LIU, Y. 1997. Immunol. Today **18:** 569–572.
6. GAO, J.-X., H. ZHANG, X.F. BAI, *et al.* 2003. J. Exp. Med. In press.

Facets of Dendritic Cell Maturation

Environmental Instructions Lead to Distinct Transcriptional Pattern in Myeloid Dendritic Cells

DAVORKA MESSMER, BRADLEY MESSMER, AND NICHOLAS CHIORAZZI

North Shore-LIJ Research Institute, Manhasset, New York 11030, USA

Dendritic cells (DCs) in the "immature" state act as environmental sentinels, recognizing components of pathogens and interpreting endogenously produced signals such as inflammatory cytokines or CD40 ligand (CD40L) on activated T cells. With sufficient stimulation, DCs mature in a process that correlates with regulated changes in the expression of various surface molecules and the release of cytokines and chemokines. Mature DCs present antigen, prime naïve T cells, and initiate an immune response. Here we address the degree to which different maturation stimuli direct the nature of the mature DC at the transcriptional level. Immature human dendritic cells were generated by culturing CD14$^+$ monocytes in 1000 U/mL GM-CSF (Immunex, Seattle, WA) and 200 U/mL IL-4 (R&D Systems, Minneapolis, MN) for 7 days. At day 7 in culture, the immature DCs were exposed to LPS (*E. coli* serotype 026:B6; Sigma), CD40L (50% of the culture medium was exchanged with CD40L containing supernatant from a J588L hybridoma cell transfected with CD40L,[1] a cocktail of inflammatory cytokines: 1000 U/mL IL-6 (R&D), 10 ng/mL TNFα (R&D), 10 ng/mL IL-1b (R&D), and 1 μg/mL PGE2 (Sigma), or left unstimulated (= immature DCs). Cells were harvested 24 and 48 h after stimulation. Total RNA was isolated using the RNeasy kit from Quiagen, labeled, and prepared for hybridization to U95 A–E oligonucleotide arrays (Affymetrix) using standard methods. Gene expression profiles were generated using all five human U95 A–E Affymetrix Genechip oligonucleotide microarrays.

Microarray image files (.cel) were analyzed using Affymetrix Microarray Suite (MAS) 5.0 and DCHIP 1.1 software. Present/absent calls were generated with MAS using the default program settings for all probesets (n = 62,981). Model-based expression values and standard errors were calculated using DCHIP,[2,3] following normalization via program defaults. Pairwise comparisons were done with a paired t-test. Genes were considered significantly different in expression if the lower bound of the 90% confidence interval of the fold change was greater than 1.5, the gene had been called present in >35% of the samples, and the P-value was ≤0.01. Hierarchical clustering was performed with Pearson correlation with average linkage.

Address for correspondence: Davorka Messmer, North Shore-LIJ Research Institute, 350 Community Dr., 4th Floor, Manhasset, NY 11030. Voice: 516-562-1080; fax: 516-562-1022. dmessmer@nshs.edu

Ann. N.Y. Acad. Sci. 987: 309–311 (2003).

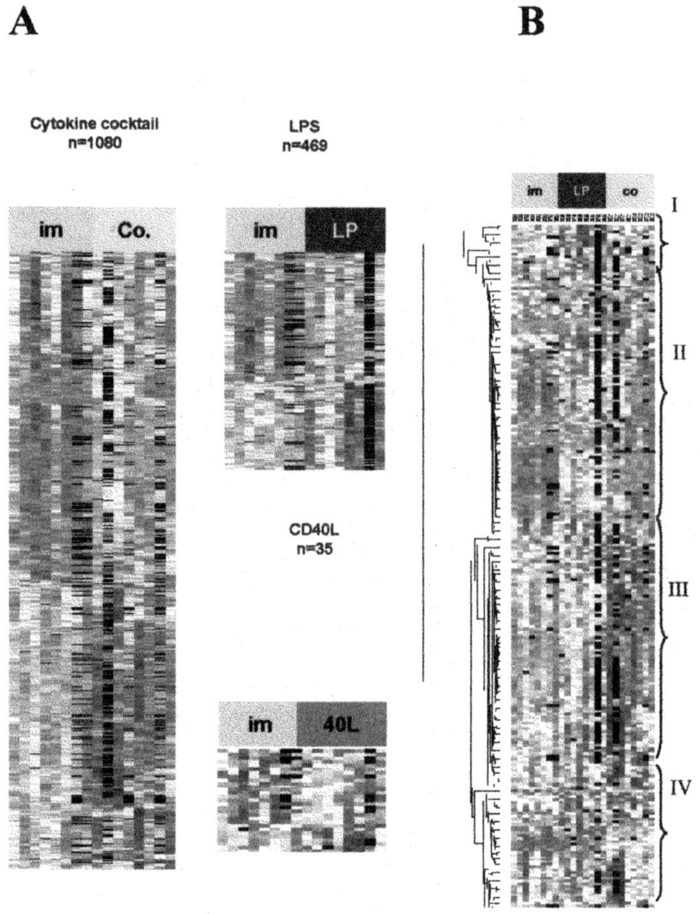

FIGURE 1. Gene expression changes in immature DCs in response to CYC, LPS, and CD40L. (**A**) Colorimetric matrix representation of genes that were significantly altered with either stimulus compared to immune DCs, with n = the number of genes altered. The *boxes* in the *top right* indicate genes that are up-regulated with the stimulus, and the *boxes* on the *bottom right* are genes down-regulated in response to the stimulus compared to immature DCs. (**B**) Hierarchical clustering of the expression patterns of LPS- and CYC-specific genes. The four principal branches of the dendrogram represent: (I) genes specifically down-regulared with LPS; (II) genes specifically up-regulated with the cytokine cocktail; (III) genes specifically down-regulated with the cytokine cocktail; (IV) genes specifically up-regulated with LPS.

Exposure to the stimuli induced the expected changes in DC surface marker expression, some of which correlated with changes in gene expression. The change in gene expression was most dramatic with a cytokine cocktail (1080 genes) and LPS (469 genes) in comparison to immature DCs. Smaller changes were observed with CD40L (FIG. 1A). The transcriptional responses to LPS or a cocktail of cytokines are to a large extent overlapping, indicating a core, stimulus-independent maturation program. However, a smaller set of genes change their expression levels in a stimulus-specific manner, suggesting distinct genetic subroutines (FIG. 1B). The genes divide into four subgroups: I, LPS down-regulated; II, cocktail up-regulated; III, cocktail down-regulated; and IV, LPS up-regulated genes. Among the genes specifically up-regulated by LPS at the tested time points were MIP-1α, MIP-1β, and RANTES. A higher expression of these chemokines could indicate enhanced immune-cell recruitment and enhanced immune response. CD40L induced a restricted response, indicating that synergistic signals might be needed for "full" maturation. Dissecting the cocktail and LPS-specific subroutines will enhance an understanding of the plasticity of DCs in response to their environment.

REFERENCES

1. LANE, P., T. BROCKER, S. HUBELE, *et al.* 1993. Soluble CD40 ligand can replace the normal T cell-derived CD40 ligand signal to B cells in T cell-dependent activation. J. Exp. Med. **177:** 1209.
2. LI, C. & W.H. WONG. 2001. Model-based analysis of oligonucleotide arrays: expression index computation and outlier detection. Proc. Natl. Acad. Sci. USA **98:** 31.
3. SCHADT, E.E., C. LI, C. SU, *et al.* 2000. Analyzing high-density oligonucleotide gene expression array data. J. Cell. Biochem. **80:** 192.

IRTA Family Proteins: Transmembrane Receptors Differentially Expressed in Normal B Cells and Involved in Lymphomagenesis

IRA MILLER,[a,b] GEORGIA HATZIVASSILIOU,[a] GIORGIO CATTORETTI,[a,b] CATHY MENDELSOHN,[b] AND RICCARDO DALLA-FAVERA[a,b,c]

[a]*Institute for Cancer Genetics,* [b]*Department of Pathology,* [c]*Department of Genetics and Development, Columbia University, New York, New York 10032, USA*

KEYWORDS: IRTA; lymphoma; Fc receptor

The IRTA (*I*mmunoglobulin superfamily *R*eceptor, *T*ranslocation *A*ssociated) genes encode a family of five type I transmembrane molecules most closely related to Fc receptors.[1,2] Whereas the number of extracellular immunoglobulin-like (Ig-like) domains (FIG. 1) varies from three to nine among the IRTA proteins, each has at least one domain situated at the amino terminal end that is similar to one of the three Ig-like domains of the Fc receptors for IgG. The more membrane proximal Ig-like domains, while lacking similarity to Fc receptors, are strikingly similar among IRTA family members, suggesting that they bind to a distinct, possibly commonly recognized ligand. The nature of the IRTA cytoplasmic domains, which contain three to five tyrosines embedded within ITIM and ITIM-like motifs, strongly suggests a signaling function.

IRTA family mRNAs are all specific to lymphoid tissues and to cell lines of B lineage, but they have distinct expression patterns revealed by *in situ* hybridization on human tonsil.[2] IRTA1 expression is exclusively within the marginal zone region,[1,2] and IRTA1+ cells, as determined by immunohistochemistry, are $CD27^+$ monocytoid B cells.[3] IRTA2 and IRTA3 are expressed in the germinal center light zone (centrocyte region) at much higher levels than the dark zone (centroblast region), and expression is maintained in interfollicular B cell areas. IRTA5, and to a lesser extent, IRTA4, show predominant expression in the mantle zone (naive B cells). Thus, the subdivision of IRTA expression suggests that these molecules serve to fine-tune the various stages of the B cell response to context-specific types of immune complexes or membrane-bound ligands.

The IRTA genes are located on chromosome 1q21,[1,2] a region frequently rearranged in many types of non-Hodgkin lymphomas, including follicular lymphomas, diffuse large B cell lymphomas, Burkitt lymphomas, and multiple myelomas. The aberrations usually involve amplification or duplication of large regions of 1q21 with translocations to nondistinct partner chromosome loci, complicating the iden-

Address for correspondence: Ira Miller, Institute for Cancer Genetics, Russ Berrie Pavilion, Room 307, 1150 St. Nicholas Avenue, New York, NY 10032. Voice: 212-851-5270; fax: 212-851-5256.

IJM8@columbia.edu

Ann. N.Y. Acad. Sci. 987: 312–313 (2003).

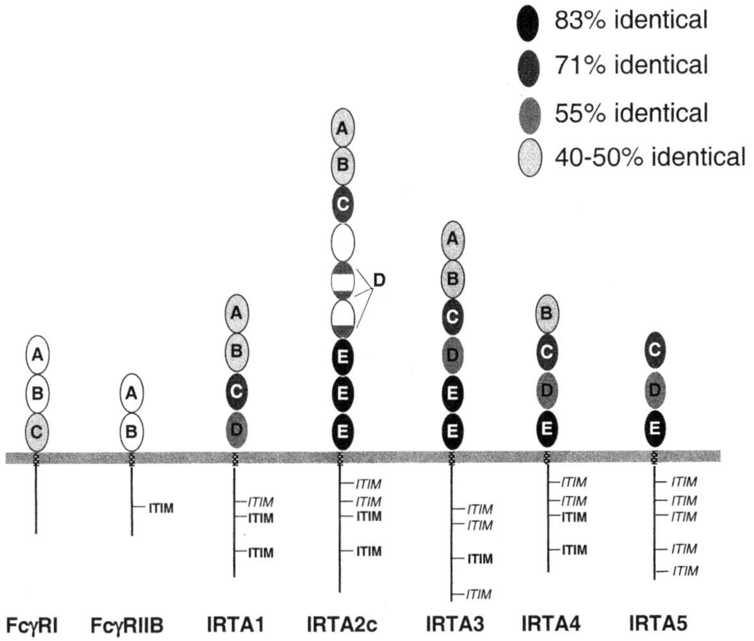

FIGURE 1. Conservation of IRTA domains and homology to Fc receptors. Extent of conservation of each Ig-like domain type (*ovals A through E*) is represented by its degree of shading, as shown. *Patterned rectangles* are transmembrane regions. *Straight lines* are cytoplasmic domains. ITIM: immunotyrosine inhibition motif; *ITIM*: immunotyrosine inhibition motif-like sequence.

tification of genes relevant to the transformation process. We identified IRTA1 as a gene disrupted in a t(1:14) in a multiple myeloma cell line,[1] resulting in expression of a fusion protein consisting almost entirely of the transmembrane and cytoplasmic portions of the IgA heavy chain. In northern blot analysis of 24 cell lines,[1] the adjacent gene, IRTA2, was found to be expressed at levels 10-fold higher in Burkitt lymphoma cell lines with cytogenetically identified aberrations of lq21 than in those with normal lq21. The other, neighboring IRTA genes were expressed indiscriminately in both groups,[2] suggesting that IRTA2 may be a specific target of deregulation by a chromosomal event, possibly contributing to lymphomagenesis.

REFERENCES

1. HATZIVASSILIOU, G., I. MILLER, *et al.* 2001. IRTA1 and IRTA2, novel immunoglobulin superfamily receptors expressed in B cells and involved in chromosome lq21 abnormalities in B cell malignancy. Immunity **14:** 277–289.
2. MILLER, I., G. HATZIVASSILIOU, *et al.* 2002. IRTAs: a new family of immunoglobulin-like receptors differentially expressed in B cells. Blood **99:** 2662–2669.
3. CATTORETTI, G., G. HATZIVASSILIOU, *et al.* 2002. IRTA-1 identifies marginal zone B cells and MZ lymphomas [abstract]. Modern Pathol. **15:** 234A.

Transcriptional Deregulation of Mutated BCL6 Alleles by Loss of Negative Autoregulation in Diffuse Large B Cell Lymphoma

LAURA PASQUALUCCI,[a] ANNA MIGLIAZZA,[a] B. HILDA YE,[b] AND RICCARDO DALLA-FAVERA[a]

[a]*Institute for Cancer Genetics, Columbia University, New York, New York 10032, USA*
[b]*Department of Cell Biology, Albert Einstein College of Medicine, Bronx, New York 10461, USA*

The BCL6 proto-oncogene encodes a POZ/zinc finger transcriptional repressor expressed in germinal center (GC) B cells and required for GC formation. In ~35% of diffuse large B cell lymphomas (DLBCL) and 5–14% of follicular lymphomas (FL), the BCL6 locus is altered by chromosomal translocations that deregulate its expression by a mechanism known as promoter substitution. In addition, the BCL6 5′ noncoding sequences are targeted by multiple somatic mutations that cluster within ~1.5 kb from the transcription initiation site and are found in 30–40% of normal GC B cells (frequency: 0.14×10^{-2}/bp) as well as in a fraction of all GC-derived B cell lymphomas.[1,2] The structural features, distribution, and specificity for GC B cells strongly suggest that BCL6 mutations are introduced by the same mechanism that targets immunoglobulin V genes in the GC. Nonetheless, the biological role of these mutations in normal and malignant B cell development remains unclear.

The present study was aimed at investigating the effect of BCL6 mutations on gene transcriptional regulation. To this end, the BCL6 5′ noncoding region (~5 kb) from 24 tumor cases [6 DLBCL, 5 Burkitt lymphoma (BL), 4 FL, and 5 chronic lymphocytic leukemia (CLL), corresponding to 35 mutated alleles] or sorted GC B cells ($N = 20$ mutated alleles) was linked to the luciferase gene, and the activity of the corresponding constructs was evaluated in transient transfection/reporter gene assays in two BCL6 permissive lymphoma lines (Ly1 and MUTUI-BL59). We found that, while mutant alleles derived from normal GC cells, or from BL, FL, and CLL cases, displayed comparable activity to that of a wild-type construct, four (33%) of the 12 DLBCL-associated alleles were significantly overexpressed (4–18-fold) in both transfected lines. Among multiple mutations scattered throughout the BCL6 5′ noncoding sequences of these four alleles, single nucleotide substitutions, all located within the first noncoding exon of the gene, were found to be responsible for the observed overexpression.

These mutations affect two adjacent BCL6 binding sites and were found to prevent BCL6 from binding to its own promoter, both *in vitro* in electrophoretic mobility shift

Address for correspondence: Laura Pasqualucci, Institute for Cancer Genetics, Columbia University, New York, NY 10032. Voice: 212-851-5273; fax: 212-851-5256.
lp171@columbia.edu

Ann. N.Y. Acad. Sci. 987: 314–315 (2003). © 2003 New York Academy of Sciences.

assays and *in vivo* in chromatin immunoprecipitation assays. Since these sites are normally used by BCL6 to negatively autoregulate its expression levels, our results suggest that DLBCL-associated mutations disrupt the negative feedback loop and can contribute to lymphomagenesis by allowing deregulated transcription of BCL6. Indeed, deregulated expression of the endogenous BCL6 gene could be documented in native lymphoma cells carrying a nucleotide substitution in one of the two exon 1 binding sites.

An expanded survey of DLBCL cases revealed that alterations in these sequences can be found in ~13% of the samples ($N = 4/32$); it is noteworthy that these samples all lack 3q27 chromosomal translocations, suggesting that exon 1 mutations and BCL6 rearrangements are mutually exclusive. Our data identify a novel mechanism for BCL6 deregulation and point to a broader involvement of this proto-oncogene in DLBCL pathogenesis, by indicating that deregulated BCL6 expression can be achieved not only by chromosomal translocations (~35% of cases) but also by mutations affecting the BCL6 exon 1 binding sites (~13% of cases).

REFERENCES

1. MIGLIAZZA, A. *et al.* 1995. Frequent somatic hypermutation of the 5′ noncoding region of the *BCL6* gene in B-cell lymphoma. Proc. Natl. Acad. Sci. USA **92:** 12520–12524.
2. PASQUALUCCI, L. *et al.* 1998. BCL-6 mutations in normal germinal center B cells: evidence of somatic hypermutation acting outside Ig loci. Proc. Natl. Acad. Sci. USA **95:** 11816–11821.

Analysis of the Structure of Proteasome-Proliferating Cell Nuclear Antigen (PCNA) Multiprotein Complex and Its Autoimmune Response in Lupus Patients

YOSHINARI TAKASAKI, KAZUHIKO KANEDA, KEN TAKEUCHI, RAN MATSUDAIRA, MASAKAZU MATSUSHITA, HIROFUMI YAMADA, MASUYUKI NAWATA, KEIGO IKEDA, AND HIROSHI HASHIMOTO

Department of Internal Medicine and Rheumatology, Juntendo University School of Medicine, Tokyo, Japan

KEYWORDS: PCNA; proteasome; SLE; autoantibody; Ki antigen

We have recently found that the PCNA multiprotein complexes (PCNA complexes) associated with cell proliferation can be purified using monoclonal antibodies (mAbs) to PCNA.[1] We attempted to analyze the structure of the complexes and found that proteasome, which had been known as an ATP-dependent proteolytic enzyme involved in antigen presentation on class I major histocompatibility molecules, was one of the elements of the PCNA complex. We also found that there was an autoimmune-response linkage between PCNA and proteasome.

MATERIAL AND METHODS

PCNA complexes were purified by affinity chromatography using mAbs to PCNA, TOB7. Murine mAbs and rabbit polyclonal antibodies (Abs) raised against various molecules associated with cell proliferation and components of proteasome were used to analyze the components of PCNA complexes in immunoblotting (IB).

RESULTS

The PCNA complexes, in turn, reacted with murine anti-DNA mAbs, as well as with Abs against p21, replication protein A (RPA), DNA helicase II (NDH II), cyclin-dependent kinase (CDK) 4, CDK 5, and Topoisomerase I (Top I) in IB

Address for correspondence: Yoshinari Takasaki, M.D., Department of Internal Medicine and Rheumatology, Juntendo University School of Medicine, Tokyo, Japan. Voice: 81-3-3813-3111; fax: 81-3-5800-4893.
tyoshi@med.juntendo.ac.jp

Ann. N.Y. Acad. Sci. 987: 316–318 (2003). © 2003 New York Academy of Sciences.

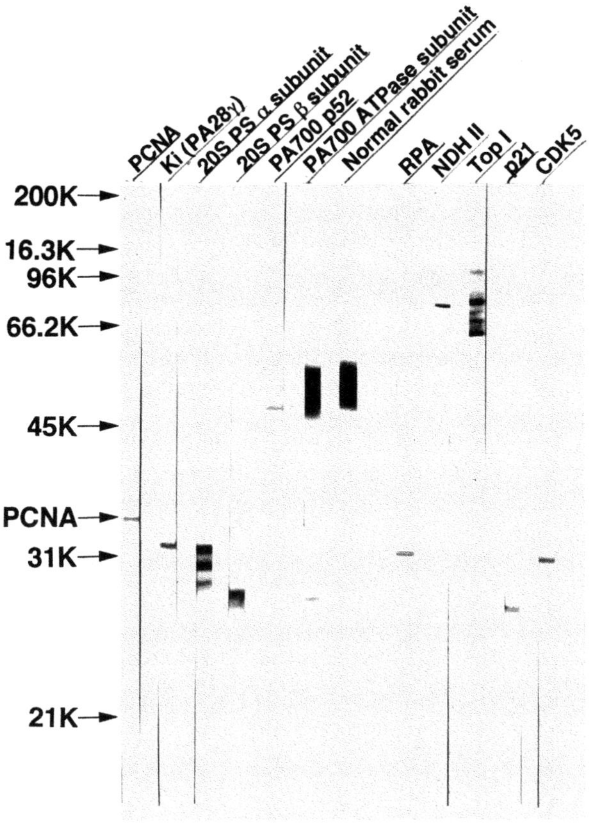

FIGURE 1. Reactivity of antibodies to various proteins associated with cell proliferation and proteasome in IB. Control is normal rabbit serum.

(FIG. 1). These findings suggested that the PCNA complexes comprise the "protein machinery" for DNA replication and cell-cycle regulation. At the same time, it was found that anti-Ki (a proteasome activator, PA28-gamma) sera from lupus patients reacted with the 32 kDa polypeptide in the PCNA complex. In addition to Ki, 20S proteasome alpha- and beta-subunits, p52 subunit, and ATPase subunit of another proteasome activator, PA700, were detected by IB (FIG. 1). The immunoprecipitation study using mAbs to beta-subunit and PA700 ATPase subunit, and the purified PCNA complexes revealed that "hybrid-type proteasome," the complex of PA700, 20S proteasome, and PA28-gamma, was interacting with PCNA.

Because it has been reported that there is an autoimmune-response linkage between PCNA and Ki, we tested the reactivity of 30 anti-Ki lupus sera to the proteasome, and found that two of them could react with the 20S proteasome alpha-subunit, and one could react with the beta-subunit, whereas 40 lupus sera, negative for anti-Ki, did not.

DISCUSSION

The proteasome is an essential component of the ATP-dependent proteolytic pathway in eukaryotic cells, and it has a role in various cellular processes.[2] These include antigen presentation on class I major histocompatibility molecules, selective proteolysis for metabolism and protein quality control, apoptosis, signal transduction, and cell-cycle regulation.[2] Among various functions of the proteasome, its association with cell-cycle regulation may explain the molecular interaction of Ki with PCNA, because recent studies have revealed that various cell-cycle regulators, such as p21, CD4, and CD5 interacting with PCNA, are degraded by the Ub-proteasome pathway. These reports and our previous findings obtained by analysis of the PCNA complex strongly suggest that Ki binding to 20S proteasome interacts with proteins involved in cell-cycle regulation, and thus is copurified with the 34-kDa PCNA polypeptide. Another possibility is that hybrid-type proteasome directly binds to PCNA and is involved in its proteolysis.

Our results suggest a novel cellular function of the hybrid-type proteasome and an important role of the molecular interaction between PCNA and proteasome to induce the autoimmune response to the complex in lupus patients.

REFERENCES

1. Takasaki, Y., T. Kogure, K. Takeuchi, *et al.* 2001. Reactivity of anti-proliferating cell nuclear antigen (PCNA) murine monoclonal antibodies to the PCNA multiprotein complexes involved in cell proliferation. J. Immunol. **166:** 4780–4787.
2. Tanaka, K. 1988. Proteasome: structure and biology. J. Biochem. **123:** 195–204.

Ethyl Pyruvate Protects against Lethal Systemic Inflammation by Preventing HMGB1 Release

LUIS ULLOA,[a] MITCHELL P. FINK,[b] AND KEVIN J. TRACEY[a]

[a]*Laboratory of Biomedical Science, North Shore-LIJ Research Institute, Manhasset, New York 11030, USA*

[b]*Department of Critical Care Medicine, Pittsburgh, Pennsylvania 15261, USA*

INTRODUCTION

Sepsis, the leading reason of death in noncoronary intensive care units, is a major cause of morbidity and mortality, affecting over 750,000 people each year in the United States. Despite the recent advances in intensive care management and antimicrobial therapies, mortality rates associated with sepsis have not declined, and remain high, exceeding 30%.[1,2] Sepsis is a lethal syndrome caused by a fatal immune response to a severe infection. Regardless of the infection source, sepsis triggers a systemic inflammatory response syndrome (SIRS) that leads to lethal multiple-organ failure, shock, and death. Experimental therapies against proinflammatory cytokines such as TNF and IL-1b have failed to improve survival in clinical trials.[3,4] One reason for this failure is because these therapies target cytokines that are released transiently during the "acute phase," long before typical clinical signs of sepsis can be recognized and treatment initiated. The recent identification of cytokines released after the acute phase provides an alternative target for delayed treatments that can be initiated after the onset of the clinical signs of sepsis. Here we review the potential of the recently discovered novel therapeutic targets.

RESULTS AND DISCUSSION

Recent studies have identified high-mobility group B1 (HMGB1) protein as a lethal cytokine released from activated macrophages during the systemic response of sepsis.[5,6] HMGB1 is released into the serum after the resolution of the acute phase, a time frame that coincides with organ injury and mortality during the "severe phase." We have analyzed the potential of different compounds to prevent the release of HMGB1 from monocyte/macropage-like RAW264.7 cells. Among these compounds, ethyl pyruvate prevented the release of HMGB1 from endotoxin-stimulated

Address for correspondence: Luis Ulloa, Laboratory of Biomedical Science, North Shore-LIJ Research Institute, 350 Community Drive, Manhasset, NY 11030. Voice: 516-562-2315; fax: 435-604-9151.

Lulloa@nshs.edu

Ann. N.Y. Acad. Sci. 987: 319–321 (2003). © 2003 New York Academy of Sciences.

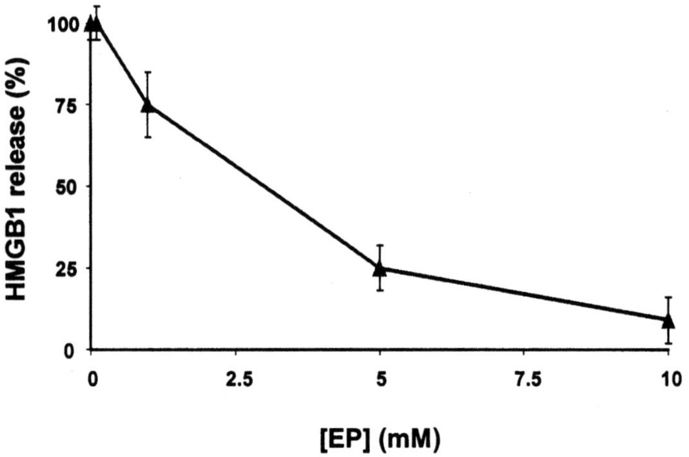

FIGURE 1. Ethyl pyruvate inhibits HMGB1 release. RAW264.7 macrophage cultures were stimulated with LPS (25 ng/mL) for 20 h. The conditioned media were analyzed by Western blot and HMGB1 levels were measured by densitometric analysis. Ethyl pyruvate (EP; 1, 5, or 10 mM) significantly inhibited release of HMGB1 from simulated macrophage cultures in a concentration-dependent manner.

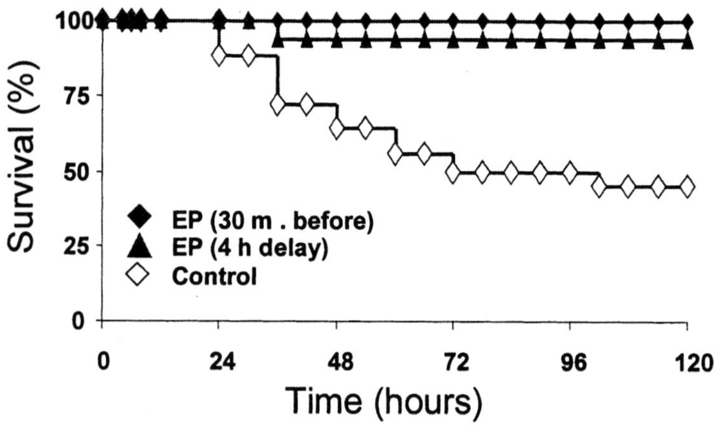

FIGURE 2. Ethyl pyruvate protects against lethal systemic inflammation. Mice ($n = 20/$ group were injected with a lethal imfusion of endotoxin (2.5 mg LPS/kg/mouse; i.p.). Ethyl pyruvate was injected (40 mg EO/kg/mouse; i.p.) in the experimental mice 30 min before or 3 h after endotoxin infusion. Ethyl pyruvate conferred significant protection against lethality ($P < 0.05$), as meaured by Fisher's exact test, even when treatment was initiated after serum TNF levels had peaked.

macrophages at clinically achievable concentrations (FIG. 1). The inhibition of HMGB1 release from macrophages was specific, and ethyl pyruvate did not significantly affect intracellular HMGB1 protein concentration or stability.

The therapeutic potential of ethyl pyruvate was evaluated after intraperitoneal injection of a lethal dose of endotoxin, an established model of endotoxemia and systemic inflammation (FIG. 2). A single administration of ethyl pyruvate (40 mg EP/kg/mouse; i.p.), 30 minutes prior to lethal injection of endotoxin (i.p.: 2.5 mg EP/kg/mouse; i.p.), prevented lethality (vehicle survival = 5/20 versus EP survival = 20/20; $P < 0.005$ Student's t-test). The therapeutic potential of ethyl pyruvate was analyzed by evaluating posttreatment, initiated after the onset of systemic TNF and the occurrence of the initial signs of sepsis. TNF reaches a maximum level in the serum of 40 ± 8 ng/mL one hour after LPS injection in control mice. Delayed treatment with EP started 3 h after endotoxin administration, prevented lethality in an established model of endotoxemia (EP survival = 18/20; $P < 0.005$ Student's t-test). Delayed administration of ethyl pyruvate attenuated the systemic release of HMGB1 measured at 20 h after the onset of endotoxemia. Ethyl pyruvate prevents mortality in both models of lethal systemic inflammation, even when treatment was initiated after serum TNF levels had peaked. These results indicate that ethyl pyruvate confers significant protection against systemic inflammation, and prevents morbidity and mortality in an established model of lethal systemic inflammation, and warrants further evaluation for the treatment of other systemic or chronic immune disorders.[7]

REFERENCES

1. BUTT, W. 2001. Septic shock. Pediatr. Clin. North Am. **48:** 601–625.
2. PERITI, P. 2000. Current treatment of sepsis and endotoxemia. Expert Opin. Pharmacother. **1:** 1203–1217.
3. REINHART, K. & W. KARZAI. 2001. Anti-tumor necrosis factor therapy in sepsis: update on clinical trials and lessons learned. Crit. Care Med. **29:** S121–S125.
4. WANG, H., O. BLOOM, M. ZHANG, *et al.* 1999. HMG-1 as a late mediator of endotoxin lethality in mice. Science **285:** 248–251.
5. OMBRELLINO, M., H. WANG, M.S. AJEMIAN, *et al.* 1999. Increased serum concentrations of high-mobility-group protein 1 in haemorrhagic shock. Lancet **354:** 1446–1447.
6. HOTCHKISS, R.S. & I.E. KARL. 2003. The pathophysiology and treatment of sepsis. N. Engl. J. Med. **348**(2): 138–150.
7. SIMS, C.A., S. WATTANASIRICHAIGOON, M.J. MENCONI, *et al.* 2001. Ringer's ethyl pyruvate solution ameliorates ischemia/reperfusion-induced intestinal mucosal injury in rats. Crit. Care Med. **29:** 1513–1518.

Central Tolerance in a Prostate Cancer Model TRAMP Mouse

PAN ZHENG,[a] XIN-CHENG ZHENG,[a] JIAN-XIN GAO,[a] HUIMING ZHANG,[a] TERRENCE GEIGER,[b] AND YANG LIU[a]

[a]*Department of Pathology and Comprehensive Cancer Center, The Ohio State University, Columbus, Ohio 43210, USA*

[b]*Department of Pathology, St. Jude Children's Research Hospital, Memphis, Tennessee 38105, USA*

During the past two decades, the development of genetic and biochemical techniques to characterize tumor antigens constitutes a major achievement for tumor immunology. The majority of the tumor antigens have the same sequences as the endogenous genes.[1] These unmutated tumor antigens are often recognized by T cells from cancer patients.[2]

However, cancers appear to progress in spite of significantly expanded CD8 T cells specific for the tumor antigens. It has been suggested that these T cells are either anergic as a result of peripheral tolerance or have low avidity for the cancer antigens. It is not clear whether the repertoire of T cells specific for the unmutated tumor antigens will be shaped by negative selection during T cell development.

The transgenic adenocarcinoma of mouse prostate (TRAMP) model is transgenic for the SV40 large T antigen (Tag) under the control of the rat probasin regulatory elements. While it has been established that T lymphocytes from TRAMP mice are tolerant to SV40 Tag,[3] the mechanism of the tolerance is largely unknown. To examine whether the T cell clonal deletion is responsible for the tolerance, we crossed the TRAMP mice with mice transgenic for a rearranged T cell receptor (TCR) specific for the SV40 Tag presented by the H-2K^k. Double transgenic TRAMP/TCR mice showed profound thymic deletion of SV40 Tag reactive T cells, including a 6- to 10-fold reduction in the total thymocyte numbers and a greater than 50-fold reduction in phenotypically mature T cells. Consistent with this finding, we observed that the SV40 Tag is expressed at low levels in the thymus. These results demonstrate that clonal deletion is a major mechanism for tolerance to antigens previously regarded as prostate-specific, and provide direct evidence that the T cell repertoire specific for an unmutated tumor antigen can be shaped by clonal deletion in the thymus.

Since the CTL epitopes for the large T antigens are among the most extensively characterized, many groups have used SV40 large T antigen transgenic mice directed by various promoters as models to study T cell tolerance to tumor-associated antigen. Self-antigen can induce tolerance by either central or peripheral tolerance.

Address for correspondence: Pan Zheng, Department of Pathology and Comprehensive Cancer Center, The Ohio State University Medical Center, 129 Hamilton Hall, 1645 Neil Avenue, Columbus, OH 43210. Voice: 614-292-2003; fax: 614-688-8152.

zheng-1@medctr.osu.edu

Ann. N.Y. Acad. Sci. 987: 322–323 (2003). © 2003 New York Academy of Sciences.

Central tolerance is characterized by clonal deletion or clonal anergy of immature T cells in the thymus, while peripheral tolerance can be mediated by a number of mechanisms, including clonal anergy, activation-induced cell death, and regulatory cells. Although significant progress has been made in understanding mechanisms of immune tolerance to SV40 Tag antigen in various tumor models, it remains unclear whether clonal deletion or functional inactivation is responsible for T cell tolerance. With regard to the tolerance in the TRAMP model, our results provide the direct evidence that tumor antigens previously perceived to be expressed exclusively in cancerous tissues can induce deletion of antigen-specific T cells in the thymus.

REFERENCES

1. ROSENBERG, S.A. 2001. Progress in human tumour immunology and immunotherapy. Nature **411:** 380–384.
2. ROMERO, P., P.R. DUNBAR, D. VALMORI, *et al.* 1998. Ex vivo staining of metastatic lymph nodes by class I major histocompatibility complex tetramers reveals high numbers of antigen-experienced tumor-specific cytolytic T lymphocytes. J. Exp. Med. **188:** 1641–1650.
3. GRANZIERO, L., S. KRAJEWSKI, P. FARNESS, *et al.* 1999. Adoptive immunotherapy prevents prostate cancer in a transgenic animal model. Eur. J. Immunol. **29:** 1127–1138.

Long-Lasting Bioactive Substance Mimicking Basic Fibroblast Growth Factor Was Associated with the Heavy Chain(s) of Immunoglobulin G in Serum from Three Patients with Breast Cancer

M.B. ZIMERING[a] AND S. THAKKER-VARIA[b]

[a]*Department of Medicine, Veterans Affairs Medical Center, Lyons, New Jersey 07939, USA, and UMDNJ-Robert Wood Johnson Medical School, Piscataway, New Jersey 08854, USA*

[b]*Department of Neuroscience and Cell Biology, UMDNJ-Robert Wood Johnson Medical School, Piscataway, New Jersey 08854, USA*

INTRODUCTION

Metastatic bone disease causes significant morbidity and mortality in cancer patients who suffer fractures, severe bone pain, hypercalcemia, or spinal cord compression.[1] Basic fibroblast growth factor (bFGF), a potent endothelial cell mitogen localized to extracellular matrix in bone[2] and other normal tissues, lacks a signal sequence and does not normally circulate. A novel kind of endothelial cell bioactivity with properties of anti-idiotype FGF-like autoantibodies was described in a subset of patients with multiple endocrine neoplasia type 1 and prolactinoma.[3] Since bFGF-like immunoreactivity (IR) was increased in some cancer sera,[4] we tested serum from patients with cancer-associated hypercalcemia for bioactive (FGF-like) autoantibodies.

SUBJECTS AND METHODS

Subjects ($n = 56$) included 19 breast cancer, 14 renal cancer, and 23 lung cancer patients who participated in a protocol evaluating intravenous bisphosphonates for the acute treatment of cancer-associated hypercalcemia. Bony metastases were present in 17/19 breast, 14/23 lung, and 8/14 renal cancer cases. Baseline serum samples (before treatment with bisphosphonates) were used in nearly all cases. A subset of three hypercalcemic breast cancer patients suffered spinal cord compression or metastases (cases 1, 2, 3; TABLE 1). These three patients were among a sub-

Address for correspondence: M.B. Zimering, Department of Medicine, Veterans Affairs Medical Center, Lyons, NJ 07939. Voice: 908-647-0180 X4426; fax: 908-604-5249.
mark.zimering@med.va.gov

Ann. N.Y. Acad. Sci. 987: 324–327 (2003). © 2003 New York Academy of Sciences.

TABLE 1. Growth-promoting activity in protein-A eluates of serum from cancer and control subjects[a]

	Mean growth activity ± 1 SD (%)
Cancer-associated hypercalcemia ($n = 56$)	
Breast cancer and spinal cord compression (cases 1, 2, 3)	
Other hyperprolactinemia ($n = 2$)	138 ± 18[b]
Normoprolactinemia	
Breast cancer ($n = 15$)	109 ± 18[c]
Renal cancer ($n = 13$)	98 ± 9
Lung cancer ($n = 23$)	98 ± 12
Normal subjects ($n = 10$)	99 ± 4
Control subjects ($n = 11$)	
Breast cancer and systemic lupus erythematosus (case 4)	191 ± 12
Primary hyperparathyroidism and skull metastases (case 5)	143 ± 15
Breast cancer (treated) ($n = 7$)	89 ± 16
Prostate cancer and diffuse bone metastases (cases 6, 7)	72 ± 9

[a]One-fiftieth dilution of the protein-A-eluted fractions of sera was added to endothelial cells. Growth activity was assessed as % change in cell number compared to cells to which no sera fractions were added.

[b]$P < 0.001$, compared to renal, lung cancer, and normal subjects.

[c]$P = 0.05$, compared to renal, lung cancer, and normal subjects.

group ($n = 5$) that had significantly increased ($P < 0.001$) mean serum prolactin (45 ± 16 ng/mL) compared to levels in other hypercalcemic (breast) cancer (17 ± 7 ng/mL) ($n = 15$) or normal subjects (6 ± 2 ng/mL) ($n = 10$) (TABLE 1). Control subjects ($n = 11$) included treated, nonrecurrent breast cancer ($n = 7$); active localized (case 4) or metastatic (case 5) breast cancer; and metastatic prostate cancer (cases 6 and 7) (TABLE 1). Normal subjects ($n = 10$) included both men and women.

In a prior study,[5] serum bFGF-IR ranged from 5 pg/mL to 27 pg/mL (normal < 4 pg/mL) in hypercalcemic cancer subjects. There was no significant difference in mean serum bFGF-IR (~5 pg/mL) in subsets of hyper- or normoprolactinemic hypercalcemic breast cancer patients. Serum bFGF-IR was undetectable in a patient with localized invasive breast cancer and systemic lupus erythematosus (case 4), and substantially elevated (44 pg/mL) in a patient with breast cancer, hyperparathyroidism, and skull metastases (case 5).

The proliferative response of bovine pulmonary artery endothelial cells to protein-A-eluted and (subsequently) hydroxyapatite-adsorbed fractions of serum was tested after 4 days incubation in medium 199 supplemented with 10% fetal calf serum. Standard recombinant human bFGF (1–10 ng/mL) was included as a positive control in all assays. Growth-promoting activity is expressed as a percentage of the control (basal) cell number for cells grown (M199 + 10% FCS) without added test factors.

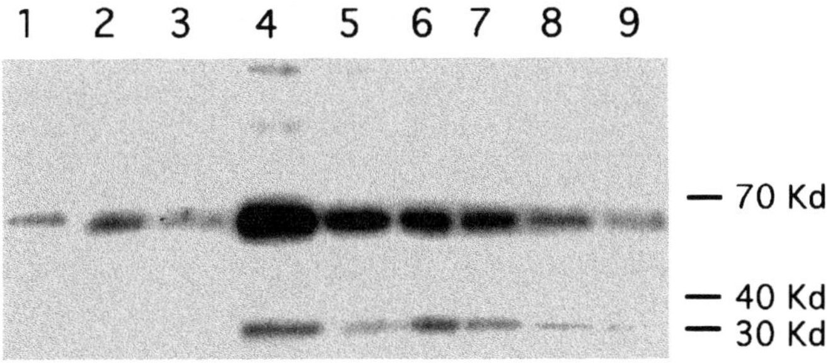

FIGURE 1. SDS-PAGE (under reducing conditions) and Western blotting [with HRP-conjugated, antihuman IgG, (1:5000)] of active hydroxyapatite-purified, protein-A-eluted fractions of serum from four breast cancer subjects (cases 1–4; lanes 1–4); and one prostate cancer subject (case 6, lane 5). Standard human IgG: 2.5, 1.25, 0.75, 0.25 μg protein (lanes 6–9), respectively. (From Zimering & Thakker-Varia.[6] Reprinted with permission from *Life Sciences.*)

RESULTS AND DISCUSSION

Mean protein-A-eluted activity (at 1:50 dilution) in serum from a hyperprolactinemic subset of hypercalcemic cancer subjects significantly exceeded ($P < 0.001$) mean activity in normal or control hypercalcemic cancer subjects (TABLE 1). Activity in three breast cancer subjects with spinal cord compression or metastases (180%, 160–310%, 150%) was significantly increased above control levels (<120%) at each of three dilutions: (1:400, 1:800, and 1:5000, respectively). Rabbit antibodies to intact recombinant bovine bFGF- (1–146) (selective for the C-terminal receptor binding domain of bFGF) completely neutralized the activity in protein-A eluates from the three breast cancer subjects. Antibodies to an amino-terminal bFGF peptide fragment (bFGF 1–24) that completely neutralized activity in 10 ng/mL bFGF had less or no significant inhibitory effect on the protein-A-eluted activity.

We purified the bioactive serum protein-A-eluted material using hydroxyapatite chromatography. Highest (peak) activity in protein-A-eluted fractions from active cancer sera (cases 1–7) was retained and eluted from hydroxyapatite (HA) columns with 0.4 M sodium phosphate. Peak inhibitory activity eluted at 0.05 M sodium phosphate ($n = 2$ prostate cancer) and 0.25 M sodium phosphate ($n = 5$ breast cancer). Mean peak activity in seven active cancer (5 breast, 2 prostate) sera (167 ± 17%) significantly exceeded ($P < 0.001$) mean peak activity in six normal sera (110 ± 14%) and seven treated nonrecurrent breast cancer sera (121 ± 7%). Activity in protein-A eluates of serum from three hypercalcemic breast cancer sera (numbers 1, 2, 3) survived storage (0–4 C) for 5 years. The HA-purified, bioactive protein-A-eluted fractions (cases 1–4, 6) were subjected to SDS-PAGE under reducing conditions. Western blotting with antihuman IgG (H and L chains) revealed bands corresponding to only heavy (cases 1–3; lanes 1–3) or heavy and light chains of IgG (cases 4, 6) (FIG. 1).

These results imply that long-lasting (endothelial cell) bioactivity in serum from three breast cancer patients with spinal cord compression or metastases likely corresponded to anti-idiotype antibodies containing an "internal image" of the active site of bFGF residing (in part) in the heavy chains of IgG. This is consistent with reports that the active combining site(s) in an anti-idiotype antibody complex was associated (in large part) with heavy-chain complementarity-determining regions.[7] Serum autoantibodies mimicking fibroblast growth factor were found in breast cancer patients in association with hyperprolactinemia, markedly elevated serum bFGF-IR (and osteoblastic metastases), or systemic lupus erythematosus (SLE). Increased affinity of the cancer sera growth stimulatory FGF-like autoantibodies for hydroxyapatite (eluted at 0.4 M sodium phosphate) suggests that the circulating autoantibodies may play a role in promoting local cell proliferation in bone, or in tissues (e.g., breast) susceptible to ectopic calcification.[8]

REFERENCES

1. JANJAN, N. 2001. Bone metastases: approaches to management. Semin. Oncol. **28:** 35–41.
2. HAUSCHKA , P.V. *et al.* 1986. Growth factors in bone matrix. Isolation of multiple types by affinity chromatography on heparin-sepharose. J. Biol. Chem. **261:** 12665–12674.
3. ZIMERING, M.B. *et al.* 1994. Circulating fibroblast growth factor-like autoantibodies in two patients with multiple endocrine neoplasia type 1 and prolactinoma. J. Clin. Endocrinol. & Metab. **79:** 1546–1552.
4. KUROBE, M. *et al.* 1993. Increased level of basic fibroblast growth factor (bFGF) in sera of patients with malignant tumors. Horm. Metab. Res. **25:** 395–396.
5. ZIMERING, M.B. 2001. Effect of intravenous bisphosphonates on release of basic fibroblast growth factor in serum of patients with cancer-associated hypercalcemia. Life Sci. **70:** 1947–1960.
6. ZIMERING, M.B. & S. THAKKER-VARIA. 2002. Increased fibroblast growth factor-like autoantibodies in serum from a subset of patients with cancer-associated hypercalcemia. Life Sci. **71:** 2939–2959.
7. GARCIA, K.C. *et al.* 1992. Three-dimensional structure of an angiotensin II-Fab complex at 3A: hormone recognition by an anti-idiotype antibody. Science **257:** 502–507.
8. THURFJELL, E. *et al.* 2001. Mammographic findings as predictor of survival in 1–9 mm invasive breast cancer. Worse prognosis for cases presenting as calcification alone. Breast Cancer Res. Treat. **67:** 177–180.

Index of Contributors

Accapezzato, D., 99–106
Adorini, L., 258–261
Albesiano, E., 302–304
Allen, S.L., 302–304
Aoufouchi, S., 158–165
Arce, S., 266–269

Bai, X.F., 307–308
Banchereau, J., 180–187
Barnaba, V., 99–106
Barrera, C.A., 285–287
Basso, K., 288–290
Belloni, A., 299–301
Bennett, L., 180–187
Berek, C., 246–248, 266–269
Blanco, P., 180–187
Bohnhorst, J.Ø., 291–294
Bolon, B., 295–298
Bonifaz, L., 15–25
Bonnyay, D., 15–25
Bouzahzah, F., 60–67
Braun, M., 158–165
Bridges, S.L., 270–273
Brot, N., 68–78
Burch, J., 230–235
Burgio, V.L., 117–124
Burmester, G.-R., 246–248
Burrows, P.D., 270–273

Califano, A., 166–172, 288–290
Callebaut, I., 150–157
Campagnuolo, G., 295–298
Capra, J.D., xi–xii
Cassese, G., 266–269
Castiglioni, P., 249–257
Cattoretti, G., 166–172, 288–290, 312–313
Cauci, S., 299–301
Ceppa, P., 117–124
Chan, T., 285–287
Chiorazzi, N., xi–xii, 117–124, 230–235, 302–304, 309–311
Choi, J.-Y., 60–67
Christensen, S.C., 38–50
Clark, M.R., 26–37
Colombo, M., 117–124
Cooper, M.D., 270–273
Craft, J.E., 60–67
Crowe, S.E., 285–287

Dalla-Favera, R., 166–172, 288–290, 312–313, 314–315
Damle, R.N., 302–304
Datta, S.K., 79–90
Davidson, A., 188–198
de Aloysio, D., 299–301
de Chasseval, R., 150–157
De Santo, D., 299–301
de Villartay, J.-P., 150–157
Debes, G.F., 266–269
Deshmukh, U.S., 91–98
Desiderio, S., 280–284
Dhodapkar, M., 15–25
Dickerson, S.K., 135–139
Doerner, T., 266–269
Dono, M., 117–124
Dörner, T., 246–248
Driussi, S., 299–301
Duryea, D., 295–298
Dustin, M.L., 51–59

Eardley, L.D., 60–67
Elkon, K.B., 68–78
Ernst, P.B., 285–287
Euler, C.W., 38–50

Faili, A., 158–165
Fay, J., 180–187
Feige, U., 295–298
Ferrarini, M., xi–xii, 117–124, 302–304
Findley, H.W., 270–273
Fink, M.P., 319–321
Fischer, A., 150–157
Fischetti, V.A., 207–214
Foell, J., 230–235
Francavilla, V., 99–106
Fu, S.M., 91–98

Gaidano, G., 117–124
Gamerdinger, K., 305–306
Gao, J.-X., 307–308, 322–323
Gaskin, F., 91–98
Geiger, T., 322–323
Gerloni, M., 249–257
Gershov, D., 68–78
Ghiotto, F., 302–304
Golfier, S., 107–116
Goronzy, J.J., 140–149
Guaschino, S., 299–301
Guéranger, Q., 158–165

Haddad, J., Jr., 166–172
Hamann, A., 266–269
Hanson, L.Å., 199–206
Harley, J.B., 215–229
Hashimoto, H., 316–318
Hatzivassiliou, G., 312–313
Hauser, A.E., 266–269
Håversen, L., 199–206
Hawiger, D., 15–25
Hiepe, F., 266–269
Hoffmann, M.K., 230–235
Honjo, T., 1–8
Höpken, U.E., 107–116
Huang, W., 188–198
Hultman, P., 236–239

Ikeda, K., 316–318
Inaba, K., 15–25
Ippolito, G.C., 262–265
Iyoda, T., 15–25

James, J.A., 215–229
Jonsson, R., 291–294
Juang, Y.-T., 240–245

Kaneda, K., 316–318
Kannapell, C.C., 91–98
Kaufman, K.M., 215–229
Keller, J.L., 166–172
Kim, S.J., 68–78
Kinoshita, K., 1–8
Kirby, M.Y., 215–229
Klein, U., 166–172, 288–290
Kong, P.L., 60–67
Kono, D.H., 236–239
Korotkova, M., 199–206
Krenn, V., 266–269
Küppers, R., 173–179

Lahita, R.G., xi–xii
Lanzafame, P., 299–301
Lewis, J.E., 91–98
Li, D.N., 9–14
Li, F., 274–279
Liew, F.Y., 274–279
Lindenau, S., 246–248, 266–269
Lipp, M., 107–116
Liu, J., 307–308
Liu, K., 15–25
Liu, Y., 307–308, 322–323
Lundin, S., 199–206

Ma, X., 68–78
Mahnke, K., 15–25
Manoury, B., 9–14
Manz, R.A., 266–269
Massara, R., 117–124
Massenburg, D., 26–37
Matsudaira, R., 316–318
Matsushita, M., 316–318
Matthews, S.P., 9–14
Mattsby-Baltzer, I., 199–206
Mazzeo, D., 9–14
McCausland, M., 230–235
Mecklenbräuker, I., 125–134
Mendelsohn, C., 312–313
Messmer, B., 309–311
Messmer, D., 309–311
Migliazza, A., 314–315
Mihara, M., 188–198
Miljkovic, V., 166–172
Miller, I., 312–313
Mittler, R.S., 230–235
Moshous, D., 150–157
Moss, C.X., 9–14
Moulon, C., 305–306
Muehlinghaus, G., 266–269
Müller, G., 107–116
Muramatsu, M., 1–8

Nagaoka, H., 1–8
Nambiar, M.P., 240–245

Natvig, J.B., 291–294
Nawata, M., 316–318
Nitschke, L., 262–265
Niu, H., 288–290
Nussenzweig, M., 15–25

O'Neil, S.P., 230–235
Odegard, J.M., 60–67
Odendahl, M., 246–248, 266–269
Okazaki, I., 1–8

Paczesny, S., 180–187
Palucka, A.K., 180–187
Papavasiliou, F.N., 135–139
Paroli, M., 99–106
Pascual, V., 180–187
Pasqualucci, L., 314–315
Pelkonen, J., 262–265
Penacchioni, P., 299–301
Poinsignon, C., 150–157
Pollard, K.M., 236–239
Posnett, D.N., 274–279
Propato, A., 99–106

Quadrifoglio, F., 299–301

Radbruch, A., 246–248, 266–269
Rai, K.R., 302–304
Rajewsky, K., 262–265
Ramanujam, M., 188–198
Ravetch, J., 15–25
Reiterer, P., 107–116
Reyes, V.E., 285–287
Reynaud, C.-A., 158–165
Riemekasten, G., 266–269

Saijo, K., 125–134
Schiffer, L., 188–198
Schmedt, C., 125–134
Schoenberger, S., 249–257
Scholze, S., 246–248
Schroeder, H.W., Jr., 262–265
Schulman, P., 302–304
Shlomchik, M.J., 38–50
Siemasko, K., 26–37
Silfverdal, S.-A., 199–206
Sinha, J., 188–198
Steinman, R.M., 15–25
Stolovitzky, G.A., 166–172, 288–290
Strandvik, B., 199–206
Suwyn, C., 230–235
Szabo, P., 274–279

Taborelli, G., 117–124
Takasaki, Y., 316–318
Takeuchi, K., 316–318
Tarakhovsky, A., 125–134
Telemo, E., 199–206
Thakker-Varia, S., 324–327
Thoen, J.E., 291–294
Thompson, K.M., 291–294
Tracey, K.J., 319–321
Tsokos, G.C., 240–245
Tu, Y., 166–172, 288–290

Ulloa, L., 319–321

Valetto, A., 302–304
Valiando, J.R., 274–279
Villey, I., 150–157
Vinciguerra, V.P., 302–304

Wang, X., 188–198
Wang, Y.-H., 270–273
Watts, C., 9–14
Weill, J.-C., 158–165
Weksler, M.E., 274–279
Weller, S., 158–165
Weltzien, H.U., 305–306
Wen, J., 307–308
West, M.A., 9–14
Weyand, C.M., 140–149
William, J., 38–50

Yamada, H., 316–318
Yan, X.-J., 230–235
Yarilin, D., 274–279
Ye, B.H., 314–315
Yoo, J.-Y., 280–284
Yoshikawa, K., 1–8

Zabriskie, J.B., xi–xii
Zack, D., 295–298

Zanetti, M., 249–257
Zemlin, M., 270–273
Zhang, H., 307–308, 322–323
Zhang, M., 26–37
Zhang, Z., 270–273
Zheng, P., 307–308, 322–323
Zheng, X., 307–308
Zheng, X.-C., 322–323
Zhu, L., 295–298
Zielinski, C.E., 60–67
Zimering, M.B., 324–327
Zupo, S., 117–124